International Review of
Cytology
A Survey of
Cell Biology

MOLECULAR BIOLOGY OF RECEPTORS AND TRANSPORTERS

PUMPS, TRANSPORTERS, AND CHANNELS

VOLUME 137C

International Review of Cytology
A Survey of Cell Biology

Guest Edited by

Martin Friedlander
Jules Stein Eye Institute
and Department of Physiology
UCLA School of Medicine
Los Angeles, California

Michael Mueckler
Department of Cell Biology
Washington University
School of Medicine
St. Louis, Missouri

MOLECULAR BIOLOGY OF RECEPTORS AND TRANSPORTERS

PUMPS, TRANSPORTERS, AND CHANNELS

VOLUME 137C

Academic Press, Inc.
Harcourt Brace Jovanovich, Publishers
San Diego New York Boston London Sydney Tokyo Toronto

This book is printed on acid-free paper. (∞)

Academic Press, Inc.
1250 Sixth Avenue, San Diego, California 92101-4311

United Kingdom Edition published by
Academic Press Limited
24–28 Oval Road, London NW1 7DX

Library of Congress Catalog Number: 52-5203

International Standard Book Number: 0-12-364539-5

PRINTED IN THE UNITED STATES OF AMERICA
93 94 95 96 97 98 EB 9 8 7 6 5 4 3 2 1

CONTENTS

Voltage-Dependent Sodium Channels

Sidney A. Cohen and Robert L. Barchi

The Vacuolar H^+-ATPases: Versatile Proton Pumps Participating in Constitutive and Specialized Functions of Eukaryotic Cells

Stephen L. Gluck

Structure of the Na,K-ATPase

Robert W. Mercer

The Mouse Multidrug Resistance Gene Family: Structural and Functional Analysis

Philippe Gros and Ellen Buschman

Molecular and Kinetic Aspects of Sodium–Calcium Exchange

Kenneth D. Philipson and Debora A. Nicoll

Molecular Analysis of the Role of Na^+/H^+ Antiporters in Bacterial Cell Physiology

Shimon Schuldiner and Etana Padan

CONTRIBUTORS

Numbers in parentheses indicate the pages on which the authors'contributions begin.

Robert L. Barchi (55), *Mahoney Institute of Neurological Sciences and the Departments of Neurology and Neuroscience, University of Pennsylvania School of Medicine, Philadelphia, Pennsylvania 19104*

Eric C. Beyer (1), *Edward Mallinckrodt Department of Pediatrics, Division of Hematology/Oncology, and the Departments of Cell Biology and Medicine and Physiology, Washington University School of Medicine, St. Louis, Missouri 63110*

Ellen Buschman (169), *Department of Biochemistry, McGill University, Montreal, Quebec, Canada H3G 1Y6*

Sidney A. Cohen (55), *Mahoney Institute of Neurological Sciences and the Department of Medicine, University of Pennsylvania School of Medicine, Philadelphia, Pennsylvania 19104*

Lynne A. Fieber (39), *Department of Cell Biology and Physiology, Washington University School of Medicine, St. Louis, Missouri 63110*

Stephen L. Gluck (105), *Departments of Medicine and Cell Biology and Physiology, Washington University School of Medicine, and the Renal Division, Jewish Hospital of St. Louis, St. Louis, Missouri 63110*

Philippe Gros (169), *Department of Biochemistry, McGill University, Montreal, Quebec, Canada H3G 1Y6*

Edwin W. McCleskey (39), *Department of Cell Biology and Physiology, Washington University School of Medicine, St. Louis, Missouri 63110*

Robert W. Mercer (139), *Department of Cell Biology and Physiology, Washington University School of Medicine, St. Louis, Missouri 63110*

Debora A. Nicoll (199), *Departments of Medicine and Physiology, and the Cardiovascular Research Laboratory, UCLA School of Medicine, Los Angeles, California 90024*

Etana Padan (229), *Division of Microbial and Molecular Ecology, The Alexander Silberman Institute of Life Sciences, The Hebrew University of Jerusalem, 91904 Jerusalem, Israel*

Kenneth D. Philipson (199), *Departments of Medicine and Physiology, and the Cardiovascular Research Laboratory, UCLA School of Medicine, Los Angeles, California 90024*

Shimon Schuldiner (229), *Division of Microbial and Molecular Ecology, The Alexander Silberman Institute of Life Sciences, The Hebrew University of Jerusalem, 91904 Jerusalem, Israel*

Mary D. Womack (39), *Department of Cell Biology and Physiology, Washington University School of Medicine, St. Louis, Missouri 63110*

FOREWORD

This volume in the Molecular Biology of Membrane Transport series is devoted to proteins engaged in the movement of ions across membranes. There are three broad categories of proteins involved in transmembrane ion movement: channels, antiporters, and pumps. The simplest of these, the ion channels, can be thought of as gated, water-filled pores in the membrane that allow the selective passage of ions of a particular size and charge. Selectivity is due to the dimensions of the pore and the chemical properties of the amino acid side chains that form the entryway and walls of the pore. Three chapters in the present volume are concerned with channels. The voltage-gated sodium channels are perhaps the most extensively studied of all channels and were the first whose complementary DNA was cloned. The subsequent blend of molecular and electrophysiological studies on these channels has resulted in an explosion of information concerning their structure and function. This channel plays a crucial role in the generation of action potentials in nerve and muscle cells. Cohen and Barchi (Chapter 3) briefly review the classic electrophysiological characterization of voltage-sensitive sodium channels and then discuss more recent data concerning the structure, function, and regulation of these molecules. The voltage-dependent calcium channels are discussed in Chapter 2 by McCleskey, Womack, and Fieber. These channels are involved in the passive flux of calcium into a variety of cell types, resulting in the elevation of cytosolic calcium levels. Changes in intracellular calcium are important in regulating a number of important cellular processes, including hormone secretion and muscle contraction. This chapter focuses on how the proposed structure of this channel relates to its gating and transport properties. Gap junctions are relatively large channels that allow the exchange of ions and small metabolites between adjacent cells. This intercellular communication is important in the propagation of action potentials, in the maintenance of metabolite homeostasis in avascular tissues, and in the regulation of hormone secretion. Communication via gap junctions has also been implicated in the regulation of cellular growth control, differentiation, and development. Beyer (Chapter 1) focuses on the biochemistry and molecular biology of the connexins, the family of proteins involved in the formation of gap junction complexes.

The dust jacket of this volume shows a schematic scale model of the cardiac gap junction ion channel based on the planar density map (shown above the model) and the edge-on density map perpendicular to the membrane plane (shown to the right of the model). The connexon has a diameter of ~65Å and is formed by a hexameric cluster of α_1 connexin subunits. The α_1 connexin subunit has a diameter of ~25Å and is highly asymmetric with an axial ratio of 4 : 1 to 5 : 1. The hexameric connexons of adjacent cells are in register to define the central channel. The schematic model is in the same orientation as the edge-on density map which is a view down the [1,1] crystal lattice plane (right). The relationship between the 50Å thick lipid bilayers and the edge-view is indicated by lines. The model shows the location of the 10Å thick lipid headgroup regions and the 30Å hydrophobic membrane interior containing the lipid alkyl chains. Based on the planar and orthogonal edge projection, the vestibule in the extracellular gap region between juxtaposed connexons accounts for the periodic stain deposition evident in original image. The detailed shape and transmembrane structure of the channel are of course not known and are depicted in a stylized fashion. The model was provided by Dr. Mark Yaeger and his colleagues at the Research Institute of Scripps Clinic.

Moving up in transporter complexity, two chapters in this volume deal with ion antiporters. Antiporters are transporters that catalyze the transmembrane movement of two different ions in the opposite direction. Transporters in general are more "enzyme-like" in their behavior than are ion channels, in that they interact more strongly with their substrates and undergo distinct conformational changes during the transport of each substrate molecule. These properties give rise to the observation that transporters in general exhibit much lower turnover values for their substrates than do ion channels (~10^2/sec versus ~10^6/sec). The sodium–calcium exchanger is one of the more extensively studied of the ion antiporters. It is present in the plasma membrane of many cell types, where it often is the major route for the efflux of calcium from the cell, thus lowering the level of the cation in the cytosol. For example, the sodium–calcium exchanger is the major route for the lowering of cytosolic calcium in cardiac myocytes during the contraction–relaxation cycle. Philipson and Nicoll (Chapter 7) discuss the physiology, purification, and cloning of this interesting protein. Sodium–proton exchange activity is present throughout the evolutionary ladder from bacteria to mammals. The sodium–hydrogen exchangers participate in a variety of cellular functions, including transmembrane signal transduction and the regulation of intracellular pH and cellular volume. Schuldiner and Padan (Chapter 8) discuss the genetics, biochemistry, and regulation of the sodium–hydrogen exchanger expressed in *Escherichia coli*.

The final class of transporters presented in this volume is the primary active transporters or "pumps," as they are commonly called. These molecules represent a further step up in complexity from the antiporters in that they harness the energy contained in a chemical bond to drive the transport of an ion against an electrochemical gradient. It is these primary active transport systems that establish the ion gradients that drive solute transport via secondary active transport systems. Although all of the pumps described in this volume use energy derived from the hydrolysis of ATP, other pumps harness energy from light or from oxidation of a reduced metabolite to generate an ion gradient. Mercer (Chapter 5) discusses molecular insights into the structure and function of the sodium–potassium ATPase. This pump is a ubiquitous constituent of animal cell plasma membranes, and is responsible for maintaining the steep gradients of sodium and potassium that exist relative to the extracellular milieu. The inward-directed sodium gradient is essential for the cellular absorption of many nutrients and is responsible for electrical signaling in excitable cell types. Gluck (Chapter 4) reviews recent work on the mammalian proton pumps. This structurally complex class of transport porteins is responsible for the acidification of intracellular organelles and the efflux of protons at the apical membrane of certain epithelial cells. Gros and Buschman (Chapter 6) review the properties of the multidrug resistant P-glycoproteins. This important family of proteins is responsible for the efflux of lipophilic xenobiotics from mammalian cells and is involved in the mechanism by which certain cancer cells acquire resistance to chemotherapeutic agents.

The proteins covered in this volume provide a small sample of the increasing number of channels, pumps, and antiporters that are being identified and studied in a wide variety of organisms. As will be abundantly evident from a perusal of the following pages, DNA cloning technology has played an indispensable role in the advancement of our understanding of the primary structures of membrane channels and transporters, how the expression of these molecules is regulated, and the mechanism of membrane protein biosynthesis. However, as was emphasized in the Forewords to the preceding volumes in this special series, additional technological breakthroughs will be required before a clear understanding of the relationship between the structure and function of membrane transporters is obtained. It is hoped that the next decade will bring much excitement as a few of these proteins are crystallized and their tertiary structures are elucidated.

Michael Mueckler

PREFACE

It has been over 6 years since we first considered putting together a volume devoted to the molecular biology of membrane proteins. We thought a single volume would provide sufficient space for several authors to describe the analysis, through molecular and cell biological approaches, of membrane protein structure and function. That was an exciting time since the dozen or so membrane protein receptors that had been cloned and sequenced appeared to fall into several large superfamilies that were related either through sequence homology or putative structural similarities. In the year it took us to prepare an outline for the volume's contents, the number of receptors had increased logarithmically; by 1988 the cloning and sequencing of several dozen receptors, transporters, and channels had been reported in the literature and we both retreated to our laboratories to try and keep up with the flood of information emerging from these studies.

By early 1990, we again began to think about putting together a multivolume treatise that summarized our knowledge of membrane protein receptors, transporters, and channels to serve a useful function for both investigators already in the field as well as those "extramembranous" students and established investigators who wanted to familiarize themselves with a rapidly expanding area of membrane biology. We invited definitive reviews from active investigators in the field; we asked prospective contributors to include a summary of their knowledge of a particular membrane protein as well as to speculate on future directions. The response was very satisfying; most of the major classes and superfamilies of membrane proteins are represented and the chapters have been written by authors whose laboratories are very active in the field. Thus, the chapters are stimulating and authoritative even though we recognize that the speed with which the field is moving makes it difficult to be current by the time a volume is published. Nevertheless, we hope these volumes will provide a useful resource for those individuals interested in the field of membrane receptors, transporters, and channels.

Assembling nearly 30 chapters from as many laboratories has required the usual cajoling, pleading, and, on occasion, threats. However, the credit for ultimate production is due to the authors themselves. We are grateful to the many scientists who contributed their efforts to writing the chapters in these volumes. For our part, the solicitation and editing of these volumes required extensive time that otherwise would have been spent with our families, and we are both very grateful to our wives, Sheila and Paula, and our children for their patience and understanding. Both of us would also like to thank Ron Kaback for his enthusiastic involvement in many of the chapters other than his own, as well as for his foreword to the first volume.

We would also like to thank the excellent editorial and production staff at Academic Press for their assistance with this project. In particular, Charlotte Brabants in editorial, Leslie Yarborough in book production, and Cathy Reynolds in the art department have been exceptionally helpful.

Each of us has our own special thanks to extend to individuals who have helped with various aspects of assembling these volumes. Martin Friedlander would like to acknowledge the support of the Heed Ophthalmic and Heed/Knapp Foundations. My chairman, Bradley Straatsma, has been firmly supportive since my arrival at UCLA and has encouraged me throughout the course of this project. I am also grateful to Allan Kreiger, Bart Mondino, Gordon Grimes, and Joe Demer for their advice throughout the preparation of these volumes. Suraj Bhat and Dean Bok provided advice and encouragement during the early phases of this project and Eileen Fallon provided excellent editorial assistance. Bernie Gilula and Bill Beers nurtured my early interest in membrane proteins and will continue to do so. I am particularly grateful to Günter Blobel for introducing me to the field of membrane protein topogenesis and sharing his infectious enthusiasm for its study with me.

Michael Mueckler would like to acknowledge the support of the Juvenile Diabetes Foundation International. I am indebted to the past and current members of my laboratory for their patience during the preparation of these volumes and for making it a (usually) pleasant experience to come into the lab in the morning. I thank my colleagues in the Department of Cell Biology and Physiology, Robert Mercer, Edwin McCleskey, Philip Stahl, and Stephen Gluck, for their contributions to this series and for many stimulating discussions on membrane proteins over the past few years. I also thank my Ph.D. thesis advisor, Henry Pitot, and my postdoctoral mentor, Harvey Lodish, for sparking my interest in membrane proteins. Lastly, I am grateful to Alan Permutt for his continued friendship, support, and encouragement since my arrival in St. Louis.

Martin Friedlander

Michael Mueckler

Gap Junctions

Eric C. Beyer
Edward Mallinckrodt Department of Pediatrics, Division of Hematology/Oncology, and the Departments of Cell Biology and Medicine and Physiology, Washington University School of Medicine, St. Louis, Missouri 63110

I. Introduction

Gap junctions are specialized plasma membrane regions that contain collections of transmembrane channels which provide a low resistance pathway directly linking adjacent cells. These channels provide for direct cell-to-cell transfer of ions and small molecules without exposure to the extracellular space (Bennett, 1966; Loewenstein, 1966). The exchange of molecules is nonspecific and includes the entire pool of metabolites and ions in each cell (Gilula *et al.*, 1972; Pitts and Simms, 1977; Simpson *et al.*, 1977; Goodenough *et al.*, 1980).

Gap junction mediated intercellular communication has been implicated in many cellular functions. Gap junctions allow action potential propagation by passage of current carrying ions in electrically excitable tissues including myocardium (Barr *et al.*, 1965; Weidman, 1966; Sperelakis and Cole, 1989) and smooth muscle (Dewey and Barr, 1962; Sperelakis and Cole, 1989; Garfield *et al.*, 1977) and at electrotonic synapses in the nervous system (Bennett, 1977; Furshpan and Potter, 1959; McMahon *et al.*, 1989). In tissue culture experiments, gap junctions allow intercellular buffering of cytoplasmic ions (Corsaro and Migeon, 1977; Ledbetter and Lubin, 1979) and the sharing of low-molecular-weight substrates, which can bypass the deleterious effects of somatic cell mutations of various enzymes (Subak-Sharpe *et al.*, 1969; Cox *et al.*, 1970; Hobbie *et al.*, 1987). Metabolic cooperation of cells through gap junctions likely maintains homeostasis in avascular systems such as lens and ovarian follicles (Goodenough, 1979; Gilula *et al.*, 1978; Brower and Schultz, 1982). Passage of second messengers through gap junctions may regulate glandular and hormonal secretion (Peterson, 1985; Meda *et al.*, 1990). A large number of studies have also suggested a role for gap junctions in growth control, embryonic development, and cellular differentiation (Atkinson *et al.*, 1981; Warner *et al.*, 1984; Mehta *et al.*, 1986; Fraser *et al.*, 1987; Lee *et al.*, 1987; Azarnia *et al.*, 1988).

New insights into the regulation of gap junctions and intercellular communication have resulted from the recent molecular cloning of a number of subunit gap junction proteins and the use of those clones to develop specific probes for gap junction nucleic acid sequences and proteins. This chapter will briefly highlight previous physiological and structural studies of gap junctions and their implications for developing a molecular understanding of intercellular communication. Most of this review will then focus on the molecular and cell biological studies of the last few years.

II. Physiology of Gap Junctional Intercellular Communication

The physiologic and pharmacologic regulation of gap junction-mediated cell–cell communication has recently been extensively reviewed (Spray and Bennett, 1985; Spray and Burt, 1990; Veenstra, 1991a). However, a number of points have particular importance in understanding the molecular structure and function of these channels. Paradigms used to study intercellular coupling have included "electrotonic" transmission of experimentally evoked electrical potentials between nonexcitable cells (Bennett, 1966; Loewenstein, 1966) and "dye transfer" of microinjected, membrane-impermeant, fluorescent dyes that can diffuse from the cytoplasm of one cell to another (Payton *et al.*, 1969; Flagg-Newton and Loewenstein, 1979). Such studies using ions or dyes of known molecular size and charge have demonstrated that the intercellular channels between mammalian cells are about 1.5 nm in diameter, with a molecular size exclusion of about 1000 Da (Flagg-Newton *et al.*, 1979; Brink and Dewey, 1980; Imanaga *et al.*, 1987). Solutes move through the channel in hydrated forms (Brink, 1983). The aqueous channel appears to contain negative charges, which may account for a mild ion-selective preference for cations over anions (Neyton and Trautman, 1985; Brink and Fan, 1989).

Although some biophysical studies of gap junctional channels have been performed on single-membrane preparations either reconstituted in artificial bilayers (Young *et al.*, 1987; Spray *et al.*, 1986a) or directly patched (Brink and Fan, 1989), the bulk of physiological data concerning these channels has been obtained from cell pairs studied by the two-cell voltage clamp technique developed by Bennett and colleagues (Spray *et al.*, 1981, 1985) and its modification, the double, whole cell, patch clamp technique (Neyton and Trautman, 1985; Veenstra and DeHaan, 1986; Burt and Spray, 1988a; Rook *et al.*, 1988; Rudisuli and Weingart, 1989;

McMahon *et al.*, 1989; Somogyi and Kolb, 1989). The patch clamp studies have allowed determination of the conductance of single gap junctional channels, which differs in different preparations. Neonatal mammalian cardiac myocytes have predominantly channels with conductances of ~50 pS (Burt and Spray, 1988a; Rook *et al.*, 1988). Embryonic chick cardiac myocytes have 160- and, rarely, 50-pS channels (Veenstra and DeHaan, 1986). Rat lacrimal gland cells and hepatocytes contain ~120-pS channels (Neyton and Trautman, 1985; Spray *et al.*, 1986b). Vascular smooth muscle cells contain 30- and 90-pS channels (Moore *et al.*, 1991). Although some of the variability observed in different laboratories using similar systems (Burt and Spray, 1988a; Rook *et al.*, 1988; Rudisuli and Weingart, 1989) can be ascribed to technical factors such as differences in pipette filling solutions, these studies clearly indicate that gap junctional channels are not all equivalent.

The patch clamp studies have also given some insight into the regulation of gap junctional current flow, by the process of "gating" (Bennett *et al.*, 1988). This very rapid control appears to occur by the transition of individual channels between the fully open and closed states. Although subconductance states of gap junctional channels have been proposed (Neyton and Trautman, 1985; Veenstra and DeHaan, 1988), they have not yet been conclusively demonstrated. There is little evidence that treatments can alter the unitary conductance of single channels. Therefore, various treatments that alter total gap junctional conductance are believed to act by altering the open state probability of the individual channels (Spray and Bennett, 1985; Spray and Burt, 1990; Burt and Spray, 1989).

A number of agents or treatments that alter gap junctional communication have been identified and are discussed below.

A. Voltage (Transjunctional and Transmembrane)

Conductance of a number of gap junctions is gated by voltage. In the rectifying synapses of crayfish (Furshpan and Potter, 1959; Giaume *et al.*, 1987) or fish (Auerbach and Bennett, 1969), depolarization only on the presynaptic side increases conductance. Symmetrical voltage dependence is exhibited by the junctions from a number of sources. In amphibian blastomeres, transjunctional voltage of either sign decreases junctional conductance (Harris *et al.*, 1981). The gap junctions between mammalian hepatocytes (Spray *et al.*, 1991a) or between chick embryo ventricular myocytes (Veenstra, 1990) exhibit a similar, but less sensitive, voltage dependence, with junctional conductance declining during prolonged transjunctional voltage steps above ±30 mV. Interestingly, this voltage dependence in chick heart becomes less prominent with maturation of the embryos (Veenstra, 1991b). Mammalian cardiac myocytes exhibit such voltage dependence only at even larger transjunctional potentials (Rook *et al.*, 1988; Veenstra, 1991c). These effects are independent of the potential between the inside of the cells and the outside. However, the gap junctions of fly salivary glands are also regulated by membrane potential (Verselis *et al.*, 1991); equal depolarization of both cells decreases conductance, and hyperpolarization increases it. These findings suggest that the gap junction structure may contain two different voltage sensors or gates and that they differ in the junctions from different sources. A gap junctional structural model should identify charged residues that may be translocated relative to the membrane or channel in response to voltage gradients.

B. Lipophiles

Several lipophilic compounds are among the pharmacologic agents that block electrical communication. The first to be investigated were octanol and heptanol, which reversibly block junctional conductance in a dose-dependent manner (Veenstra and DeHaan, 1986; Burt and Spray, 1988b; Johnston *et al.*, 1980; White *et al.*, 1985; Niggli *et al.*, 1989). Volatile anesthetics such as halothane and ethrane (in mM concentrations) reversibly block cardiac gap junctions (Burt and Spray, 1989). Arachidonic acid, other unsaturated fatty acids, and doxyl stearic acids block junctional conductance at relatively low (μM) concentrations (Giaume *et al.*, 1987; Fluri *et al.*, 1990; Aylsworth *et al.*, 1986; Burt, 1989). All of these lipophilic agents are likely to act by incorporation into the lipid bilayer. Different potencies of fatty acids of different structures suggest that ionic interactions at the surface and disruption of the bilayer are both important in the inhibitory activity of these compounds (Burt, 1989). However, the specificity of activity of these agents on gap junctional versus other membrane channels is not established.

C. pH and Ca^{2+}

Cytoplasmic acidification or elevation of cytoplasmic calcium ion concentrations leads to reductions of junctional conductance in many systems (Noma and Tsuboi, 1987; Reber and Weingart, 1982; Maurer and Weingart, 1987; Rose and Rick, 1978; Peracchia, 1990; DeMello, 1989; White *et al.*, 1990). A quantitative analysis of the relation of junctional conductance to pH and Ca^{2+} concentration in amphibian blastomeres (Spray *et al.*, 1982) suggested that intracellular pH was the more physiologically relevant variable; Ca^{2+}-dependent channel closure might only be relevant at concentrations seen under pathologic conditions (DeMello, 1983). Further study suggested that pH-dependent gating and voltage dependence might occur by different mechanisms (Spray *et al.*, 1986b). pH sensitivities of hepatocyte or cardiac myocyte junctions differ (Spray and Burt, 1990; White *et al.*, 1985; Reber and Weingart, 1982; Spray *et al.*, 1986c). Chick lens epithelial–fiber cell junctions and mature fiber–fiber cell junctions are quite insensitive to cytoplasmic acidification (Miller and Goodenough, 1986; Schuetze and Goodenough, 1982). Recent studies have suggested that Ca^{2+} can enhance the sensitivity of junctional channels to acidification, suggesting a synergistic action (White *et al.*, 1990; Burt, 1987).

D. Cyclic Nucleotides and Tumor Promoters

In rat hepatocytes, membrane-permeant cAMP derivatives increase junctional conductance, whereas cGMP derivatives have no effect (Saez *et al.*, 1986). In heart, cAMP derivatives and β-adrenergic agonists increase junctional conductance (DeMello, 1983, 1989; Burt and Spray, 1988b), but cGMP derivatives decrease conductance (Spray and Burt, 1990). Carbachol has an effect similar to that of cGMP in heart cells (Burt and Spray, 1988b). These effects are all seen over a time course of minutes. In retinal horizontal cells, dopamine modulates conductance via a cAMP-dependent pathway (McMahon *et al.*, 1989). In rat lacrimal gland cells, acetylcholine induces the closure of gap junctions by a protein kinase C dependent mechanism (Randriamampita *et al.*, 1988). Activators of protein kinase C (such as diacylglycerol and phorbol esters) also reduce junctional conductance in a variety of cells (Gainer and Murray, 1985; Yada *et al.*, 1985; Chanson *et al.*, 1988; Somogyi *et al.*, 1989).

E. Cellular Transformation and Viral Carcinogenesis

The early observation that many transformed cells are deficient in cell–cell communication (Loewenstein, 1979) has been investigated in many systems. Studies have shown that the *src* oncogene of Rous sarcoma virus (RSV) induces a profound loss of cellular coupling in cultured fibroblasts. In cells infected with temperature-sensitive RSV, communication was disrupted rapidly upon a shift to the permissive temperature, which correlated with activation of $pp60^{v\text{-}src}$ kinase activity (Azarnia and Loewenstein, 1984; Atkinson *et al.*, 1981). Cells transfected with *v-src* or *c-src* gene constructs containing kinase activating mutations also exhibited reduced communication (Azarnia *et al.*, 1988). Similarly, transformation by polyomavirus middle T antigen (which activates $pp60^{c\text{-}src}$ and other cellular kinases) produced decreased junctional permeability (Azarnia and Loewenstein, 1987). Activation of the receptors for epidermal growth factor and platelet-derived growth factor (receptor protein kinases) in nontransformed cells caused decreased cell–cell communication (Maldonado *et al.*, 1988). Finally, transformation with other viral oncogenes has also been implicated in blocking cellular coupling (Atkinson and Sheridan, 1988).

III. Gap Junction Structure

The gap junction was originally characterized by its appearance in thin section electron micrographs as a pair of membranes of variable area separated by a 2-nm "gap" (Revel and Karnovsky, 1967; Robertson, 1963). Freeze fracture replicas showed that the structure is characterized by a plaque-shaped, differentiated region of the plasma membrane containing a dense array of intramembrane particles (connexons) on the P-fracture face and a complementary array of depressions or pits on the E-fracture face. (See Figs. 1 and 2).

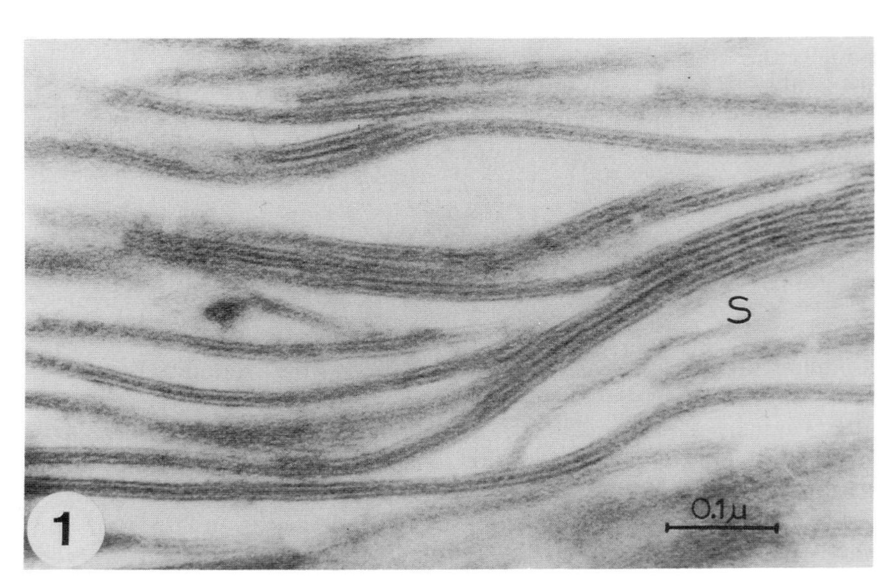

FIG. 1 Thin section electron micrograph of isolated rat liver gap junctions. S indicates "stacks" of junctions in these isolated preparations. (From Goodenough, 1976, reprinted with permission.)

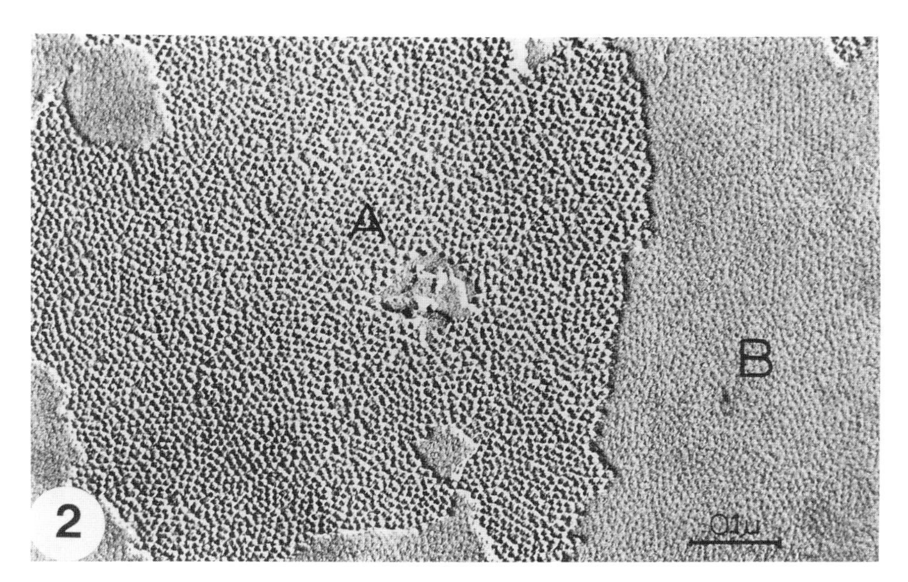

FIG. 2 Freeze fracture electron micrograph of a gap junction in rat liver. A and B refer to the P and E fracture faces, respectively. (From Goodenough, 1976, reprinted with permission.)

The notion that the gap junction structures were the sites of intercellular coupling originally came from correlation of visualized structures with function(s), but has subsequently been proven by immunological and expression studies (see below).

Procedures for the isolation of gap junctions from liver were developed (Goodenough and Stoeckenius, 1972; Evans and Gurd, 1972). These isolated gap junctions retained their *in vivo* appearance in thin sections, but negatively stained images revealed that following isolation the arrays of particles in junctional plaques formed regular two-dimensional crystals. This crystalline nature of the connexons in isolated liver junctions facilitated study of some of their molecular structure. Makowski and colleagues (Caspar *et al.*, 1977; Makowski *et al.*, 1977; Makowski, 1988) used X-ray diffraction and electron microscopy to develop a low-resolution (25 Å) structural model (Fig. 3). The model shows that a gap junction plaque is composed of from tens to thousands of channels. Each channel is composed of a hexameric structure (connexon) composed of six apparently identical integral membrane subunits (connexins) that surround a central pore. The connexon joins in mirror symmetry with a connexon in the plasma membrane of the adjacent cell. The channel appears to have an inside radius of 15- to 20-Å and may taper, being wider near the extracellular end. Studies of junctions imaged with cationic and anionic dyes suggest

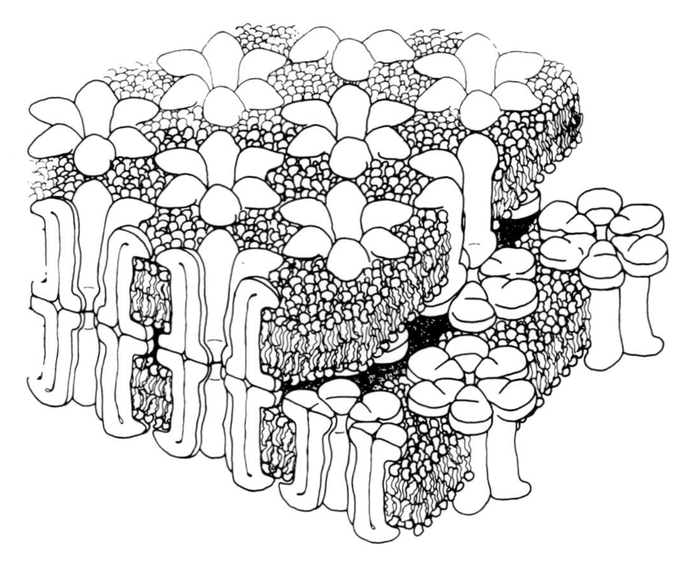

FIG. 3 Structural model of a gap junction, based on X-ray diffraction data. (From Makowski *et al.*, 1977, reproduced from *J. Cell Biol.* **74,** 643, by copyright permission of the Rockefeller University Press.)

a fixed negative charge on the channel wall (Baker *et al.,* 1983). Low-irradiation microscopy studies have given evidence for a bulbous gate closing the mouth of the opening and for a domain at the threefold axis that links trimers of connexons into the hexagonal lattice. Unwin and Zampighi (1980) have proposed that the connexin subunits are relatively rigid, rodlike proteins, with much α-helical transmembrane structure. Makowski (1988) originally proposed that much of the transmembrane region was formed of a barrel of β-sheet structure, but recently Sosinsky *et al.* (1990) have reinterpreted the data to predict parallel bundles of α-helices that are not perpendicular, but are tilted relative to the plane of the membrane. Circular dichroism spectroscopy has confirmed the presence of 40–50% α-helical structure within rat liver gap junction proteins (Cascio *et al.,* 1990).

There have been numerous attempts to understand the regulation/gating of gap junctions by structural studies. Morphologic studies initially suggested that the transitions of connexons from the disordered to an ordered state might be correlated with physiological closure of junctional channels (Peracchia, 1977). But, subsequent studies showed no common structural alterations in connexon packing as visualized by quick freeze fracture that correlated with a block in intercellular dye transfer (Miller and

Goodenough, 1985; Hanna *et al.*, 1985). Therefore, it is much more likely that gap junction gating occurs on a molecular scale. Unwin and colleagues (Unwin and Zampighi, 1980; Unwin and Ennis, 1984) pioneered the study of gap junction structure by Fourier low-dose microscopy. They provided evidence for structural changes accompanying changes in Ca^{+2} concentration that appear as constriction of the pore by tilting and twisting the subunits and could serve as a channel gating mechanism. Caspar *et al.* (1988) and Makowski (1988), in contrast, proposed that channel gating might be achieved by a hinged six-segmented barrier that might open or block the cytoplasmic surface of the channel.

The general structural features of gap junctions in other tissues are believed to be similar to those determined from rodent hepatocyte junctions, but less information is available (Zampighi *et al.*, 1982; Green and Severs, 1984; Fujimoto *et al.*, 1985; Yeager, 1987). The connexons in lens fiber junctions appear less "condensed" or uniformly packed than those from liver (Schuetze and Goodenough, 1982; Goodenough, 1979; Miller and Goodenough, 1985). Cardiac gap junctions contain a "fuzzy" coat along their cytoplasmic sides as seen in thin section or freeze fracture (Kensler and Goodenough, 1980; Manjunath *et al.*, 1984; Shibata and Yamamoto, 1986). These cytoplasmic domains have been suggested to have a regulatory function.

IV. Biochemistry of Gap Junction Proteins

In addition to the liver junctions, methods have been developed for the isolation of myocardial gap junctions (Kensler and Goodenough, 1980; Manjunath *et al.*, 1984) and of lens fiber plasma membranes which contain 5–10% gap junction profiles (Goodenough, 1979; Bloemendal *et al.*, 1972; Alcala *et al.*, 1975; Broekhuyse *et al.*, 1976; Paul and Goodenough, 1983). Sodium dodecylsulfate–polyacrylamide gel electrophoresis (SDS–PAGE) of these preparations has shown that the isolated liver gap junctions are composed primarily of a 27-kDa polypeptide, accompanied by proteolysis fragments, aggregates, and a 21-kDa polypeptide (Henderson *et al.*, 1979; Hertzberg and Gilula, 1979; Finbow *et al.*, 1980). In the mouse, the 21 kDa is present at 50% the abundance of the 27 kDa, whereas in the rat its relative abundance is only 10% (Nicholson *et al.*, 1987). Isolated myocardial gap junctions contain a 43- to 47-kDa polypeptide, cleaved by endogenous proteases to 34-, 32-, and 29-kDa bands (Kensler and Goodenough, 1980; Manjunath *et al.*, 1987). Isolated bovine and ovine lens fiber plasma membranes contain a predominant polypeptide of 26 kDa, called MP26 or MIP26, and numerous other polypeptides, notably one of 70 kDa (MP70) (Kistler *et al.*, 1985). N-terminal sequencing of these proteins by Edman degradation has shown that the liver 27 and 21 kDa (Nicholson *et al.*, 1987), the heart 43–47 kDa and its degradation products (Gros *et al.*, 1983; Nicholson *et al.*, 1985), and the lens 70 kDa (Kistler *et al.*, 1988) are homologous proteins, whereas the lens 26 kDa appears unrelated (Nicholson *et al.*, 1983).

Polyclonal and monoclonal antisera have been raised to the liver 27 and 21 kDa, the lens MP26, and the lens MP70. Immunocytochemical studies have shown that anti-27 kDa and anti-21 kDa bind directly to the cytoplasmic surfaces of gap junctions from liver and anti-MP70 similarly binds to cytoplasmic surfaces of lens fiber junctions (Janssen-Timmen *et al.*, 1983; Paul, 1985, 1986; Stevenson *et al.*, 1986; Gruijters *et al.*, 1987; Milks *et al.*, 1988; Goodenough *et al.*, 1988; Traub *et al.*, 1989). Indeed, Traub *et al.* (1989) demonstrated that the 21 and 27 kDa colocalize in the same hepatocyte junctional plaque. In some laboratories, anti-MP26 has been localized to both the junctional and the nonjunctional membranes of lens fibers (Bok *et al.*, 1982; Fitzgerald *et al.*, 1983; Sas *et al.*, 1985), whereas in others, the antisera can only be localized to the nonjunctional membranes (Paul and Goodenough, 1983; Zampighi *et al.*, 1989). In a novel study, Gruijters (1989) reports that MP26 is associated only briefly with lens fiber junctions during their assembly. The antisera have also been used to show that although MP26 and MP70 appear to be lens-specific proteins, the 27-kDa molecule is not unique to the liver, localizing to junctions in other tissues as well, including stomach, exocrine pancreas, renal tubules, and brain (Dermietzel *et al.*, 1984; Paul, 1985). On Western blots, some antisera have proven to be specific for the 27-kDa molecule, whereas others will cross-react with junctional proteins in other tissues (Goodenough *et al.*, 1988; Hertzberg and Skibbens, 1984). Taken together, these antibody data corroborate the results from the Edman degradation studies: gap junctions between different cell types may be made from different members of a protein family that share some structure, but also contain unique domains.

The isolated junction protein preparations and the antibody probes have also been utilized in a variety of functional assays to demonstrate that the liver 27-kDa protein is indeed a channel-forming protein. Isolated rat liver 27-kDa polypeptide, when reconstituted into artificial lipid bilayers, produces single-channel activity similar to that seen in studies of isolated cell pairs (Young *et al.*, 1987; Spray *et al.*, 1986a; Harris, 1991). Intracellularly injected anti-liver 27-kDa antibodies resulted in electrical and dye uncoupling in different cell systems (Warner *et al.*, 1984; Hertzberg *et al.*, 1985).

V. Molecular Cloning of Gap Junction Proteins: The Connexins

In 1986, Paul (1986) cloned a cDNA for the rat liver 27-kDa protein by antibody screening of a bacteriophage expression library, and Kumar and Gilula (1986) cloned a cDNA for its human counterpart by hybridization screening with an amino-terminal oligonucleotide. Both cDNAs encode a polypeptide of 32 kDa. Paul demonstrated by Northern blotting that mRNA corresponding to this cDNA was expressed in rat liver, brain, stomach, and kidney, but was not detectable in heart or lens. That this protein was indeed a component of gap junction structures was confirmed by demonstrating that an antiserum raised against a bacterial fusion protein reacts with gap junctions by electron microscopic immunocytochemistry (Paul, 1986). Expression of the cloned protein in paired *Xenopus* oocytes demonstrated its ability to form functional cell-to-cell channels (Dahl *et al.*, 1987).

Beyer *et al.* (1987) isolated a homologous sequence from a rat heart cDNA library by screening with the rat liver cDNA at reduced stringency. The rat heart cDNA codes for a polypeptide of 43 kDa that contains 43% identical amino acids to the protein cloned from rat liver, but it contains two regions with many more identical residues (see below). The amino-terminal sequence predicted from the heart clone matches that determined by Edman degradation of the major protein in isolated rat heart gap junctions (Manjunath *et al.*, 1987; Nicholson *et al.*, 1985). Morphological and functional proof that this protein forms cardiac gap junctional channels was provided by immunocytochemistry using specific anti-peptide antisera (Beyer *et al.*, 1989; Yancey *et al.*, 1989) and by expression in *Xenopus* oocytes (Swenson *et al.*, 1989; Werner *et al.*, 1989).

Cloning of this second gap junction protein confirmed previous suggestions that there was a family of related gap junction proteins. Many of these proteins are not uniquely expressed in a single tissue (Paul, 1986; Beyer *et al.*, 1987). The mobilities of these proteins on SDS–PAGE may vary with electrophoresis conditions (Green *et al.*, 1988). Therefore, we abandoned previous descriptions of gap junction proteins based on tissue of origin or electrophoretic mobility and proposed a new operational nomenclature using the generic term "connexin" for the protein family, with an indication of species (as necessary) and a numeric suffix designating the molecular mass in kilodaltons (Beyer *et al.*, 1987, 1988, 1990). Thus, the 27-kDa protein from rat liver is termed rat connexin32, the 43-kDa protein from rat heart is termed rat connexin43. An alternative, less widely accepted, nomenclature system that assigns Greek letters and numbers to different gap junction proteins to emphasize structural differences has been suggested (Gimlich *et al.*, 1990).

Further investigations have allowed the cloning of other sequences that may explain some of the gap junctional diversity in different systems predicted by previous biochemical and physiological experiments. A second cDNA from rat liver that codes for the 21-kDa protein has been cloned (Zhang and Nicholson, 1989); this cDNA codes for a 26-kDa protein, which is thus termed connexin26. *Xenopus* connexin38, which was cloned from an embryo library (Ebihara *et al.*, 1989; Gimlich *et al.*, 1990) and is expressed only in oocytes and early embryos, appears to correspond to the voltage-dependent gap junction of amphibian blastomeres (Harris *et al.*, 1981). The 70-kDa lens protein MP70 has not been cloned, but a cDNA that predicts a 46-kDa polypeptide (connexin46) has been isolated from a rat lens cDNA library that shares exact amino acid sequence with the N-terminus of MP70 (Beyer *et al.*, 1988; Goodenough *et al.*, 1990). The exact relationship of MP70 to connexin46 is unclear; the anti-MP70 monoclonal antibody and a polyclonal antiserum directed against a connexin46 synthetic peptide react with different polypeptides on Western blots of lens proteins (Goodenough *et al.*, 1990). Northern analysis of lens RNA probed with connexin46 reveals a single 3.0-kb mRNA. Recently, Rup and Beyer (1991) cloned a related sequence, connexin56, which is expressed in chick lens.

The search for other homologous sequences has identified further members of the connexin gene family. Beyer (1990) used low stringency hybridization screening of a chick embryo cDNA library with a connexin43 probe to clone the novel sequences connexin42 and connexin45, which are expressed in the developing heart and other tissues. Kanter *et al.* (1992) used the polymerase chain reaction (PCR) and genomic cloning to isolate mammalian counterparts of these connexins, dog connexin40, and dog connexin45. Hoh *et al.* (1991) cloned connexin31 from rat genomic DNA and demonstrated that it is most abundantly expressed in placenta. Genomic cloning and PCR amplification of genomic DNA have led to the identification of further expressed connexins, including connexin31.1, connexin33, and connexin37 (Haefliger *et al.*, 1992; Willecke *et al.*, 1990; K. E. Reed and E. C. Beyer, unpublished observations).

Some connexins have been cloned from multiple different species. Connexin43 or its close homolog has been cloned from *Xenopus* (Gimlich *et al.*, 1990), chick (Musil *et al.*, 1990a), human (Fishman *et al.*, 1990), mouse (Beyer and Steinberg, 1991), and cow (Lash *et al.*, 1990) as well as rat (Beyer *et al.*, 1987). The sequences are extremely similar: the mammalian connexin43 proteins show ≥97% amino acid identity; the chick protein has 92% identical amino acids to the rat; the *Xenopus* protein has 87%

identical amino acids to the rat. Many of the substitutions are conservative. The amphibian homolog of rat connexin32 (*Xenopus* connexin30) contains 71% identical amino acids to the rat or human proteins (Gimlich *et al.*, 1988).

Availability of specific DNA (and antibody) probes for different connexins has allowed determination of which gap junction sequence is expressed by different cells and tissues. Nearly all connexins are expressed in multiple locations. Connexin43 appears to be the most ubiquitous connexin: it is expressed by cardiac myocytes, myometrial and other smooth muscle cells, ovarian granulosa cells, endothelial cells, lens epithelial cells, pancreatic acinar cells, astrocytes, renal tubular epithelial cells, macrophages, connective tissue cells, fibroblasts and many other tissue culture cell lines, and hepatocellular carcinoma lines (Moore *et al.*, 1991; Beyer *et al.*, 1989; Dermietzel *et al.*, 1989a, 1991; Larson *et al.*, 1990; Crow *et al.*, 1990; Meda *et al.*, 1991; Spray *et al.*, 1991b; Che *et al.*, 1989; Luke *et al.*, 1989; Giaume *et al.*, 1991). Other connexins like connexin26 and connexin32 are widely expressed (Paul, 1986; Zhang and Nicholson, 1989). Thus far, there do not seem to be any clear rules that explain connexin distribution. Interestingly, some cells express multiple connexins, potentially providing an additional complexity to the regulation of intercellular coupling by those cells. In the liver, connexin32 and connexin26 are both found together in the same junctional plaques, but show different distributions throughout the hepatic lobule, connexin26 being more concentrated in the periportal zones (Nicholson *et al.*, 1987; Traub *et al.*, 1989). Cardiac myocytes appear to contain three connexins in similar locations (Kanter *et al.*, 1992). Connexin32 and connexin43 are both found in renal proximal tubule cells, but in different plaques (Beyer *et al.*, 1989). A7r5 smooth muscle cells express both connexin40 and connexin43 (Beyer *et al.*, 1992), which may explain the multiple distinct channels observed in these cells (Moore *et al.*, 1991). A compilation of the different cloned connexins is shown in Table I.

Figure 4 shows a representation comparing the primary sequences of several different connexin polypeptides. All identified connexins contain two homologous regions, one at the beginning of the protein and one in the middle. The other portions of the connexins (marked A and B) are unique in each connexin, varying in both length and sequence. Gilula and colleagues (Gimlich *et al.*, 1990; Risek *et al.*, 1990) and Bennett *et al.* (1991) have tried to devise classification schemes that emphasize the relatedness of different connexins by grouping together connexins that have a short A region like connexin32 (Group I in Bennett's terminology, β in Gilula's) as opposed to a longer A region like connexin43 (group II in Bennett's terminology, α in Gilula's). However, as seen in Fig. 4, there is

TABLE I
Cloned Connexins

Connexin	Species from which cloned	Tissue/cell type distribution of mRNA expression
Cx26	Rat	Hepatocytes, pineal, brain, leptomeninges, kidney, intestine, lung, spleen, stomach, testes, exocrine pancreas, endometrium
Cx31	Rat	Placenta, skin, Harderian gland
Cx31.1	Rat	Skin
Cx32	Rat, human, *Xenopus*[a]	Hepatocytes, stomach, brain, kidney, exocrine pancreas, postgastrulation embryo
Cx33	Rat	Testis
Cx37	Rat, human	Lung, endothelium
Cx38	*Xenopus*	Oocyte, early embryo
Cx40/Cx42[b]	Rat, dog, chick	Lung, vascular smooth muscle, myocardium
Cx43	Rat, mouse, human, cow, chick, *Xenopus*	Myocardium, myometrium, testis, ovarian granulosa cells, lens epithelium, endothelium, smooth muscle, fibroblasts, astrocytes, immature oocyte, macrophages, pancreatic islet cells
Cx45	Chick, dog	Myocardium
Cx46[c]	Rat	Lens fibers
Cx56[c]	Chick	Lens

[a] *Xenopus* homolog of rat/human connexin32 is *Xenopus* connexin30.
[b] Connexin40 in rat, dog, and mouse appear to be homologous to chick connexin42.
[c] The relation between rat connexin46, chick connexin56, and the lens membrane protein MP70 has not been clarified.

a wide range of different sizes, and it has not yet been demonstrated that this is a physiologically important distinguishing feature among gap junction proteins.

VI. Connexin Topological Structure

The amino acid sequences derived from the cloned cDNAs have been used to predict the structures of the connexins. Hydropathy plots of connexin32 predict four hydrophobic domains, with a large carboxyl-terminal hydrophilic tail. Three smaller hydrophilic domains separate the hydrophobic regions (Paul, 1986). These data, together with that from

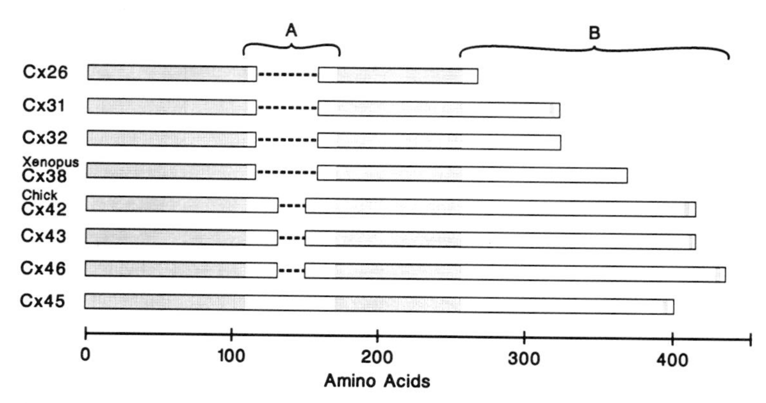

FIG. 4 Comparison of the primary amino acid sequences of several cloned connexins. Regions containing many identical amino acids are shaded, those that are unique are not shaded. Dashes represent spaces added to optimize alignment. Each connexin contains two stretches of ~100 amino acids that are highly conserved and two unique regions marked A and B. Several connexins also contain a short serine-rich region near the carboxyl terminus.

proteolysis studies of isolated junctions, have been used to construct topology models for the relation of this polypeptide to the junctional plasma membrane, assuming that each hydrophobic domain represents a transmembrane segment of the molecule (Beyer *et al.*, 1987; Zimmer *et al.*, 1987). This model has been tested by examining the protease sensitivity of isolated liver gap junctions and by the mapping of site-specific antisera by immunocytochemistry. The controlled proteolytic cleavage has demonstrated that both the N- and C-termini of connexin32 face the cytoplasm and that an additional cytoplasmically accessible proteolytic site is located between the second and the third transmembrane segments (Zimmer *et al.*, 1987; Hertzberg *et al.*, 1988). Antisera against synthetic oligopeptides representing various segments of connexin32 have been raised and used to map the topology in the electron microscope (Goodenough *et al.*, 1988; Milks *et al.*, 1988). These studies have confirmed the proteolysis studies. They have shown that the amino terminus, the carboxyl terminus, and a loop in the middle of the protein are all located on the cytoplasmic face of the junctional membrane. And, to the degree that the harsh experimental conditions do not alter the protein topology, they also demonstrate that both predicted extracellular domains of connexin32 can be detected on the extracellular surfaces of the junctional membranes. This connexin model is illustrated in Fig. 5.

Studies have suggested that connexin43 has a similar structure. Hydropathy analysis reveals a set of four hydrophobic domains similar to those in connexin32. Modeling of connexin43 in a manner comparable to that of

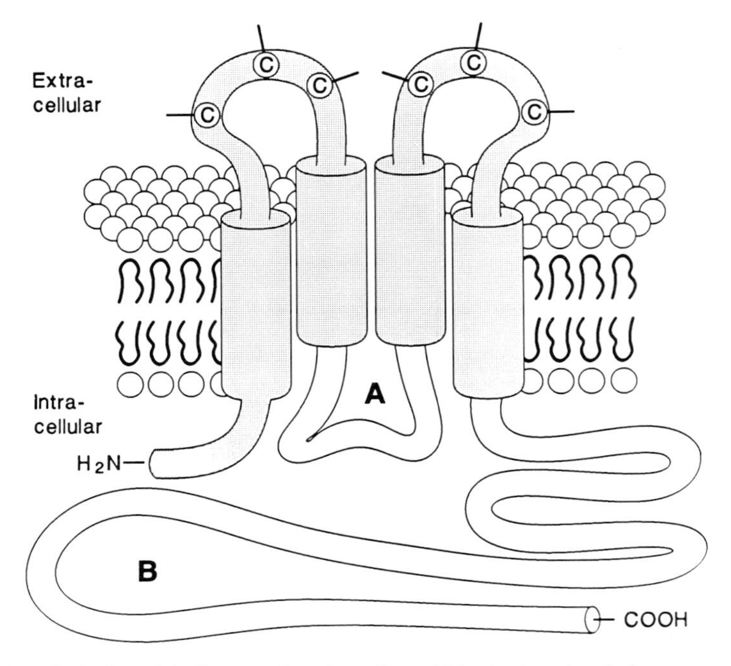

FIG. 5 Topological model of connexin orientation within the junctional plasma membrane. Shaded regions represent those sequences shared among all connexins. Each extracellular loop contains three cysteines (C) that are conserved in all connexins. The unshaded, predicted cytoplasmic domains (A and B), correspond to the unique, connexin-specific regions shown in Fig. 4.

connexin32 suggests that those hydrophobic/transmembrane domains and the extracellular regions correspond to the homologous regions between connexin43 and -32 (Beyer *et al.,* 1987). These are represented as the shaded regions in Figs. 4 and 5. Proteolysis studies of isolated cardiac gap junctions combined with the localization of anti-synthetic peptide antisera have confirmed the three separate cytoplasmic domains and an overall membrane topology similar to that of connexin32 (Beyer *et al.,* 1989; Yancey *et al.,* 1989).

Such studies mapping sites by proteolysis and immunocytochemistry have not been conducted for the other connexins, but sequence comparison and analogy to the connexin32/43 studies allow some prediction of their structure. All of the sequenced connexins share the four hydrophobic domains corresponding to the transmembrane regions in connexin32. Comparison of the primary sequences of all the known connexins reveals that the two predicted extracellular domains are the most conserved regions, each containing three invariant cysteines. The transmembrane domains are somewhat less well conserved. The cytoplasmic domains, with the exception of the short N-terminal region, differ markedly between connexins both in sequence and in length. This is summarized in Figs. 4 and 5, in which the two cytoplasmic domains are labeled A and B. *Xenopus* connexin38 has a unique feature: it contains an additional predicted hydrophobic domain in the C-terminal tail (Ebihara *et al.,* 1989). It is not known whether this fifth hydrophobic segment represents an additional transmembrane span placing the C-terminus of connexin38 extracellularly.

VII. Connexin Gene Structure

Relatively little is known about the genomic organization of connexin genes. Some characterization and genomic sequence have been published concerning the genes for rat connexin32 (Miller *et al.,* 1988) and human connexin43 (Fishman *et al.,* 1991c). These studies show that both connexin genes have a rather similar structure, which is illustrated in Fig. 6. Each is composed of two exons: the first is small and contains only 5′-untranslated sequence; the second contains the entire coding sequence. Although the genes for other connexins have not been as well characterized, they may have similar structures, since Southern blots, PCR experiments, and genomic DNA sequencing suggest that their coding sequences are not interrupted by introns (Zhang and Nicholson, 1989; Beyer, 1990; Rup and Beyer, 1991; Kanter *et al.,* 1992; Hoh *et al.,* 1991; Beyer and Steinberg, 1991). The 5'-ends of the connexin32 and connexin43 genes have not yet revealed interesting or important regulatory elements, but analysis using reporter gene constructs has not yet been performed. One interesting result of the genomic studies has been the demonstration of the presence of a processed pseudogene for human connexin43 (Fishman *et al.,* 1990, 1991c).

Some chromosomal mapping of connexin genes has been performed (Willecke *et al.,* 1990; Hsieh *et al.,* 1991; Fishman *et al.,* 1991c). These studies, which produce consistent results, place human connexin26 and connexin46 both on chromosome 13, connexin32 on chromosome X, connexin43 on chromosome 6, connexin 37 and connexin40 both on chromo-

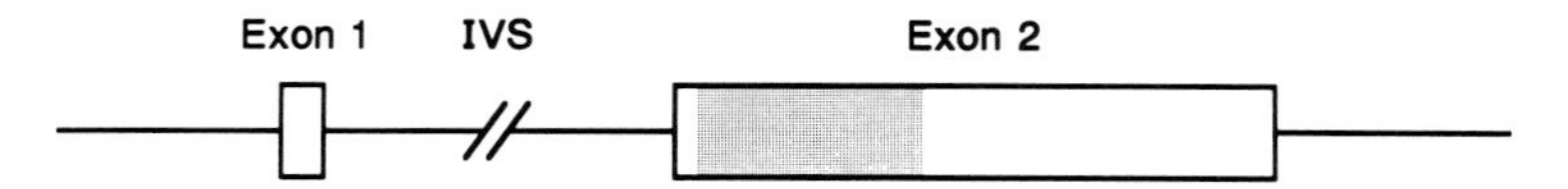

FIG. 6 Connexin gene structure, based on studies of rat connexin32 and human connexin43.

some 1, and the connexin43 pseudogene on chromosome 5. Hsieh *et al.* (1991) report that the homologous mouse loci were assigned to regions of known conserved synteinic groups. Relatively crude subchromosomal assignments have been made. The gap junction genes have not yet been linked to any known genetic disease loci.

VIII. Regulation of Connexin Gene Expression

The availability of specific connexin DNA and antibody probes has made possible new molecular approaches to understanding gap junction regulation. Although there is little mechanistic comprehension of controls of gap junction expression, several systems have been explored where modulation of junctional communication may have biological importance and where it may be amenable to further study. Alterations of connexin gene expression have been examined in uterine myometrium, in developing embryos and tissues, in response to tissue injury (partial hepatectomy), and in response to chemically induced carcinogenesis.

A. Connexin43 in Uterine Smooth Muscle.

The onset of labor in mammals is associated with the development of intense, coordinated electrical and contractile activity in the uterine myometrium, from a previously relatively inactive state. This period is marked by the abrupt development of large numbers of morphologically (Dahl and Berger, 1978; Garfield *et al.,* 1977, 1978) and physiologically (Sims *et al.,* 1982) recognizable gap junctions, which likely coordinate the electrical communication between these smooth muscle cells. This production of gap junctions is dependent on estrogens and may be blocked by progesterone and requires synthesis of both RNA and protein (MacKenzie and Garfield, 1985a, b; Burghardt *et al.,* 1984a,b; Cole and Garfield, 1985). Beyer *et al.* (1987, 1989) have identified connexin43 as a component of myometrial gap junctions in estradiol-treated rats. Risek *et al.* (1990) have now provided evidence that increased expression of connexin43 may be responsible for this burst of gap junction synthesis. They report an increase of connexin43 mRNA levels in the myometrium of term pregnant rats, which reaches a maximum of 5.5 times that of control nonpregnant animals on the day prior to parturition. Immunoreactive connexin43 protein in frozen sections of myometrium peaked 24 hr later. The myometrial connexin43 mRNA and protein declined rapidly following delivery. Risek *et al.* (1990) found similar changes in connexin43 mRNA and protein expression in the ovarian stroma, but not in myocardium, suggesting tissue-specific and potentially steroid-responsive regulation. Winterhager *et al.* (1991) have found very similar increases in connexin43 in myometrium. Lye *et al.* (1991) have extended these observations by investigating hormonal effects on connexin43 expression. They find that, in pregnant animals, administration of progesterone just prior to term inhibits the expected increase in connexin43 transcripts and administration of the progesterone receptor antagonist RU486 on Day 15 resulted in a premature increase in connexin43 transcripts. They also found modulation of myometrial connexin43 expression by estrogen and progesterone in ovariectomized, nonpregnant rats. In contrast to these studies, Lang *et al.* (1991) found high levels of connexin43 in the uterine tissue of nonpregnant rats or early in pregnancy, with only very modest increases in connexin43 mRNA at term. Although some of the discrepancies may be due to differences in mRNA quantitation or tissue procurement, these results are difficult to reconcile. All studies suggest that there may also be post-transcriptional controls of connexin43 gap junction synthesis, since even with a 5-fold increase in mRNA, it is hard to explain an increase of 50- to 100-fold in morphometrically quantitatable gap junctions. Hendrix *et al.* (1991) have provided some suggestion that there may be pools of cytoplasmic connexin43 protein at a time prior to when it is detectable in gap junction structures.

B. Connexin26 in Endometrium

The studies of Risek *et al.* (1990) and Winterhager *et al.* (1991) also demonstrated that connexin26 is a component of endometrial gap junctions. Its expression was modulated in endometrium both during pregnancy and during the estrus cycle, with peak levels during the estrogen-dominated phase, suggesting steroid hormone regulation of expression of this connexin as well. Risek and Gilula (1991) have now expanded their examination of connexin expression in the rat implantation chamber, suggesting exquisite regulation of these proteins involved in feto-maternal communication.

C. Ovarian Follicle Gap Junctions.

Prior to ovulation, intercellular coupling and gap junction structures between follicular cells and the oocyte are lost (Gilula *et al.*, 1978). This uncoupling might be involved in releasing the oocyte from meiotic arrest (Larsen *et al.*, 1987). In mammals, this downregulation of gap junctions results from the preovulatory luteinizing hormone surge and can be induced in immature animals by gonatrophin injection (Gilula *et al.*, 1978). In *Xenopus*, this event may be controlled by progesterone; Gimlich *et al.* (1990) showed that *in vitro* maturation of *Xenopus* oocytes with progesterone leads to a decrease of connexin43 mRNA levels to less than 9% of control values within 3 hr. *In vivo* injection of human chorionic gonadotrophin produced similar results. They suggest that the loss of connexin43 gap junctions between follicular cells and oocyte is responsible for the observed uncoupling.

D. Gap Junctions in Early Embryos.

It has been frequently suggested that gap junctions might play an important role in early embryonic development by the formation of communication compartments, which would lead to pattern formation established by concentration gradients of diffusible morphogens (Guthrie and Gilula, 1989; Lo, 1985). This has lead to investigations of connexin expression in early embryos. In the *Xenopus* embryo, connexin38 mRNA (a maternal product) is plentiful in the oocyte, then declines to an undetectable level by the neurula stage; *Xenopus* connexin30 (the homolog of mammalian connexin32) is first detected at the midgastrula stage, then increases with further development; connexin43 expression begins with organogenesis (Ebihara *et al.*, 1989; Gimlich *et al.*, 1988, 1990). In the mouse, initial activation of the zygotic genome results in transcription of connexin43, but not connexin32 (Barron *et al.*, 1989; Valdimarsson *et al.*, 1991), yet preimplantation embryos develop gap junctions with connexin32 (Lee *et al.*, 1987; Bevilacqua *et al.*, 1989), indicating that these early gap junctions are derived from maternal stores.

E. Gap Junctions in the Developing Heart.

Although electrical coupling through gap junctions is crucial to normal impulse propagation in the mature heart, only a little is known about the assembly of these intercellular junctions during development. There have been some descriptive morphological studies (Gros *et al.*, 1978, 1979). Beyer (1990) demonstrated that in the developing chick heart, in addition to connexin43, there are two additional gap junction proteins, connexin42 and connexin45. Connexin45 mRNA is present primarily in early embryos and falls to 10-fold lower levels in the adult heart. Three studies have examined expression of connexin43 in the developing mammalian heart (Fishman *et al.*, 1991a; van Kempen *et al.* 1991; Fromaget *et al.*, 1990).

Connexin43 mRNA is present at the earliest cardiac stage tested (11 days postcoitum in the mouse), but rises 5- to 8-fold by about 1 week postpartum, before falling to lower levels in juvenile and adult hearts. Fishman *et al.* (1991a) used immunoblots to demonstrate that connexin43 protein showed parallel, but temporally delayed, changes. Van Kempen *et al.* (1991) demonstrate that gap junctions demonstrable by immunostaining with anti-connexin43 are assembled with a similar time course. However, it is a bit hard to reconcile this peak of connexin expression in neonatal animals with the previous observations that adult myocytes have a much greater total junctional conductance than neonatal myocytes. Does this represent organization of these junctions into the intercalated disc?

F. Response to Tissue Injury (Partial Hepatectomy).

Revel and colleagues (Yee and Revel, 1978; Meyer *et al.,* 1981) have extensively examined the changes in gap junction structures and connexin expression that follow partial hepatectomy in the rat. Beginning about 20 hr following surgery, there is a dramatic loss of gap junctions and intercellular communication between hepatocytes, but by 48 hr following partial hepatectomy, they return to control levels. Using quantitative immunoblots, Traub and co-workers (Traub *et al.,* 1983, 1989; Dermietzel *et al.,* 1987) demonstrated that connexin32 and connexin26 protein levels fell and then rose in a time course consistent with these morphological changes. They found a similar loss and reacquisition of connexin mRNA and protein in wounded primary cultures of primary hepatocytes, which showed an inverse correlation with BrdU-labeling (Dermietzel *et al.*, 1987), suggesting a relation between the loss of gap junctions and cell proliferation. However, multiple factors must control connexin expression in the cultured hepatocytes; Spray *et al.* (1986c) demonstrated that they could modulate connexin32 expression by treatment with different extracellular matrix components.

G. Altered Connexin Expression in Carcinogenesis.

The previous studies linking transformation and loss of gap junctional communication (Loewenstein, 1979) have lead several workers to investigate connexin expression in several models of carcinogenesis. These results are largely consistent with previous functional and morphologic observations. In several models of chemical-induced liver carcinogenesis, steady-state connexin32 mRNA levels and connexin32 immunoreactive protein are reduced (Beer *et al.,* 1988; Janssen-Timmen *et al.,* 1986;

Fitzgerald *et al.*, 1989). It has also been observed that some tumors no longer synthesize the connexin(s) made in their normal cellular counterparts, but switch to connexin43 synthesis (Oyamada *et al.*, 1990). In a breast cancer system, Sager and colleagues used subtractive hybridization to demonstrate that connexin26 and connexin43 were expressed in normal mammary epithelium, but not in the malignant counterparts (Lee *et al.*, 1991). Bertram and colleagues have been investigating the effects of retinoids on transformed cells and have demonstrated that in 10T1/2 cells retinoid-enhanced cellular coupling corresponds to increased connexin43 mRNA and protein levels (Rogers *et al.*, 1990).

IX. Connexin Protein Biosynthesis and Post-Translational Modifications

Very little is known about the complex biosynthetic pathway by which connexins are synthesized, inserted into the membrane, assembled into connexons (hexamers), connected to connexons from the adjacent cell, and aggregated into gap junctional plaque. In 1981, Fallon and Goodenough (1981) used *in vivo* labeling with [^{14}C]bicarbonate to demonstrate that the protein subunits of gap junctions are very dynamic, having a half-life of approximately 5 hr. Since then, the development of specific anti-connexin antibody probes have made possible *in vitro* studies of connexin biosynthesis by metabolic labeling and immunoprecipitation. These studies have demonstrated similar short half-lives of 1–4 hr for connexin32 and connexin43 in tissue culture cells of different sources (Musil *et al.*, 1990a; Crow *et al.*, 1990; Traub *et al.*, 1989; Laird *et al.*, 1991; Hertzberg *et al.*, 1989; Larson *et al.*, 1991). They have also examined covalent modifications of the connexins during biosynthesis. Glycosylation has never been demonstrated for any of the gap junction proteins; indeed, the connexins contain no predicted extracellular N-linked glycosylation sites. No definitive evidence for lipid-linked modifications has been published, although it has been suggested (Traub *et al.*, 1989). The most detailed studies have been of connexin phosphorylation, probably in response to the physiologic experiments suggesting roles for protein kinases in regulating communication.

A. Connexin32 Phosphorylation.

Saez *et al.* (1986) demonstrated by metabolic labeling with [^{32}P]orthophosphate in primary hepatocyte cultures that connexin32 is a phosphoprotein and that treatment of the cells with 8-bromo-cAMP produced a 1.6-fold increase in ^{32}P incorporation. This phosphorylation of connexin32 occurred on serine residues. Isolated liver gap junctions could also be phosphorylated on serines by purified cAMP-dependent kinase, but with low stoichiometry (0.025–0.07 mol phosphate/mol connexin protein) (Saez *et al.*, 1986; Takeda *et al.*, 1989). Takeda *et al.* (1989) have shown that protein kinase C also produces a stimulation of low level phosphorylation of connexin32. Using synthetic peptide substrates, Saez *et al.* (1990) have suggested that Ser233 is the phosphorylated residue in connexin32. However, it is still not demonstrated that any of this connexin32 phosphorylation directly affects gap junctional intercellular coupling.

B. Serine Phosphorylation of Connexin43

Pulse–chase studies have demonstrated that connexin43 is initially synthesized as a 42-kDa polypeptide that subsequently is post-translationally modified to forms with slightly slower mobility on SDS–PAGE by the addition of phosphate on serine residues (Musil *et al.*, 1990a; Crow *et al.*, 1990; Filson *et al.*, 1990). The exact phosphorylated residues have not been identified, but there apparently are multiple phosphates added to each mole of connexin43, and there is a serine-rich sequence near the carboxyl terminus with multiple potential sites (Beyer *et al.*, 1987). (Indeed, other connexins have similar sequences, suggesting that they might be similarly modified. See Fig. 4.) The kinase(s) responsible for connexin43 phosphorylation or the exact phosphorylated serines have not been determined. Modulation of connexin43 phosphorylation by agents that alter intercellular coupling has not been examined.

Musil *et al.* (1990b) have further investigated connexin43 serine phosphorylation. They showed that the noncommunicating cell lines L929 and S180 constitutively synthesized connexin43 mRNA and protein. But, the connexin43 protein in these cells was not present in cell surface gap junctional plaques, but rather it accumulated intercellularly. The connexin43 was incompletely phosphorylated (only to their "P_1" form). However, transfection of the S180 cells with the cell adhesion molecule LCAM (E cadherin) restored gap junctional communication, full phosphorylation of connexin43, and expression in cell surface gap junctions. These findings suggest a relation between the ability of cells to fully phosphorylate connexin43 and the ability to form communicating junctions. They also suggest a hierarchy of events in the formation of intercellular junctions: a primary cell adhesion event is required prior to formation of gap junctions. Similar observations regarding the requirement of cadherin-mediated cell adhesion for development of gap junctions have also been made in epidermal cells (Jongen *et al.*, 1991).

C. Tyrosine Phosphorylation of Connexin43

The large body of data demonstrating that intercellular communication is abolished in fibroblasts infected with RSV (Atkinson *et al.*, 1981; Azarnia *et al.*, 1988) has lead several investigators to investigate the effects of $pp60^{src}$ on connexin43-expressing cells. Crow *et al.* (1990) demonstrated that RSV infection uncoupled vole fibroblasts and led to the incorporation of phosphate in connexin43 tyrosine residues. Filson *et al.* (1990) extended these observations by showing that each molecule of phosphorylated connexin43 contained both phosphoserine and phosphotyrosine and by showing that the ability of *src* variants to abolish cellular communication correlated with tyrosine phosphorylation of connexin43. Swenson *et al.* (1990) showed that in a *Xenopus* oocyte expression system coexpression of $pp60^{v\text{-}src}$ with connexin43 reduced cell–cell coupling and led to tyrosine phosphorylation of connexin43. But, site-directed mutagenesis of Tyr265 in connexin43 eliminated the tyrosine phosphorylation and the depression of communication.

D. Connexin/Connexon Biosynthesis and Assembly

Little is known about assembly of gap junction plaques. There is a large body of electron microscopic data describing structures called ''formation plaque'' and annular gap junctions that have been proposed to be intermediates in the assembly and degradation of gap junctions, respectively (Larsen, 1983; Mazet *et al.*, 1985). Their relation to biochemically assayed connexins is not determined. Biosynthetic studies utilizing drugs such as monensin and Brifeldin A suggest that the intercellular trafficking of connexin43 is similar to that of other membrane proteins (Larson *et al.*, 1991; Puranam *et al.*, 1990). Cellular sites of phosphorylation events have not been determined. Experiments in a Novikoff hepatoma system (Meyer *et al.*, 1990) in which gap junction synthesis can be synchronized demonstrate that application of an antibody to extracellular domains in connexin43 can block gap junction formation, demonstrating that at least transient exposure of connexins or connexons free in the plasma membrane does occur.

X. Functional Expression Studies Using Cloned Connexin Sequences

The availability of the cloned connexin DNAs has facilitated more detailed investigation of their functional properties by application of various expression systems. A powerful system has been developed (Werner *et al.*, 1985; Dahl *et al.*, 1987) involving the expression of injected gap junction mRNAs in paired *Xenopus* oocytes and the subsequent study of the channels induced between the two cells by voltage clamp techniques. In this system, homologous injection of two oocytes with mRNA for rat connexin32 (Dahl *et al.*, 1987), connexin43 (Swenson *et al.*, 1989; Werner *et al.*, 1989), or connexin26 (Barrio *et al.*, 1990) or *Xenopus* connexin38 (Ebihara *et al.*, 1989) induces development of high junctional conductance between the paired cells up to three orders of magnitude above background levels. This demonstrates that each of these four connexins is individually sufficient to form communicating channels. Junctional conductance induced by the exogenous mRNA can be distinguished from the endogenous background channels by differences in voltage sensitivity or by injection of antisense oligonucleotides to *Xenopus* connexin38 sequences.

This system has been used to begin to determine different biophysical properties of the channels produced from different connexins. For instance, *Xenopus* connexin38 forms channels that show a pronounced voltage-dependent inactivation, similar to that previously observed in amphibian blastomeres (Ebihara *et al.*, 1989). Expression of rat connexin26 in the oocytes has demonstrated that it forms channels sensitive to both transmembrane and transjunctional potentials (Barrio *et al.*, 1990). Rat connexin46, when expressed in the *Xenopus* oocytes, does not make communicating channels, but rather appears to make transmembrane channels between the cytoplasm and the extracellular space (Goodenough *et al.*, 1990). It is unclear whether this is an artifact of this expression system or whether it implies that this is not a cell–cell communication channel.

The oocyte system has also been exploited to investigate the possible formation of hybrid junctions composed of two different connexins. In culture, junctions between heterologous cells have been documented a number of times (Gilula *et al.*, 1972; Subak-Sharpe *et al.*, 1969; Michalke and Loewenstein, 1971; Fentiman *et al.*, 1976; Flagg-Newton and Loewenstein, 1980), but in most cases the connexins expressed by these cells are not known. Because of the high degree of conservation of their extracellular domains, it seemed possible that heterologous junctions could form (containing two different connexins located on opposite sides of the junction). To test this hypothesis, oocytes were injected with either connexin32 or connexin43 and then paired as connexin32/43; these pairs of oocytes displayed voltage-insensitive high junctional conductances similar to oocytes injected with the same connexin, indicating that heterologous channels had formed (Swenson *et al.*, 1989; Werner *et al.*, 1989). Connexin43 can also form high conductances with the oocyte's endogenous channels in heterologous pairs. In this case, the resultant hybrid channels are asymmetrically voltage sensitive, showing physiological properties of both the endogenous and the connexin43 channels. This hybrid channel has been proposed as a molecular model of the rectifying electrical synapse.

The oocyte system is also being exploited to examine mutagenized connexin constructs to determine the functional properties of different portions of the connexin sequences. The studies of Swenson *et al.* (1990) have pinpointed a tyrosine in connexin43 that is the target for $pp60^{v\text{-}src}$. Mutants of connexin32 have shed some light on the importance of extracellular disulfide bonds in cell–cell channel formation (Werner *et al.*, 1991; Dahl *et al.*, 1991). Preparation of chimeric molecules with portions of connexin32 and connexin43 suggest that an interaction between the cytoplasmic loop and tail of a single connexin type may be important for functional channel formation (Bruzzone *et al.*, 1990).

Functional properties of the cloned connexins have also been investigated by the stable transfection of a communication-deficient cell line with connexin-containing constructs, which will restore intercellular coupling (Eghbali *et al.*, 1990). Although this system is less amenable to screening multiple connexin mutants, it does make possible measurements of single gap junctional channels and assay of junctional coupling in a less artificial state, and avoids some of the problems of the endogenous amphibian channels. Such experiments have confirmed the differences in biophysical properties of different connexins: connexin32 has a unitary conductance of ~120 pS and is mildly voltage sensitive (Eghbali *et al.*, 1990); connexin43 has a single channel conductance of ~50 pS (Fishman *et al.*, 1990). Veenstra *et al.* (1992a,b) have examined the properties of three transfected chick connexins, demonstrated that they make channels with different conductances and voltage-dependence properties, and shown that they correspond to the properties previously observed in chick embryo ventricular myocytes. Fishman *et al.* (1991b) produced mutants of connexin43 with progressive deletions of the carboxyl tail with different unitary conductances, suggesting that this portion of the protein contributes to determining this property. Certainly, many more structure–function studies are yet to come.

XI. Other Roles for Connexins and Nonconnexin Components of Gap Junctions

This review has concentrated on the connexin family of proteins, several of which have been definitively demonstrated to be cell–cell channel–forming components of gap junctions. However, it is important to consider briefly two other concepts. (*a*) Connexins might also play other roles. (*b*) Other proteins may also be components of gap junctions.

It has not been proven that any connexin has another function than as a gap junctional channel protein, but there are several interesting observations. Goodenough *et al.* (1990) demonstrated that connexin46 when expressed in *Xenopus* oocytes will form transmembrane, but not cell–cell, channels. Harris (1991) demonstrated that connexin32 when reconstituted into single lipid bilayers formed single membrane channels. Beyer and Steinberg (1991) have suggested that connexin43 is the ATP-induced plasma membrane pore of mouse macrophages.

Several studies have suggested that plants might contain connexin-related proteins. Anti-connexin32 and anti-connexin43 antisera have been shown to react with plant polypeptides, and the anti-connexin43 immunoreactivity localizes to plasmodesmata, structures specialized for cellular communication (Meiners and Schindler, 1987; Yaholom *et al.*, 1991). The protein antigenically related to connexin32 has been cloned from *Arabidopsis* (Meiners *et al.*, 1991). Its DNA sequence shows no relation to the connexins; its amino acid sequence shares a number of identical amino acids within the cytoplasmic loop region of connexin32, whereas there is little similarity to the extracellular or transmembrane domains shared in all connexins.

It seems likely that the connexin components of gap junctions interact with other membrane, cytoskeletal, or cytoplasmic components. However, because the connexin polypeptides have been difficult to solubilize without harsh, denaturing conditions, no such proteins have been demonstrated to coisolate with a connexin. Several indirect experiments have suggested that calmodulin can interact with gap junctions and with connexin32 (Zimmer *et al.*, 1987; Peracchia, 1985). The experiments with LCAM suggest a possible interaction between this adhesion molecule and connexin43 (Musil *et al.*, 1990b; Jongen *et al.*, 1991).

Other, nonconnexin, proteins have been proposed to be gap junction proteins. The lens protein MP26, which shows no sequence similarity to the connexins (Gorin *et al.*, 1984), was once widely believed to be a lens gap junction protein, because its abundance in lens membranes parallels the abundance of gap junction structures in that tissue (Bloemendal *et al.*,

1972). However, as discussed above, recent immunocytochemical studies have suggested that it is not localized to classical gap junction structures. However, this protein does form transmembrane channels (Ehring and Hall, 1991; Zampighi *et al.*, 1982). One suggested role has been that of volume regulation, but it is also possible that by forming transmembrane channels in adjacent cells that are narrowly separated, it could still function to allow cell–cell communication. Another nonconnexin lens protein, MP19, that has been isolated and cloned also localizes to gap junctions (Louis *et al.*, 1989; Galvan *et al.*, 1989; Guntekunst *et al.*, 1990). No functional information about the ability of this protein to form channels has yet been obtained; it could be an additional channel protein, or could be an accessory junctional component. Finally, Finbow *et al.* (1983, 1984) have studied a 16-kDa polypeptide that they find coenriches with gap junction structures; however, its protein sequence suggests that it is indeed a component of the vacuolar ATPase (Dermietzel *et al.*, 1989b).

XII. Future Directions: Use of Molecular Tools to Investigate Gap Junction Biology

With the explosive influence of connexin-specific antibody and DNA probes, recombinant expression techniques, and elegant electrophysiological techniques, we shall soon have a detailed molecular understanding of the process of gap junction-mediated intercellular communication. But, the coming challenge for this field is to address the important biological roles of gap junctions. Can we use these tools to elucidate the contribution of gap junctional channels to such processes as embryonic development and cardiac electrical conduction? Are gap junctions of importance in any disease processes?

A. Gap Junctions and Development

Several experiments have utilized anti-connexin antibodies or antisense oligonucleotides to interfere with embryonic development and cellular differentiation (Warner *et al.*, 1984; Lee *et al.*, 1987; Fraser *et al.*, 1987; Bevilacqua *et al.*, 1989). With a better understanding of which connexins are expressed and their spatial and temporal patterns, it should be possible to design more specific experiments, possibly using a transgenic mouse model. Although no insect or invertebrate gap junction sequences have yet been cloned, it will be desirable to pursue developmental questions in such systems that are more amenable to genetic manipulation.

B. Gap Junctions and Electrical Conduction

Alterations in intercellular coupling have been implicated in the development of reentrant ventricular arrhythmias (Spach and Dobler, 1986; Ursell *et al.*, 1985; Luke and Saffitz, 1991). Investigations may uncover alterations of connexin expression in diseased hearts, but it should also be possible to develop testable models of the perturbation of cardiac gap junctions. Use of information about the molecular structure of connexins may lead to the development of more specific pharmacologic agents to modulate cellular electrical coupling. Such drugs would have profound implications for the treatment of arrhythmias or preterm labor.

Acknowledgments

The author thanks Dr.Daniel Goodenough for contributing the electron micrographs of liver gap junctions and Dr. David Paul for discussing recent cloning of novel connexin sequences. Drs. H. Lee Kanter and Diane Rup provided valuable discussions. Research in the author's laboratory is supported by NIH Grants HL45466 and EY08368.

References

Alcala, J., Lieska, N., and Maisel, H. (1975). *Exp. Eye Res.* **21,** 581–589.

Atkinson, M. M., and Sheridan, J. D. (1988). *Am. J. Physiol.* **255,** C674–C683.

Atkinson, M. M., Menko, A. S., Johnson, R. G., Sheppard, J. R., and Sheridan, J. R. (1981). *J. Cell Biol.* **91,** 573–578.

Auerbach, A. A., and Bennett, M. V. L. (1969). *J. Gen. Physiol.* **53,** 211–237.

Aylsworth, C. F., Trosko, J. E., and Welsch, C. W. (1986). *Cancer Res.* **46,** 4527–4533.

Azarnia, R., and Loewenstein, W. R. (1984). *J. Membr. Biol.* **82,** 191–205.

Azarnia, R., and Loewenstein, W. R. (1987). *Mol. Cell Biol.* **7,** 946–950.

Azarnia, R., Reddy, S., Kmiecik, T. E., Shalloway, D., and Loewenstein, W. R. (1988). *Science* **239,** 398–401.

Baker, T. S., Caspar, D. L. D., Hollingshead, C. J., and Goodenough, D. A. (1983). *J. Cell Biol.* **96,** 204–216.

Barr, L., Dewey, M. M., and Berger, W. (1965). *J. Gen. Physiol.* **48,** 797–823.

Barrio, L. C., Suchyna, T., Bargiello, T. A., Roginski, R., Zukin, R. S., Nicholson, B. J., and Bennett, M. V. L. (1990). *Soc. Neurosci. Abstr.* **16,** 185.

Barron, D. J., Valdimarsson, G., Paul, D. L., and Kidder, G. M. (1989). *Dev. Genet.* **10,** 318–323.

Beer, D. G., Neveu, M. J., Paul, D. L., Rapp, U. R., and Pitot, H. C. (1988). *Cancer Res.* **48,** 1610–1617.

Bennett, M. V. L. (1966). *Ann. N.Y. Acad. Sci.* **137,** 509–539.

Bennett, M. V. L. (1977). *In* "Handbook of Physiology, Section 1: The Nervous System" (E. Kandel, ed.), pp. 357–416. Williams & Wilkins, Baltimore.

Bennett, M. V. L., Verselis, V., White, R. L., and Spray, D. C. (1988). *Mol. Cell. Biol.* **7,** 287–304.

Bennett, M. V. L., Barrio, L. C., Bargiello, T. A., Spray, D. C., Hertzberg, E., and Saez, J. C. (1991). *Neuron* **6,** 305–320.
Bevilacqua, A., Loch-Caruso, R., and Erickson, R. P. (1989). *Proc. Natl. Acad. Sci. U.S.A.* **86,** 5444–5448.
Beyer, E. C. (1990). *J. Biol. Chem.* **265,** 14439–14443.
Beyer, E. C., and Steinberg, T. H. (1991). *J. Biol. Chem.* **266,** 7971–7974.
Beyer, E. C., Paul, D. L., and Goodenough, D. A. (1987). *J. Cell Biol.* **105,** 2621–2629.
Beyer, E. C., Goodenough, D. A., and Paul, D. L. (1988). *Mol. Cell. Biol.* **7,** 167–175.
Beyer, E. C., Kistler, J., Paul, D. L., and Goodenough, D. A. (1989). *J. Cell Biol.* **108,** 595–605.
Beyer, E. C., Paul, D. L., and Goodenough, D. A. (1990). *J. Membr. Biol.* **116,** 187–194.
Beyer, E. C., Reed, K. E., Westphale, E. M., Kanter, H. L., and Larson, D. M. (1992). *J. Membr. Biol.* **127,** 69–76.
Bloemendal, H., Zweers, A., Dunia, I., and Benedetti, E. L. (1972). *Cell Differ.* **1,** 91–10.
Bok, D., Dockstader, J., and Horwitz, J. (1982). *J. Cell Biol.* **92,** 213–220.
Brink, P. R. (1983). *J. Membr. Biol.* **71,** 79–87.
Brink, P. R., and Dewey, M. M. (1980). *Nature (London)* **285,** 101–102.
Brink, P. R., and Fan, S.-f. (1989). *Biophys. J.* **56,** 579–594.
Broekhuyse, R., Kuhlman, E., and Stols, A. (1976). *Exp. Eye Res.* **23,** 365–371.
Brower, P. T., and Schultz, R. M. (1982). *Dev. Biol.* **90,** 144–153.
Bruzzone, R., White, T., and Paul, D. L. (1990). *J. Cell Biol.* **111,** 154a.
Burghardt, R. C., Matheson, R. L., and Gaddy, D. (1984a). *Biol. Reprod.* **30,** 249–255.
Burghardt, R. C., Mitchell, P. A., and Kurten, R. (1984b). *Biol. Reprod.* **30,** 249–255.
Burt, J. M. (1987). *Am. J. Physiol.* **253,** C607–C612.
Burt, J. M. (1989). *Am. J. Physiol.* **256,** C913–C924.
Burt, J. M., and Spray, D. C. (1988a). *Proc. Natl. Acad. Sci. U.S.A.* **85,** 3431–3434.
Burt, J. M., and Spray, D. C. (1988b). *Am. J. Physiol.* **254,** H1206–H1210.
Burt, J. M., and Spray, D. C. (1989). *Circ. Res.* **65,** 829–837.
Cascio, M., Gogol, E., and Wallace, B. A. (1990). *J. Biol. Chem.* **265,** 2358–2364.
Caspar, D. L. D., Goodenough, D. A., Makowski, L., and Phillips, W. C. (1977). *J. Cell Biol.* **74,** 605–628.
Caspar, D. L. D., Soskinsky, G. E., Tibbitts, T. T., Phillips, W. C., and Goodenough, D. A. (1988). *Mol. Cell. Biol.* **7,** 117–133.
Chanson, M., Bruzzone, R., Spray, D. C., Regazzi, R., and Meda, P. (1988). *Am. J. Physiol.* **255,** C699–C704.
Che, M., Saez, J. C., and Risley, M. D. (1989). *Biol. Reprod.* **40,** 143.
Cole, W. C., and Garfield, R. E. (1985). *In* "Gap Junctions" (M. V. L. Bennett and D. C. Spray, eds.), pp. 215–230. Cold Spring Harbor Press, Cold Spring Harbor, New York.
Corsaro, C. M., and Migeon, B. R. (1977). *Nature (London)* **268,** 737–739.
Cox, R. P., Krauss, M. R., Balis, M. E., and Dancis, J. (1970). *Proc. Natl. Acad. Sci. U.S.A.* **67,** 1573–1579.
Crow, D. S., Beyer, E. C., Paul, D., Kobe, S. S., and Lau, A. F. (1990). *Mol. Cell. Biol.* **10,** 1754–1763.
Dahl, G., and Berger, W. (1978). *Cell Biol. Int. Rep.* **2,** 381–387.
Dahl, G., Miller, T., Paul, D., Voellmy, R. and Werner, R. (1987). *Science* **236,** 1290–1293.
Dahl, G., Levine, E., Rabadan-Diehl, C., and Werner, R. (1991). *Eur. J. Biochem.* **197,** 141–144.
DeMello, W. C. (1983). *J. Physiol. (London)* **339,** 299–307.
DeMello, W. C. (1989). *Biochim. Biophys. Acta* **1012,** 291–298.
Dermietzel, R., Leibstein, A., Rrixen, U., Janssen-Timmen, U., Traub, O., and Willeke, K. (1984). *EMBO J.* **3,** 2261–2270.

Dermietzel, R., Yancey, S. B., Traub, O., Willecke, K., and Revel, J.-P. (1987). *J. Cell Biol.* **105,** 1925–1934.

Dermietzel, R., Traub, O., Hwang, T. K., Beyer, E. C., Bennett M. V. L., Spray, D. C., and Willecke, K. (1989a). *Proc. Natl. Acad. U.S.A.* **86,** 10148–10152.

Dermietzel, R., Volker, M., Hwang, T.-K., Berzborn, R. J., and Meyer, H. E. (1989b). *FEBS Lett.* **253,** 1–5.

Dermietzel, R., Hertzberg, E. L., Kessler, J. A., and Spray, D. C. (1991). *J. Neurosci.* **11,** 1421–1432.

Dewey, M. M., and Barr, L. (1962). *Science* **137,** 670–672.

Ebihara, L., Beyer, E. C., Swenson, K. I., Paul, D. L., and Goodenough, D. A. (1989). *Science* **243,** 1194–1195.

Eghbali, B., Kessler, J. A., and Spray, D. C. (1990). *Proc. Natl. Acad. Sci. U.S.A.* **87,** 1328–1331.

Ehring, G. R., and Hall, J. E. (1991). *In* "Biophysics of Gap Junction Channels" (C. Perrachia, ed.), pp. 333–351. CRC Press, Boca Raton, Florida.

Evans, W. H., and Gurd, J. W. (1972). *Biochem. J.* **128,** 691–700.

Fallon, E. F., and Goodenough, D. A. (1981). *J. Cell Biol.* **90,** 521–526.

Fentiman, I., Taylor-Papadimitriou, J., and Stoker, M. (1976). *Nature (London)* **264,** 760–762.

Filson, A. J., Azarnia, R., Beyer, E. C., Loewenstein, W. R., and Brugge, J. S. (1990). *Cell Growth Differ.* **1,** 661–668.

Finbow, M., Yancey, S. B., Johnson, R., and Revel, J.-P. (1980). *Proc. Natl. Acad. Sci. U.S.A.* **77,** 970–974.

Finbow, M. E., Shuttleworth, J., Hamilton, A. E., and Pitts, J. D. (1983). *EMBO J.* **2,** 1479–1486.

Finbow, M. E., Buultjens, T. E. J., Lane, N. J., Shuttleworth, J., and Pitts, J. D. (1984). *EMBO J.* **3,** 2271–2278.

Fishman, G. I., Spray, D. C., and Leinwand, L. A. (1990). *J. Cell Biol.* **111,** 589.

Fishman, G. I., Hertzberg, E. L., Spray, D. C., and Leinward, L. A. (1991a). *Circ. Res.* **68,** 782–787.

Fishman, G. I., Moreno, A. P., Spray, D. C., and Leinwand, L. A. (1991b). *Proc. Natl. Acad. Sci. U.S.A.* **88,** 3525–3529.

Fishman, G. I., Eddy, R. L., Shows, T. B., Rosenthal, L., and Leinwand, L. A. (1991c). *Genomics* **10,** 250–256.

Fitzgerald, D. J., Mesnil, M., Oyamada, M., Tsuda, H., Ito, N., and Yamasaki, H. (1989). *J. Cell. Biochem.* **41,** 97–102.

Fitzgerald, P. G., Bok, D., and Horwitz, J. (1983). *J. Cell Biol.* **97,** 1491–1499.

Flagg-Newton, J., and Loewenstein, W. R. (1979). *J. Membr. Biol.* **50,** 65–100.

Flagg-Newton, J. L., and Loewenstein, W. R. (1980). *Science* **207,** 771–773.

Flagg-Newton, J., Simpson, I., and Loewenstein, W. R. (1979). *Science* **205,** 404–407.

Fluri, G. S., Rudisuli, A., Willi, M., Rohr, S., and Weingart, R. (1990). *Pflugers Arch.* **417,** 149–156.

Fraser, S. E., Green, C. R., Bode, H. R., and Gilula, N. B. (1987). *Science* **237,** 49–55.

Fromaget, C., el Aoumari, A., Dupont, E., Briand, J. P., and Gros, D. (1990). *J. Mol. Cell. Cardiol.* **22,** 1245–1258.

Fujimuto, K., Narimoto, K., Ogawa, K. S., Kondo, S., and Ogawa, K. (1985). *Acta Histochem. Cytochem.* **18,** 467–482.

Furshpan, E. J., and Potter, D. D. (1959). *J. Physiol. (London)* **145,** 289–325.

Gainer, H. S. C., and Murray, A. M. (1985). *Biochem. Biophys. Res. Commun.* **126,** 1109–1113.

Galvan, A. C., Lampe, P. D., Hur, K. C., Howard, J. B., Eccleston, E. D., Arneson, M., and Louis, C. F. (1989). *J. Biol. Chem.* **264,** 19974–19978.

Garfield, R. E., Sims, S., and Daniel, E. E. (1977). *Science* **198,** 958–960.

Garfield, R. E., Sims, S., Kannan, M. S., and Daniel, E. E. (1978). *Am. J. Physiol.* **235,** C168–C179.

Giaume, C., Kado, R. T., and Korn, H. (1987). *J. Physiol. (London)* **386,** 91–112.

Giaume, C., Fromaget, C., el-Aoumari, A., Cordier, J., Glowinski, J., and Gros, D. (1991). *Neuron* **6,** 133–143.

Gilula, N. B., Reeves, O. R., and Steinbach, A. (1972). *Nature (London)* **235,** 262–265.

Gilula, N. B., Epstein, M. L., and Beers, W. H. (1978). *J. Cell Biol.* **78,** 58–75.

Gimlich, R. L., Kumar, N. M., and Gilula, N. B. (1988). *J. Cell Biol.* **107,** 1065–1073.

Gimlich, R. L., Kumar, N. M., and Gilula, N. B. (1990). *J. Cell Biol.* **110,** 597–605.

Goodenough, D. A. (1976). *Cold Spring Harbor Symp. Quant. Biol.* **40,** 37–43.

Goodenough, D. A. (1979). *Invest. Ophthalmol. Visual Sci.***18,** 1104–1122.

Goodenough, D. A., and Stoeckenius, W. (1972). *J. Cell Biol.* **94,** 646–656.

Goodenough, D. A., Dick, J. S. B., II, and Lyons, J. E. (1980). *J. Cell Biol.* **86,** 576–589.

Goodenough, D. A., Paul, D. L., and Jesaitis, L. A. (1988). *J. Cell Biol.* **107,** 1817–1824.

Goodenough, D. A., Paul, D. L., and Takemoto, L. (1990). *J. Cell Biol.* **111,** 153A.

Gorin, M. B., Yancey, S. B., Cline, J., Revel, J. P., and Horwitz, J. (1984). *Cell* **39,** 49–59.

Green, C. R., and Severs, N. J. (1984). *J. Cell Biol.* **99,** 453–463.

Green, C. R., Harfst, E., Gourdie, R. G., and Severs, N. J. (1988). *Proc. R. Soc. London* **233,** 165–174.

Gros, D., Mocquard, J. P., Challice, C. E., and Schrevel, J. (1978). *J. Cell Sci.* **30,** 45–61.

Gros, D., Mocquard, J. P., Challice, C. E., and Schrevel, J. (1979). *J. Mol. Cell. Cardiol.* **11,** 543–554.

Gros, D. B., Nicholson, B. J., and Revel, J.-P. (1983). *Cell* **35,** 539–549.

Gruijters, W. T. M. (1989). *J. Cell Sci.* **93,** 509–513.

Gruijters, W. T. M., Kistler, J., Bullivant, S., and Goodenough, D. A. (1987). *J. Cell Biol.* **104,** 565–572.

Gutekunst, K. A., Rao, G. N., and Church, R. L. (1990). *Curr. Eye Res.* **9,** 955–961.

Guthrie, S. C., and Gilula, N. B. (1989). *Trends Neurosci.* **12,** 12–16.

Haefliger, J.-A., Bruzzone, R., Jenkens, N. A., Gilbert, D. J., Copeland, N. G., and Paul, D. L. (1992). *J. Biol. Chem.* **267,** 2057–2064.

Hanna, R. B., Ornberg, R. L., and Reese, T. S. (1985). *In* "Gap Junctions" (M. V. L. Bennett and D. C. Spray, eds.), pp. 23–32. Cold Spring Harbor Lab. Press, Cold Spring Harbor, New York.

Harris, A. L. (1991). *In* "Biophysics of Gap Junction Channels" (C. Peracchia, ed.), pp. 373–389. CRC Press, Boca Raton, Florida.

Harris, A. L., Spray, D. C., and Bennett, M. V. L. (1981). *J. Gen. Physiol.* **77,** 95–117.

Henderson, D., Eibl, H., and Weber, K. (1979). *J. Mol. Biol.* **132,** 193–218.

Hendrix, E. M., Mao, S. J., Everson. W., Myatt, L., and Larsen, W. J. (1991). *Biol. Reprod.* **44,** 151a.

Hertzberg, E. L., and Gilula, N. B. (1979). *J. Biol. Chem.* **254,** 2138–2147.

Hertzberg, E. L., and Skibbens, R. V. (1984). *Cell* **39,** 61–69.

Hertzberg, E. L., Spray, D. C., and Bennet, M. V. L. (1985). *Proc. Natl. Acad. Sci. U.S.A.* **82,** 2412–2416.

Hertzberg, E. L., Disher, R. M., Tiller, A. A., Zhou, Y., and Cook, R. G. (1988). *J. Biol. Chem.* **263,** 19105–19111.

Hertzberg, E. L., Corpina, R., Roy, C., Dougherty, M. J., and Kessler, J. A. (1989). *J. Cell Biol.* **109,** 47a.

Hobbie, L., Kingsley, D. M., Kozarsky, K. F., Jackman, R. W., and Krieger, M. (1987). *Science* **235,** 69–73.

Hoh, J. H., John, S. A., and Revel, J.-P. (1991). *J. Biol. Chem.* **266,** 6524–6531.

Hsieh, C.-L., Kumar, N. M., Gilula, N. B., and Francke, U. (1991). *Somatic Cell Mol. Genet.* **17,** 191–200.
Imanaga, I., Kameyama, M., and Irasawa, H. (1987). *Am. J. Physiol.* **252,** H223–H232.
Janssen-Timmen, U., Dermietzel, R., Frixen, U., Leibstein, A., Traub, O., and Willeke, K. (1983). *EMBO J.* **2,** 295–302.
Janssen-Timmen, U., Traub, O., Dermietzel, R., Rabes, H. M., and Willecke, K. (1986). *Carcinogenesis* **7,** 1475–1482.
Johnston, M. F., Simon, S. A., and Ramon, F. (1980). *Nature (London)* **286,** 498–500.
Jongen, W. M. F., Fitzgerald, D. J., Asamoto, M., Picolli, C., Slaga, T. J., Gros, D., Takeichi, M., and Yamasaki, H. (1991). *J. Cell Biol.* **114,** 545–555.
Kanter, H. L., Saffitz, J. E., and Beyer, E. C. (1992). *Circ. Res.* **70,** 438–444.
Kensler, R. W., and Goodenough, D. A. (1980). *J. Cell Biol.* **86,** 755–764.
Kistler, J., Kirkland, B., and Bullivant, S. (1985). *J. Cell Biol.* **101,** 28–35.
Kistler, J., Christie, D., and Bullivant, S. (1988). *Nature (London)* **331,** 721–723.
Kumar, N. M., and Gilula, N. B. (1986). *J. Cell Biol.* **103,** 767–776.
Laird, D. M., Puranam, K. L., and Revel, J. P. (1991). *Biochem. J.* **273,** 67–72.
Lang, L., Beyer, E. C., Schwartz, A. L., and Gitlin, J. D. (1991). *Am. J. Physiol.* **260,** E787–793.
Larsen, W. J. (1983). *Tissue Cell* **15,** 645–671.
Larsen, W. J., Wert, S. E., and Brunner, G. D. (1987). *Dev. Biol.* **122,** 61–71.
Larson, D. M., Haudenschild, C. C., and Beyer, E. C. (1990). *Circ. Res.* **66,** 1074–1080.
Larson, D. M., Beyer, E. C., and Haudenschild, C. C. (1991). *FASEB J.* **5,** A528.
Lash, J. A., Critser, E. S., and Pressler, M. L. (1990). *J. Biol. Chem.* **265,** 13113.
Ledbetter, M. L. S., and Lubin, M. (1979). *J. Cell Biol.* **80,** 150–165.
Lee, S., Gilula, N. B., and Warner, A. E. (1987). *Cell* **51,** 851–860.
Lee, S. W., Tomasetto, C., and Sager, R. (1991). *Proc. Natl. Acad. Sci. U.S.A.* **88,** 2825–2829.
Lo, C. W. (1985). *In* "Gap Junctions" (M. V. L. Bennett and D. C. Spray, eds.), pp. 251–263. Cold Spring Harbor Press, Cold Spring Harbor, New York.
Loewenstein, W. R. (1966). *Ann. N.Y. Acad. Sci.* **137,** 441–472.
Loewenstein, W. R. (1979). *Biochim. Biophys. Acta* **560,** 1–56.
Louis, C. F., Hur, K. C., Galvan, A. C., TenBroek, E., Jarvis, L. J., Eccleston, E. D., and Howard, J. B. (1989). *J. Biol. Chem.* **264,** 19967–19978.
Luke, R. A., and Saffitz, J. E. (1991). *J. Clin. Invest.* **87,** 1594–1602.
Luke, R. A., Beyer, E. C., Hoyt, R. H., and Saffitz, J. E. (1989). *Circ. Res.* **65,** 1450–1457.
Lye, S. J., Mascarenhas, M., Zander, D., McKenzie, L., and Petrocelli, T. (1991). Abstract presented at the International Gap Junction Conference. Asilomar, California.
MacKenzie, L. W., and Garfield, R. E. (1985a). *Am. J. Physiol.* **248,** C296–C308.
Mackenzie, L. W., and Garfield, R. E. (1985b). *Can. J. Physiol.* **64,** 703–706.
Makowski, L. (1988). *Adv. Cell Biol.* **2,** 119–158.
Makowski, L., Caspar, D. L. D., Phillips, W. C., and Goodenough, D. A. (1977). *J. Cell Biol.* **74,** 629–645.
Maldonado, P. E., Rose, B., and Loewenstein, W. R. (1988). *J. Membr. Biol.* **106,** 203–210.
Manjunath, C. K., Goings, G. E., and Page, E. (1984). *Am. J. Physiol.* **246,** H865–H875.
Manjunath, C. K., Nicholson, B. J., Teplow, D., Hood, L., Page, E., and Revel, J.-P. (1987). *Biochem. Biophys. Res. Commun.* **142,** 228–234.
Maurer, P., and Weingart, R. (1987). *Pflugers Arch.* **409,** 394–402.
Mazet, F., Wittenberg, B. A., and Spray, D. C. (1985). *Circ. Res.* **56,** 195–204.
McMahon, D. G., Knapp, A. G., and Dowling, J. E. (1989). *Proc. Natl. Acad. Sci. U.S.A.* **86,** 7639–7643.
Meda, P., Bosco, D., Chanson, M., Giordano, E., Vallar, L., Wollheim, C., and Orci, L. (1990). *J. Clin. Invest.* **86,** 759–768.

Meda, P., Chanson, M., Pepper, M., Giordano, E., Bosco, D., Traub, O., Willeke, K., El Aoumari, A., Gros, D., Beyer, E. C., Orci, L., and Spray, D. C. (1991). *Exp. Cell Res.* **192,** 469–480.

Mehta, P. P., Bertram, J. S., and Loewenstein, W. R. (1986). *Cell* **44,** 187–196.

Meiners, S., and Schindler, M. (1987). *J. Biol. Chem.* **262,** 951–953.

Meiners, S., Xu, A., and Schindler, M. (1991). *Proc. Natl. Acad. Sci. U.S.A.* **88,** 4119–4122.

Meyer, D. J., Yancey, S. B., Revel, J.-P., and Peskoff, A. (1981). *J. Cell Biol.* **91,** 503–523.

Meyer, R. A., Laird, D. L., Revel, J.-P., and Johnson, R. G. (1990). *J. Cell Biol.* **111,** 154a.

Michalke, W., and Loewenstein, W. R. (1971). *Nature (London)* **232,** 121–123.

Milks, L. C., Kumar, N. M., Houghten, R., Unwin, N., and Gilula, N.B. (1988). *J. Biol. Chem.* **260,** 6514–6517.

Miller, T. M., and Goodenough, D. A. (1985). *J. Cell Biol.* **101,** 1741–1748.

Miller, T. M., and Goodenough, D. A. (1986). *J. Cell Biol.* **102,** 194–199.

Miller, T., Dahl, G., and Werner, R. (1988). *Biosci. Rep.* **8,** 455–464.

Moore, L., Beyer, E. C., and Burt, J. M. (1991). *Am. J. Physiol.* **260,** C975–C981.

Musil, L. S., Beyer, E. C., and Goodenough, D. A. (1990a). *J. Membr. Biol.* **116,** 163–175.

Musil, L. S., Cunningham, B. A., Edelman, G. M., and Goodenough, D. A. (1990b). *J. Cell Biol.* **111,** 2077–2088.

Neyton, J., and Trautman, A. (1985). *Nature (London)* **317,** 331–335.

Nicholson, B. J., Takemoto, L. J., Hunkapillar, L. E., Hood, L. E., and Revel, J.-P. (1983). *Cell* **32,** 967–978.

Nicholson, B. J., Gros, D. B., Kent, S. B. H., Hood, L. E., and Revel, J.-P. (1985). *J. Biol. Chem.* **260,** 6514–6517.

Nicholson, B. J., Dermietzel, R., Teplow, D., Traub, O., Willecke, K., and Revel, J.-P. (1987). *Nature (London)* **329,** 732–734.

Niggli, E., Rudisuli, A., Maurer, P., and Weingart, R. (1989). *Am. J. Physiol.* **256,** C273–C281.

Noma, A., and Tsuboi, N. (1987). *J. Physiol. (London)* **382,** 193–211.

Obaid, A. L., Socolar, S. J., and Rose, B. (1983). *J. Membr. Biol.* **73,** 69–89.

Oyamada, M., Krutovskikh, V. A., Mesnil, M., Partensky, C., Berger, F., and Yamasaki, H. (1990). *Mol. Carcinog.* **3,** 273–278.

Paul, D. L. (1985). *In* "Gap Junctions" (M. V. L. Bennett and D. C. Spray, eds.), pp. 107–122. Cold Spring Harbor Press, Cold Spring Harbor, New York.

Paul, D. L. (1986). *J. Cell Biol.* **103,** 123–134.

Paul, D. L., and Goodenough, D. A. (1983). *J. Cell Biol.* **96,** 625–632.

Payton, B. W., Bennett, M. V. L., and Pappas, G. D. (1969). *Science* **165,** 594–597.

Peracchia, C. (1977). *J. Cell Biol.* **72,** 628–641.

Peracchia, C. (1985). *Mol. Cell. Biol.* **7,** 267–282.

Peracchia, C. (1990). *J. Membr. Biol.* **113,** 75–92.

Peterson, O.H. (1985). *In* "Gap Junctions" (M. V. L. Bennett and D. C. Spray, eds.), pp. 315–324. Cold Spring Harbor Press, Cold Spring Harbor, New York.

Pitts, J. D., and Simms, J. W. (1977). *Exp. Cell Res.* **104,** 153–163.

Puranam, K. L., Laird, D. W., and Revel, J.-P. (1990). *J. Cell Biol.* **111,** 155a.

Randriamampita, C., Giaume, C., Neyton, J., and Trautman, A. (1988). *Pflugers Arch.* **412,** 462–468.

Reber, W., and Weingart, R. (1982). *J. Physiol. (London)* **328,** 87–104.

Revel, J.-P., and Karnovsky, M. J. (1967). *J. Cell Biol.* **33,** C7–C12.

Risek, B., and Gilula, N. B. (1991). *Development* **113,** 165–181.

Risek, B., Guthrie, S., Kumar, N., and Gilula, N. B. (1990). *J. Cell Biol.* **110,** 269–282.

Robertson, J. D. (1963). *J. Cell Biol.* **19,** 201–221.

Rogers, M., Berestecky, J. M., Hossain, M. Z., Guo, H. M., Kadle, R., Nicholson, B. J., and Bertram, J. S. (1990). *Mol. Carcinog.* **3,** 553–564.

Rook, M. B., Jongsma, H. J., and van Ginneken, A. C. G. (1988). *Am. J. Physiol.* **414,** H770–H782.

Rose, B., and Rick, R. (1978). *J. Membr. Biol.* **44,** 377–415.

Rudisuli, A., and Weingart, R. (1989). *Pflugers Arch.* **415,** 12–21.

Rup, D., and Beyer, E. C. (1991). *J. Cell Biol.* **115** (in press).

Saez, J. C., Spray, D. C., Nairn, A., Hertzberg, E. L., Greengard, P., and Bennett, M. V. L. (1986). *Proc. Natl. Acad. Sci. U.S.A.* **83,** 2473–2477.

Saez, J. C., Nairn, A. C., Czernick, A., Spray, D. C., Hertzberg, E. L., Greengard, P., and Bennett, M. V. L. (1990). *Eur. J. Biochem.* **192,** 263–273.

Sas, D. F., Sas, M. J., Johnson, K. R., Menko, A. S., and Johnson, R. G. (1985). *J. Cell Biol.* **100,** 216–225.

Schuetze, S. M., and Goodenough, D. A. (1982). *J. Cell Biol.* **92,** 694–705.

Shibata, Y., and Yamamoto, T. (1986). *Anat. Rec.* **214,** 107–112.

Simpson, I., Rose, B., and Loewenstein, W. R. (1977). *Science* **195,** 294–297.

Sims, S. M., Daniel, E. E., and Garfield, R. E. (1982). *J. Gen. Physiol.* **80,** 353–375.

Somogyi, R., and Kolb, H. A. (1989). *Pflugers Arch.* **412,** 54–65.

Somogyi, R., Batzer, A., and Kolb, H. A. (1989). *J. Membr. Biol.* **108,** 273–282.

Sosinsky, G. E., Baker, T. S., Caspar, D. L., and Goodenough, D. A. (1990). *Biophys. J.* **58,** 1213–1226.

Spach, M. S., and Dobler, P. C. (1986). *Circ. Res.* **58,** 356–371.

Sperelakis, N., and Cole, W. C. (1989). "Cell Interactions and Gap Junctions," Vol. 2. CRC Press, Boca Raton, Florida.

Spray, D. C., and Bennett, M. V. L. (1985). *Annu. Rev. Physiol.* **47,** 281–303.

Spray, D. C., and Burt, J. M. (1990). *Am. J. Physiol.* **258,** C195–C205.

Spray, D. C., Harris, A. L., and Bennett, M. V. L. (1981). *J. Gen. Physiol.* **77,** 77–93.

Spray, D. C., Stern, J. H., Harris, A. L., and Bennett, M. V. L. (1982). *Proc. Natl. Acad. Aci. U.S.A.* **79,** 441–445.

Spray, D. C., White, R. I., Verselis, V., and Bennett, M. V. L. (1985). *In* "Gap Junctions" (M. V. L. Bennett and D. C. Spray, eds.), pp. 139–154. Cold Spring Harbor Press, Cold Spring Harbor, New York.

Spray, D. C., Saez, J. C., Brosius, D., Bennett, M. V. L., and Hertzberg, E. L. (1986a). *Proc. Natl. Acad. Sci. U.S.A.* **83,** 5494–5497.

Spray, D. C., Campos de Carvallho, A. C., and Bennett, M. V. L. (1986b). *Proc. Natl. Acad. Aci. U.S.A.* **83,** 3533–3536.

Spray, D. C., Ginzberg, R. D., Morales, E. A., Gatmaitan, Z., and Arias, I. M. (1986c). *J. Cell Biol.* **103,** 135–144.

Spray, D. C., Fujita, M., Saez, J. C., Choi, H., Watanabe, T., Hertzberg, E., Rosenberg, L. C., and Reid, L. M. (1987). *J. Cell Biol.* **105,** 541–551.

Spray, D. C., Bennett, M. V. L., Campos de Carvallho, A. C., Eghbali, B., Moreno, A. P., and Verselis, V. K. (1991a). *In* "Biophysics of Gap Junction Channels" (C. Peracchia, ed.), pp. 97–116. CRC Press, Boca Raton, Florida.

Spray, D. C., Chanson, M., Moreno, A. P., Dermietzel, R., and Meda, P. (1991b). *Am. J. Physiol.* **260,** C513–527.

Stevenson, B. R., Siliciano, J. D., Mooseker, M. S., and Goodenough, D. A. (1986). *J. Cell Biol.* **104,** 565–572.

Subak-Sharpe, H., Burk, R. R., and Pitts, J. D. (1969). *J. Cell Sci.* **4,** 353–367.

Swenson, K. I., Jordan, J. R., Beyer, E. C., and Paul, D. L. (1989). *Cell* **57,** 145–155.

Swenson, K. I., Piwnicka-Worms, H., McNamee, J., and Paul, D. L. (1990). *Cell Regul.* **1,** 989–1002.

Takeda, A., Saheki, S., Shimazu, T., and Takeuchi, N. (1989). *J. Biochem. (Tokyo)* **106,** 723–727.
Traub, O., Druege, M., and Willecke, K. (1983). *Proc. Natl. Acad. Sci. U.S.A.* **80,** 755–759.
Traub, O., Look, J., Dermietzel, R., Brummer, F., Hulser, D., and Willecke, K. (1989). *J. Cell Biol.* **108,** 1039–1051.
Unwin, P. N. T., and Ennis, P. D. (1984). *Nature (London)* **307,** 609–613.
Unwin, P. N. T., and Zampighi, G. (1980). *Nature (London)* **283,** 545–549.
Ursell, P. C., Gardner, P. I., Albala, A., Fenoglio, J. J., Jr., and Wit, A. L. (1985). *Circ. Res.* **56,** 436–451.
Valdimarsson, G., de Sousa, P. A., Beyer, E. C., Paul, D. L., and Kidder, G. M. (1991). *Mol. Reprod. Dev.* **30,** 18–26.
van Kempen, M. J. A., Fromaget, C., Gros, D., Moorman, A. F. M., and Lamers, W. H. (1991). *Circ. Res.* **68,** 1638–1651.
Veenstra, R. D. (1990). *Am. J. Physiol.* **258,** C662–C672.
Veenstra, R. D. (1991a). *J. Cardiovasc. Electrophysiol.* **2,** 168–189.
Veenstra, R. D. (1991b). *J. Membr. Biol.* **119,** 253–265.
Veenstra, R. D. (1991c). *In* "Biophysics of Gap Junction Channels" (C. Peracchia, ed.), pp. 131–144. CRC Press, Boca Raton, Florida.
Veenstra, R. D., and DeHaan, R. L. (1986). *Science* **233,** 972–974.
Veenstra, R. D., and DeHaan, R. L. (1988). *Am. J. Physiol.* **254,** H170–H180.
Veenstra, R. D., Wang, H. Z., Berg, K., Westphale, E. M., and Beyer, E. C. (1993). *In* "Gap Junctions" (G. Zampighi and J. Hall, eds.). Elsevier, Amsterdam. In press.
Veenstra, R. D., Wang, H. Z., Westphale, E. M., and Beyer, E. C. (1992b). *J. Cell Biol.* (in press).
Verselis, V., Bennett, M. V. L., and Bargiello, T. A. (1991). *Biophys. J.* **59,** 114–126.
Warner, A. E. Guthrie, S. C., and Gilula, N. B. (1984). *Nature (London)* **311,** 127–131.
Weidman, S. (1966). *J. Physiol. (London)* **187,** 323–342.
Werner, R., Miller, T., Azarnia, R., and Dahl, G. (1985). *J. Membr. Biol.* **87,** 253–268.
Werner, R., Levine, E., Rabadan-Diehl, C., and Dahl, G. (1989). *Proc. Natl. Acad. Sci. U.S.A.* **86,** 5380–5384.
Werner, R., Levine, E., Rabadan-Diehl, C., and Dahl, G. (1991). *Proc. R. Soc. London* **243,** 5–12.
White, R. L., Spray, D. C., Campos de Carvallho, A. C., Wittenberg, B. A., and Bennett, M. V. L. (1985). *Am. J. Physiol.* **249,** C447–C455.
White, R. L., Doeller, J. E., Verselis, V. K., and Wittenberg, B. A. (1990). *J. Gen. Physiol.* **95,** 1061–1075.
Willecke, K., Jungbluth, S., Dahl, E., Hennemann, H., Heyukes, R., and Grzeschik, K.-H. (1990). *Eur. J. Cell Biol.* **53,** 275–280.
Winterhager, E., Stutenkemper, R., Traub, O., Beyer, E., and Willecke, K. (1991). *Eur. J. Cell Biol.* **55,** 133–142.
Yada, T., Rose, B., and Loewenstein, W. R. (1985). *J. Membr. Biol.* **88,** 217–232.
Yaholom, A., Warmbrodt, R. D., Laird, D. W., Traub, O., Revel, J.-P., Willecke, K., and Epel, B. L. (1991). *Plant Cell* **3,** 407–417.
Yancey, S. B., John, S. A., Lal, R., Austin, B. J., and Revel, J.-P. (1989). *J. Cell Biol.* **108,** 2241–2254.
Yeager, M. (1987). *J. Mol. Cell. Cardiol.* 19, Suppl. IV, S.54.
Yee, A. G., and Revel, J.-P. (1978). *J. Cell Biol.* **78,** 554–564.
Young, J. D.-E., Cohn, Z. A., and Gilula, N. B. (1987). *Cell* **48,** 733–743.
Zampighi, G., Simon, S. A., Robertson, J. D., McIntosh, T. J., and Costello, M. J. (1982). *J. Cell Biol.* **93,** 175–189.

Zampighi, G. A., Hall, J. E., and Kreman, M. (1985). *Proc. Natl. Acad. Sci. U.S.A.* **82,** 8468.

Zampighi, G., Hall, J. E., Ehring, G. R., and Simon, S. A. (1989). *J. Cell Biol.* **108,** 2255–2275.

Zhang, J.-T., and Nicholson, B. (1989). *J. Cell Biol.* **109,** 3391–3401.

Zimmer, D. B., Green, C. R., Evans, W. H., and Gilula, N. B. (1987). *J. Biol. Chem.* **262,** 7751–7763.

Structural Properties of Voltage-Dependent Calcium Channels

Edwin W. McCleskey, Mary D. Womack, and Lynne A. Fieber
Department of Cell Biology and Physiology, Washington University School of Medicine, St. Louis, Missouri 63110

I. Introduction

Voltage-dependent Ca^{2+} channels are gated pores that open in response to depolarization of the plasma membrane and selectively allow the entry of Ca^{2+} into the cell. The inward flow of Ca^{2+} ions through these channels further depolarizes the membrane potential, contributing to the depolarizing electrical signal. But a more important role for Ca^{2+} channels is to serve as translators that decode the surface membrane electrical signal into an intracellular chemical signal. When the channels open, the accumulation of Ca^{2+} within the cytoplasm activates enzymes that control processes such as secretion and contraction. Structural studies of voltage-dependent Ca^{2+} channels must explain how the channels sense changes in membrane potential and how the pores allow Ca^{2+} to pass through at a high rate while excluding other cations. Further goals are to identify sites where activation of protein kinases modify Ca^{2+} channel activity and to explain the differences among various types of Ca^{2+} channels.

The study of Ca^{2+} channels is now at a turning point. Biochemical and genetic investigations of structure are beginning to converge with functional studies carried out by electrophysiologists. Functional studies of single Ca^{2+} channels and of Ca^{2+} currents in whole cells have provided clear predictions about the structure of the voltage sensor, the diameter of the pore, the existence of intrapore binding sites, and the presence of intracellular phosphorylation sites. These predictions have been further refined by structural studies of other voltage gated channels. Now that Ca^{2+} channels have been cloned and successfully expressed, it becomes possible to test, extend, prove, or disprove such predictions.

The purpose of this review is to briefly summarize what is known about Ca^{2+} channel structure from biochemical and molecular biological studies and then to explain the predictions about channel structure that have been derived from purely functional studies. There are a variety of recent reviews that are less specific in their aim and broader in their coverage of the literature (Tsien *et al.*, 1987; Bean, 1989; Hess, 1990; McCleskey and Schroeder, 1992).

II. Protein Biochemistry: The Subunit of the Skeletal Muscle Channel

Purification and cloning of Ca^{2+} channels have relied on the use of dihydropyridines (DHPs), drugs that have been used to treat angina and other cardiovascular diseases (Triggle and Janis, 1987). These ligands bind to the class of Ca^{2+} channels predominant in cardiac, skeletal, and smooth muscle and modify their activity. Most of the biochemical work has been done on the Ca^{2+} channel in mammalian skeletal muscle because transverse tubules, a system of invaginations of the plasma membrane, contain a very high density of DHP binding sites. Radiolabeled DHPs have been used to follow transverse tubule Ca^{2+} channels through a series of conventional purification steps involving, first, lectin–Sepharose chromatography, then ion exchange chromatography, and, finally, sucrose gradient sedimentation (Curtis and Catterall, 1984; Borsotto *et al.*, 1985; Flockerzi *et al.*, 1986a). Following purification in the presence of the DHP Ca^{2+} channel activator Bay k 8644, Ca^{2+} channels reconstituted in planar lipid bilayers exhibit a single channel conductance (20 pS with 90mM Ba^{2+} as the charge carrier; Flockerzi *et al.*, 1986b) similar to that found for Ca^{2+} channels in intact ventricular muscle cells (Reuter *et al.*, 1982). Purified channels have also been shown to exhibit subconductance states lower in value than 20 pS (Talvenheimo *et al.*, 1987; Hymel *et al.*, 1988). The activity of purified channels is increased following phosphorylation by cAMP-dependent protein kinase (Flockerzi *et al.*, 1986b), as are the channels *in vivo* (Kameyama *et al.*, 1985).

The proposed subunit structure of the skeletal muscle Ca^{2+} channel is illustrated in Fig. 1 (from Catterall, 1988). Five subunits, α_1, α_2, δ, β and γ, cosediment during centrifugation or coprecipitate with monoclonal antibodies (Leung *et al.*, 1987; Takahashi *et al.*, 1987; Sieber *et al.*, 1987). The α_1 subunit is largest (175 kDa), is the binding site for DHPs, and, judging from its amino acid sequence (see Section III), is certain to be both the pore-and the voltage-sensing subunit. The α_2 (143 kDa) and δ (27 kDa) subunits are linked by disulfide bonds. Under nonreducing conditions, the α_2–δ combination is similar in molecular weight to α_1, leading to an initial confusion about whether α_1 or α_2 was the DHP binding site.

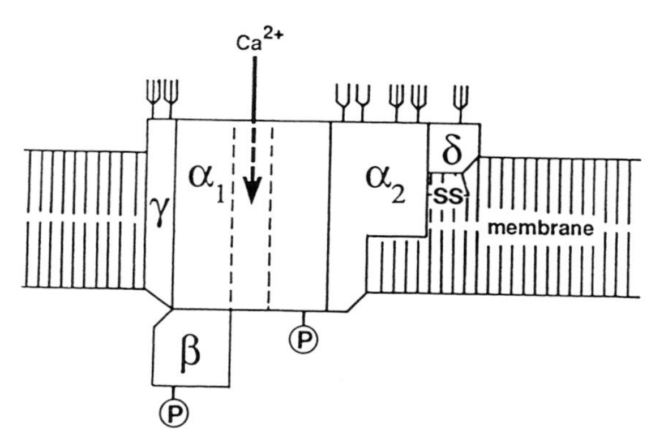

FIG. 1 Subunit structure of the Ca^{2+} channel in skeletal muscle. (From Catterall, 1988.)

This issue was resolved by the observation that a DHP photoaffinity label only incorporated into α_1 (Leung *et al.*, 1987; Takahashi *et al.*, 1987; Sieber *et al.*, 1987). The β subunit (54 kDa) and γ subunit (30 kDa) remain associated with α_1 in the presence of 0.5% Triton X-100, whereas the α_2–δ subunit complex readily dissociates from the other three subunits (Takahashi *et al.*, 1987). All subunits except β have some hydrophobic domains and, therefore, are expected to be transmembrane- or membrane-associated. The α_2, δ, and γ subunits are heavily glycosylated, indicating that they have an extracellular face (Takahashi *et al.*, 1987). Cyclic AMP-dependent protein kinase phosphorylates both the purified α_1 and the β subunits. However, in the crude membrane preparations containing functional Ca^{2+} channels, only the β subunit is phosphorylated (Curtis and Catterall, 1985); this suggests that the β subunit mediates the cAMP-dependent increase in channel activity.

Ligands other than DHPs bind to different classes of Ca^{2+} channels. Toxins that suppress activity of other types of Ca^{2+} channels have been isolated from the venoms of funnel web spiders (Llinas *et al.*, 1989) and from marine snails (Olivera *et al.*, 1985). Purification of different classes of Ca^{2+} channels based on toxin binding is currently being pursued.

III. Molecular Biology: Cloning and Expression of the Subunits

Four of the five subunits of the skeletal muscle channel have been independently cloned and sequenced: α_1 (Tanabe *et al.*, 1987; Ellis *et al.*, 1988),

α_2 (Ellis *et al.*, 1988), β (Ruth *et al.*, 1989), and γ (Jay *et al.*, 1990). The fifth, δ, is encoded on the same gene as α_2 (De Jongh *et al.*, 1990). Synthesis as part of the same protein precursor may be important for formation of correct disulfide bonds between these two subunits. The α_1 sequence predicts a protein of 1873 amino acids that is homologous to the α subunit of the voltage-dependent Na^+ channel (55% identical or conservative substitutions) and has four internally homologous domains (Tanabe *et al.*, 1987) (Fig. 2). Each domain has six putative transmembrane regions and itself is homologous to voltage-dependent potassium channels (Jan and Jan, 1989). Since expression of the α subunit of the Na^+ channel generates functioning channels in frog oocytes (Noda *et al.*, 1986), the homologous α_1 subunit of the Ca^{2+} channel is considered to be the pore and voltage sensor, with the other subunits serving unknown auxiliary roles. The amino acid sequence of α_1 predicts a protein with 212-kDa molecular mass rather than the 175 kDa found in biochemical studies. Using a monoclonal antibody directed against a synthetic peptide representing the 18 C-terminal amino acids of the α_1 sequence, De Jongh *et al.* (1989) have

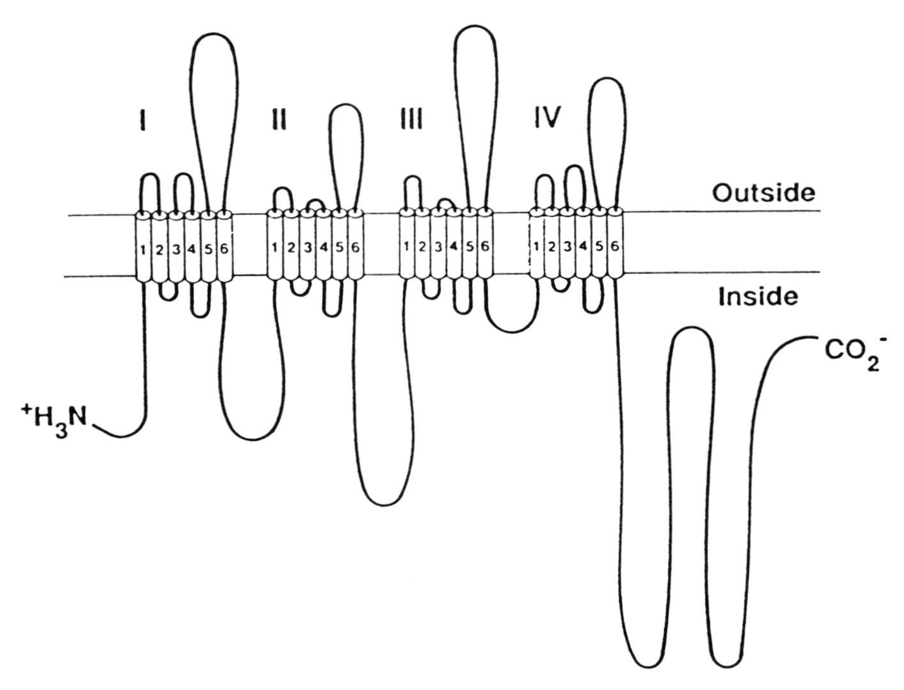

FIG. 2 Proposed transmembrane arrangement of the α_1 subunit. I, II, III, IV refer to the four internally homologous domains. Each domain includes six putative membrane-spanning α helices indicated by the cylinders. (From Catterall, 1988.)

discovered a 212-kDa protein as predicted; presumably, the 175-kDa α_1 protein is created by posttranslational proteolytic cleavage of the full-length sequence.

Sequences homologous to the skeletal muscle α_1 have been obtained from a variety of different tissues (heart: Mikami *et al.*, 1989; aorta: Koch *et al.*, 1990; brain: Mori *et al.*, 1991; Snutch *et al.*, 1991; Hui *et al.*, 1991; lung: Biel *et al.*, 1990; ovary and a neuroendocrine cell line: Perez-Reyes *et al.*, 1990). Analysis of the sequences indicates that alternative splicing generates several channel variants from three distinct genes originally found in muscle, heart, and brain/endocrine tissues (Perez-Reyes *et al.*, 1990; Snutch *et al.*, 1991). Moreover, a single tissue can express multiple α_1 genes as well as alternatively spliced products of these genes. Clearly the machinery exists to create a diversity of Ca^{2+} channel types.

It has been difficult to express functional Ca^{2+} channels in oocytes; success appears to involve coexpression of the α_1 subunit with other subunits. A small whole-cell Ca^{2+} channel current is seen in oocytes injected with the mRNA for the cardiac α_1 subunit and this current is greatly enhanced if mRNA for the skeletal muscle α_2 subunit is coinjected (Mikami *et al.*, 1989). The brain α_1 clone is functionally expressed only weakly in oocytes, but Ca^{2+} channel current increases by two orders of magnitude when the α_2 and β subunits from muscle are coinjected with α_1 from brain (Mori *et al.*, 1991).

Expression of the α_1 subunit has been obtained in murine L cells, a fibroblast-like cell. When L cells are transfected with only the muscle α_1 subunit cDNA, active Ca^{2+} channels result despite the fact that L cells contain neither endogenous Ca^{2+} current nor subunits analogous to α_2 or β (Perez-Reyes *et al.*, 1989; Kim *et al.*, 1990). The number of DHP binding sites increases 10-fold if the β subunit is coexpressed but, interestingly, the density of active Ca^{2+} channels is not increased (Varadi *et al.*, 1991). This supports the hypothesis that DHP binding sights and functional Ca^{2+} channels are not always identical (Schwartz *et al.*, 1985).

The presence of the β subunit also affects the kinetics of the transfected channels. When expressed alone, α_1 channels have very slow activation and inactivation rates. The presence of β dramatically speeds the kinetics of the expressed channels in L cells (Varadi *et al.*, 1991; Lacerda *et al.*, 1991).

IV. Structure–Function Studies of Excitation–Contraction Coupling

Recent experiments with a skeletal muscle model have demonstrated the importance of Ca^{2+} channels in excitation–contraction (EC) coupling.

Excitation–contraction coupling is the process linking changes in membrane potential to muscle contraction. In both skeletal and cardiac muscle the Ca^{2+} required for movement of contractile proteins is provided by release from the sarcoplasmic reticulum (SR), a Ca^{2+}-rich intracellular organelle. But cardiac muscle contraction also requires Ca^{2+} influx, whereas contraction of skeletal muscle persists in the absence of extracellular Ca^{2+} or in the presence of inorganic Ca^{2+} channel blockers. What, then, is the role of plasma membrane Ca^{2+} channels in skeletal muscle? Skeletal muscle transverse tubules contain a high concentration of DHP binding sites and, by implication, Ca^{2+} channels. They are also the site of the voltage sensor for EC coupling. Rios and Brum (1987) proposed that Ca^{2+} channels in the transverse tubules might act as voltage sensors during EC coupling, even though Ca^{2+} flux through the channels is not necessary for muscle contraction. They showed that a low concentration of DHPs inhibits Ca^{2+} release from skeletal muscle SR by diminishing the activity of the voltage sensor in the transverse tubules. They suggested that DHP receptors perform two tasks in skeletal muscle: to act as Ca^{2+} channels and as the voltage sensor responsible for triggering Ca^{2+} release from the SR.

Support for this hypothesis was provided in a series of experiments carried out on skeletal muscles of dysgenic mice. Muscular dysgenesis results from a single recessive defect in an inbred colony of mice that asphyxiate at birth because their skeletal muscles cannot contract. The failure to contract was correlated with the absence of DHP-sensitive Ca^{2+} channels (Beam *et al.*, 1986; Rieger *et al.*, 1987). When the nuclei of dysgenic skeletal muscle cells in primary tissue culture were injected with muscle α_1 cDNA, both EC coupling and Ca^{2+} current were restored (Tanabe *et al.*, 1988). Though the influx of Ca^{2+} is unimportant, Ca^{2+} channels themselves are critical for EC coupling in skeletal muscle.

What are the structural differences between cardiac and skeletal muscle DHP receptors that are responsible for differences in EC coupling? Tanabe *et al.* (1990a) showed that the answer lies in the composition of the α_1 subunit. When the skeletal muscle clone for α_1 was injected into dysgenic muscle (Fig. 3A, top), contraction did not require extracellular Ca^{2+} (trace 2) and occurred in the presence of the Ca^{2+} channel blocker, Cd^{2+} (trace 3). Thus the skeletal clone elicits skeletal-type EC coupling. Injection of the cardiac α_1 clone caused cardiac-like EC coupling, which requires Ca^{2+} influx (Fig. 3A, bottom). The part of the α_1 amino acid sequence that determines the nature of EC coupling has been addressed through the use of chimeras of skeletal and cardiac α_1 subunits (Tanabe *et al.*, 1990b). In Fig. 3B, contractile responses are plotted on the left for the corresponding chimeras indicated on the right (heavy lines represent sequences from the skeletal clone and fine lines represent sequences from the cardiac clone).

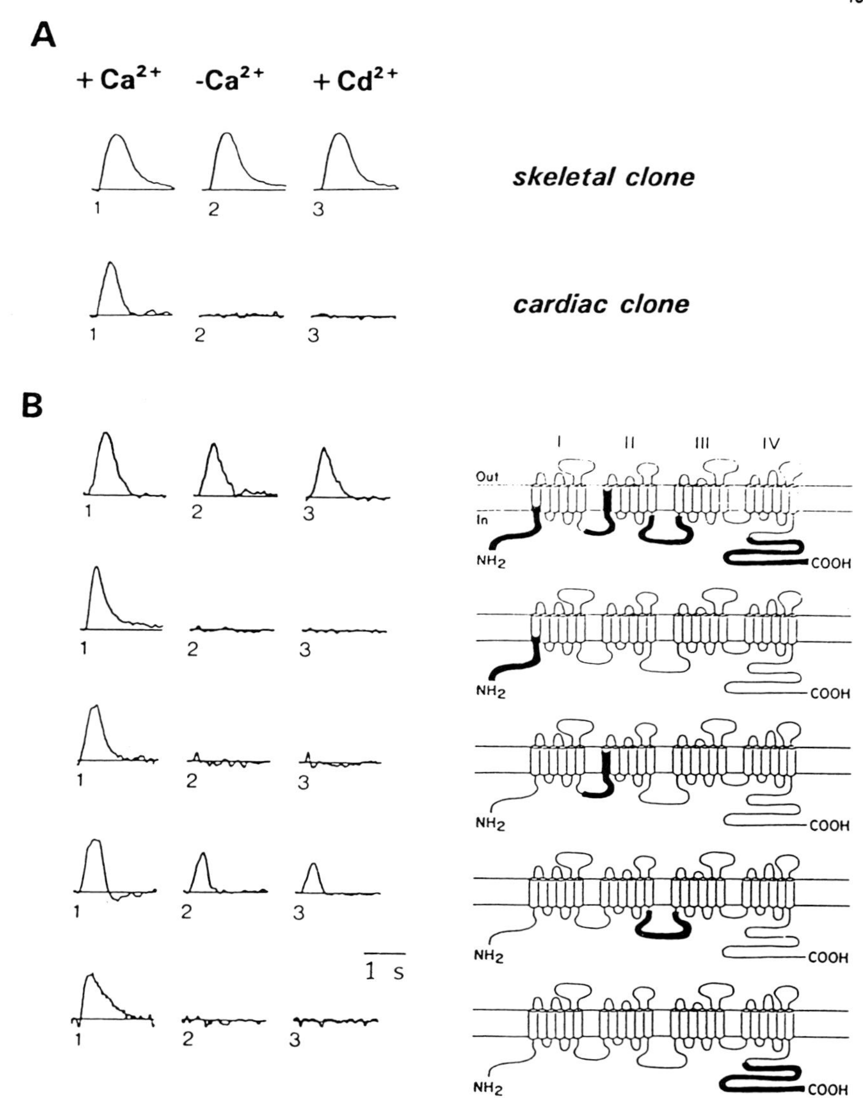

FIG. 3 Magnitude of contractions of dysgenic muscle cells expressing skeletal, cardiac, or chimeric α_1 subunits. In each row, recording 1 is in the presence of Ca^{2+}, 2 is in the absence of Ca^{2+}, and 3 is in the presence of Cd^{2+}, a Ca^{2+} channel blocker. Compositions of the chimeras are indicated on the right of (B). Thick lines indicate the skeletal portions of the chimeras and thin lines indicate the cardiac portions. (From Tanabe *et al.*, 1990b.)

If all putative cytoplasmic regions are skeletal (top chimera), but all extracellular and transmembrane segments are cardiac, the contraction is like that in skeletal muscle (i.e., Ca^{2+}-independent). Thus the cytoplasmic regions, rather than the transmembrane regions, determine whether EC coupling will be like skeletal muscle or like cardiac muscle. Moreover, substitution of only one of the α_1 cytoplasmic loops into the cardiac gene is sufficient to generate skeletal-type EC coupling: the loop between the second and the third homologous domains.

Although two of the chimeras in Fig. 3B induce skeletal-type EC coupling, each generates a Ca^{2+} current like that in heart. Cardiac Ca^{2+} current activates about 100 times faster than that in skeletal muscle. Clearly, the activation kinetics are not controlled by the cytoplasmic domains. Another set of chimeras have been generated with transmembrane domains from the skeletal α_1 subunit and cytoplasmic loops from the cardiac α_1 clone. Changing the transmembrane segments of domains II, III, and IV from cardiac to skeletal does not alter channel kinetics. However, a skeletal-type domain I results in a slow, skeletal-type Ca^{2+} current even if all other parts of the channel are from the cardiac clone (Tanabe *et al.,* 1992). The results can be interpreted by assuming that each domain acts as an independent gate, all of which must be open for the channel to pass Ca^{2+}. The gate corresponding to domain I is especially slow in skeletal muscle. This is analogous to the multiple gate model for Na^+ channel gating proposed by Hodgkin and Huxley (1952).

V. What Tertiary Structures Are Predicted by Functional Studies?

A. Gating Structures

1. The S4 Helix Is Probably the Voltage Sensor

Movement of a charged residue(s) in the channel protein is the most obvious way to sense a changing membrane voltage. By considering the voltage dependence of Na^+ channel gating, Hodgkin and Huxley (1952) deduced that six charges must move all the way across the membrane (or a larger number partway across) for the channel to open. A similar number of charges are required to gate Ca^{2+} channels (Kostyuk *et al.,* 1981). It now seems clear that the general structure for this highly charged voltage sensor is shared by all voltage-dependent channels.

A practical structure was proposed by Clay Armstrong (1981) following a decade of study of Na^+ channel gating currents, the small signals generated by movement of the voltage sensor. The evidence showed that the sensor was a discrete structure extending from the intracellular to the extracellular face of the channel. Armstrong proposed that a linear array of linked charges interacted with a similar array of countercharges (Fig. 4A). The arrays move relative to one another like a ratchet; a single step

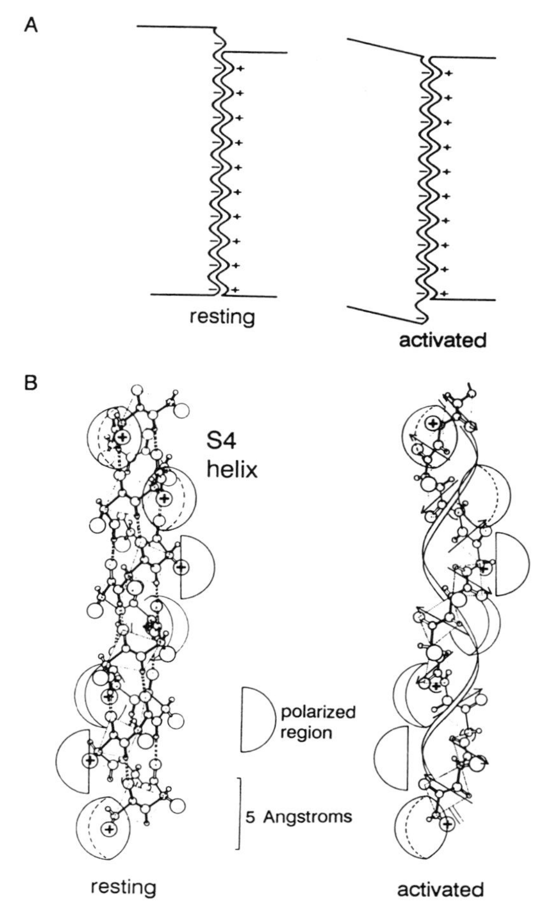

FIG. 4 (A) Clay Armstrong's ratchet model for the voltage sensor of voltage-dependent ion channels. In going from the resting to activated states, the two charge arrays ratchet one step relative to one another. The result is the net movement of a negative charge all the way from top to bottom. (From Armstrong, 1981.) (B) The S4 helix. The charged amino acids are indicated by the hemispheres that snake around the helix. In Klaus Benndorf's model, the resting-to-activated transition involves a relaxation of the helical structure due to breaking of the H bonds between neighboring residues when there is an input of electrical energy. (From Benndorf, 1989.)

results in the net movement of one charge across the membrane, though each individual charge moves only a small fraction of the distance. The movement of several such structures would be necessary to open the channel.

Cloning of the Na^+ channel revealed a highly charged region (the "S4" helix) that could correspond to the positive half of Armstrong's ratchet (Guy and Seetharamulu, 1986; Catterall, 1986; Noda *et al.*, 1986). The structure consists of a 22-amino acid sequence with a positively charged residue at every third position. The sequence is predicted to form an α-helix with the charges arranged to form a ratchet-like structure (Fig. 4B). There are four S4 helices in both Ca^{2+} and Na^+ channels, one in each of the homologous domains; voltage-dependent K^+ channels each have one S4 region (Jan and Jan, 1989). Figure 4B shows a model in which the S4 helix converts from the resting to activated states by undergoing a "melting" of its H bonds upon input of electrical energy (Benndorf, 1989), providing a thermodynamically plausible mechanism for the conformation change.

Site-directed mutations within the S4 region cause dramatic shifts in voltage-dependent gating of both Na^+ (Stuhmer *et al.*, 1989) and K^+ (Papazian *et al.*, 1991; Lopez *et al.*, 1991) channels. These structure–function studies, its high charge density, and its dramatic conservation among voltage-dependent channels all point to the S4 helix as a key element of the voltage sensor—it appears to be the positive half of Armstrong's ratchet.

2. Multiple Ways to Inactivate Ca^{2+} Channels

Most voltage-dependent ion channels close soon after they open even though the stimulus is maintained; the process is called "inactivation." This can happen by two mechanisms in Ca^{2+} channels. In one, the channel closes soon after it opens, independent of how much Ca^{2+} has passed through it. Many Ca^{2+} channels also exhibit Ca^{2+}-dependent inactivation, in which accumulation of intracellular Ca^{2+} closes the pore (Fig. 5). What structures are expected to mediate these forms of inactivation?

The structure underlying one mechanism of inactivation has been deduced in a K^+ channel and agrees with a proposal based upon prior functional studies with Na^+ channels. Armstrong and Bezanilla (1977) proposed that the Na^+ channel, shortly after it has opened, is blocked from the inside when a flexible cytoplasmic domain of the channel swings in and plugs the pore. Among the evidence for the "ball and chain model" was the observation that lysine- and arginine-specific proteases remove inactivation when perfused into the cytoplasm of the cell (Armstrong *et al.*, 1973). Hoshi *et al.* (1990) have shown that the first 20 amino acids of

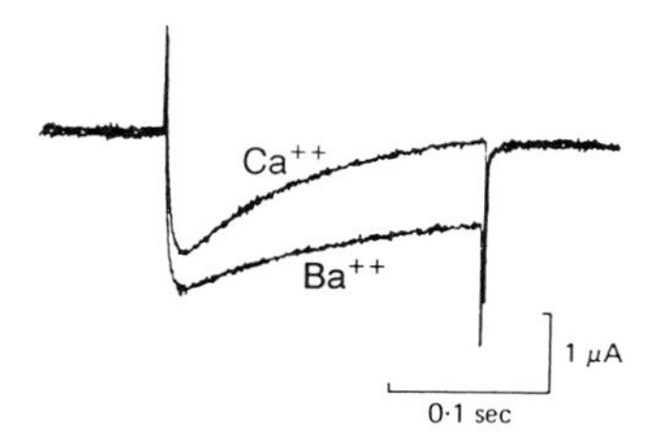

FIG. 5 Ca^{2+}-dependent inactivation. Inactivation of Ca^{2+} channels is faster if Ca^{2+} carries the current than if Ba^{2+} carries the current. Thus, there is a chemical component to the inactivation process. (From Eckert and Tillotson, 1981.)

a rapidly inactivating K^+ channel correspond to the predicted inactivation ball. When this sequence of amino acids is deleted, fast inactivation is lost in exact analogy to the action of the proteases. Moreover, inactivation is recovered by perfusing the interior of the cell with a synthetic peptide with the sequence of the first 20 residues (Zagotta *et al.*, 1990). Whether a similar structure is responsible for the rapid inactivation that occurs in some types of Ca^{2+} channels remains to be investigated.

Ca^{2+}-dependent inactivation is less well studied. The one concrete proposal is that Ca^{2+} accumulation activates a phosphatase that diminishes channel activity by dephosphorylation (Chad and Eckert, 1986).

B. Properties of the Pore

Ca^{2+} channels choose Ca^{2+} in preference to Na^+ at a ratio of 1000:1, allowing them to specifically pass Ca^{2+} even though the extracellular Na^+ concentration is 100-fold higher than that of Ca^{2+} (Lee and Tsien, 1982; Hess *et al.*, 1986). The extraordinary specificity is surprising because Ca^{2+} and Na^+ are so similar: the naked ions are identical in size (2 Å in diameter) and are both positively charged. What are the structures within the pore that confer such high selectivity between such similar ions?

A key property is the diameter of the pore. It is possible to measure the pore diameter in the absence of extracellular Ca^{2+} because the pore becomes permeable to monovalent cations, which are available in a variety of sizes. Figure 6A shows how the permeability of the Ca^{2+} channel to different monovalent cations diminishes with increasing ion size (McCleskey and Almers, 1985). The largest permeant ion is tetramethylammonium. Since tetramethylammonium is about 5.5 Å in diameter, the pore can be no less at its narrowest point (Fig. 6A,right). This diameter is nearly three times wider than the Ca^{2+} and Na^+ ions that the channel so perfectly

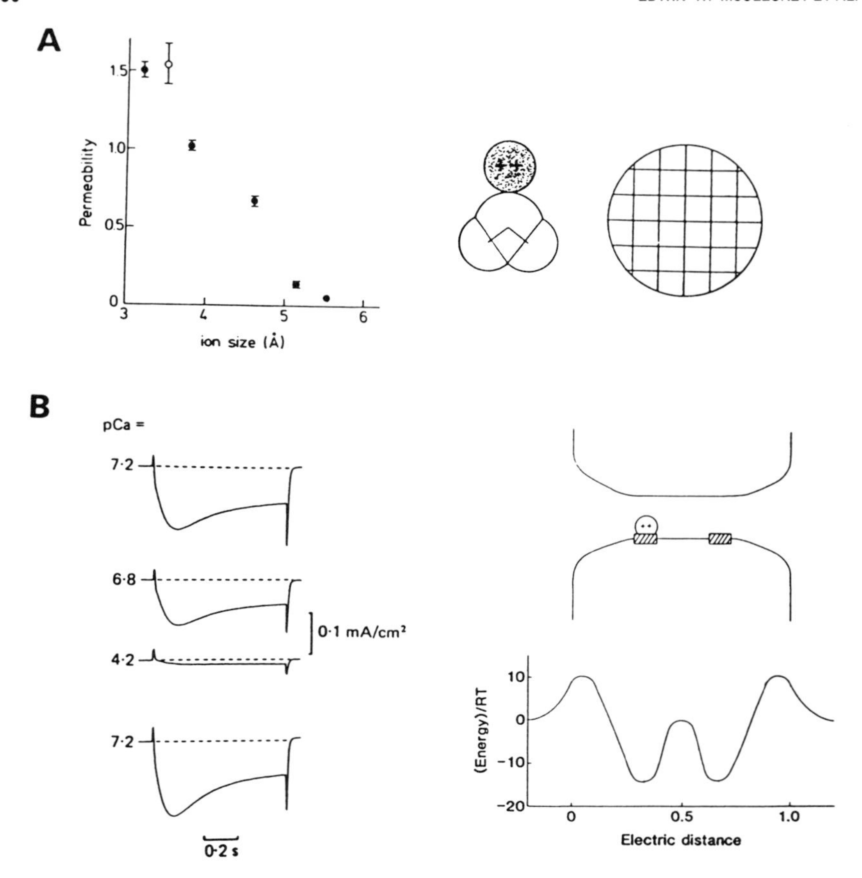

FIG. 6 (A) Pore size of the Ca^{2+} channel in the absence of Ca^{2+}. (Left) Permeability of the Ca^{2+} channel to hydrazinium (open circles) and to ammonium and its four methylated derivatives (filled circles) plotted against ion size. (Right) Scale drawing of a Ca^{2+} ion with an attached water and of a 6-Å pore. (From McCleskey and Almers, 1985.) (B) (Left) Ca^{2+} reversibly blocks Na^+ currents through Ca^{2+} channels as extracellular Ca^{2+} rises in the micromolar range. Ca^{2+} concentration is indicated in negative log units. (From Almers *et al.*, 1984). (Right) A pore with two binding sites and the corresponding potential energy diagram. (From Almers and McCleskey, 1984.)

distinguishes, so the pore appears not to choose among physiological ions by acting as a molecular sieve.

The alternative to a molecular sieve is selection by affinity; evidence points to a high-affinity Ca^{2+} binding site within the pore of the channel. Na^+ current through the channel is blocked with increasing Ca^{2+} concentration in the micromolar range (Fig. 6B, left) (Kostyuk *et al.*, 1983; Almers *et al.*, 1984). Thus, the high selectivity for Ca^{2+} over Na^{+} is conferred upon the channel by Ca^{2+} itself. This high affinity site was shown to be located within the pore by demonstrating that access to and exit from the site were modified by changes in the membrane voltage (Lansman *et al.*, 1986; Fukushima and Hagiwara, 1985).

The presence of a high affinity site within the pore severely limits flux through the channel. The maximum off-rate from a binding site with a 1 μM dissociation constant is about 1000/sec, whereas flux rates through Ca^{2+} channels are about 100,000/sec. How can the flux rate through the pore be greater than the maximum off-rate of a binding site within the pore? There are several theoretical solutions. One idea is that there are actually a pair of such sites, represented by deep energy wells in Fig. 6B, within the pore (Hess and Tsien, 1984; Almers and McCleskey, 1984). In this model, double occupancy is relatively unstable because of electrostatic repulsion between similarly charged ions brought into such close proximity. When two Ca^{2+} ions occupy the pore, one rapidly pops out, thereby speeding the flux rate. As appealing as this simple idea may be, it has not been proven; site-directed mutation studies might prove useful in this regard.

C. Different Types of Channels

There are several functionally distinct Ca^{2+} channels (Bean, 1989). Activation range and inactivation rates distinguish "T-type" channels from high-threshold channels. High-threshold channels are further subdivided by pharmacology and tissue location: L channels, present on both neurons and muscle, are sensitive to dihydropyridine drugs; N channels are present only on neurons and are blocked by ω-conotoxin; other neuronal channels, similar in kinetics to L and N, are insensitive to either conotoxin or DHPs. The short history of Ca^{2+} channel molecular biology has already generated many more clones than there are functionally distinct channels. The molecular basis of the pharmacological and kinetic differences among Ca^{2+} channels and the issue of the seemingly redundant array of Ca^{2+} channel clones are being addressed in a number of studies.

VI. Summary

Following purification using Ca^{2+} channel drugs as ligands, the skeletal muscle Ca^{2+} channel was shown to be a five-subunit structure containing one large (175 kDa) protein that is the pore and four auxiliary subunits.

Each subunit has been cloned and expression studies are proceeding rapidly. Particular success has been made in structure–function studies of excitation–contraction coupling using a Ca^{2+} channel-free mutant muscle. The work confirmed the suggestion made from physiological studies that muscle Ca^{2+} channels serve dual roles: passing Ca^{2+} and triggering Ca^{2+} release from an intracellular organelle. A variety of other predictions about the structure of Ca^{2+} channels have been reviewed here and these may soon be possible to test. Such concrete predictions along with analogies to studies on other voltage-dependent ion channels should speed progress in structure–function studies of Ca^{2+} channels.

References

Almers, W., and McCleskey, E. W. (1984). *J. Physiol. (London)* **353,** 585–608.
Almers, W., McCleskey, E. W., and Palade, P. T. (1984). *J. Physiol. (London)* **353,** 565–583.
Armstrong, C. M. (1981). *Physiol. Rev.* **61,** 644–683.
Armstrong, C. M., and Bezanilla, F. (1977). *J. Gen. Physiol.* **70,** 567–590.
Armstrong, C. M., Bezanilla, F., and Rojas, E. (1973). *J. Gen. Physiol.* **62,** 375–391.
Beam, K. G., Knudson, C. M., and Powell, J. A. (1986). *Nature (London)* **320,** 168–170.
Bean, B. P. (1989). *Annu. Rev. Physiol.* **51,** 367–385.
Benndorf, K. (1989). *Eur. Biophys. J.* **17,** 257–271.
Biel, M., Ruth, P., Bosse, E., Hullin, R., Stuhmer, W., Flockerzi, V., and Hofmann, F. (1990). *FEBS Lett.* **269,** 409–412.
Borsotto, M., Barhanin, J., Fosset, M., and Lazdunski, M. (1985). *J. Biol. Chem.* **260,** 14255–14263.
Catterall, W. A. (1986). *Trends Neurosci.* **9,** 7–10.
Catterall, W. A. (1988). *Science* **242,** 50–61.
Chad, J. E., and Eckert, R. (1986). *J. Physiol. (London)* **378,** 31–51.
Curtis, B. M., and Catterall, W. A. (1984). *Biochemistry* **23,** 2113.
Curtis, B. M., and Catterall, W. A. (1985). *Proc. Natl. Acad. Sci. U.S.A.* **82,** 2528–2532.
De Jongh, K. S., Merrick, D. K., and Catterall, W. A. (1989). *Proc. Natl. Acad. Sci. U.S.A.* **86,** 8585–8589.
De Jongh, K. S., Warner, C., and Catterall, W. A. (1990). *J. Biol. Chem.* **265,** 14738–14741.
Eckert, R., and Tillotson, D. L. (1981). *J. Physiol. (London)* **314,** 265–280.
Ellis, S. B., Williams, M. E., Ways, N. R., Brenner, R., Sharp, A. H., Leung, A. T., Campbell, K. P., McKenna, E., Koch, W. J., Hui, A., Schwartz, A., and Harpold, M. M. (1988). *Science* **241,** 1661–1664.
Flockerzi, V., Oeken, H. J., and Hoffmann, F. (1986a). *Eur. J. Biochem.* **161,** 217–224.
Flockerzi, V., Oeken, H. J., Hofmann, F., Pelzer, D., Cavalie, A., and Trautwein, W. (1986b). *Nature (London)* **323,** 66–68.
Fukushima, Y., and Hagiwara, S. (1985). *J. Physiol. (London)* **358,** 255–284.
Guy, H. R., and Seetharamulu, P. (1986). *Proc. Natl. Acad. Sci. U.S.A.* **83,** 508–512.
Hess, P. (1990). *Annu. Rev. Neurosci.* **13,** 337–356.
Hess, P., and Tsien, R. W. (1984). *Nature (London)* **309,** 453–456.
Hess, P., Lansman, J. B., and Tsien, R. W. (1986). *J. Gen. Physiol.* **88,** 293–319.
Hodgkin, A. L., and Huxley, A. F. (1952). *J. Physiol. (London)* **117,** 500–544.
Hoshi, T., Zagotta, W. N., and Aldrich, R. W. (1990). *Science* **250,** 533–538.

Hui, A., Ellinor, P. T., Krizanova, O., Wang, J.-J., Diebold, R. J., and Scwartz, A. (1991). *Neuron* **7,** 35–44.

Hymel, L., Striessnig, J., Glossmann, H., and Schindler, H. (1988). *Proc. Natl. Acad. Sci. U.S.A.* **85,** 4290–4294.

Jan, L. Y., and Jan, Y. N. (1989). *Cell* **56,** 13–25.

Jay, S. D., Ellis, S. B., McCue, A. F., Williams, M. E., Vedvick, T. S., Harpold, M. M., and Campbell, K. P. (1990). *Science* **248,** 490–492.

Kameyama, M., Hofmann, F., and Trautwein, W. (1985). *Pfluegers Arch.* **405,** 285–293.

Kim, H. S., Wei, X., Ruth, P., Perez-Reyes, E., Flockerzi, V., Hofmann, F., and Birnbaumer, L. (1990). *J. Biol. Chem.* **265,** 11858–11863.

Koch, W. J., Ellinor, P. T., and Schwartz, A. (1990). *J. Biol. Chem.* **265,** 17786–17791.

Kostyuk, P. G., Krishtal, O. A., and Pidoplichko, V. I. (1981). *J. Physiol. (London)* **310,** 403–421.

Kostyuk, P. G., Mironov, S. L., and Shuba, Y. A. (1983). *J. Membr. Biol.* **76,** 83–93.

Lacerda, A. E., Haeyoung, S. K., Ruth, P., Perez-Reyes, E., Flockerzi, V., Hofman, F., Birnbaumer, L., and Brown, A. M. (1991). *Nature (London)* **352,** 527–530.

Lansman, J. B., Hess, P., and Tsien, R. W. (1986). *J. Gen. Physiol.* **88,** 321–347.

Lee, K. S., and Tsien, R. W. (1982). *Nature (London)* **297,** 498–501.

Leung, A. T., Imagawa, T., and Campbell, K. P. (1987). *J. Biol. Chem.* **262,** 7943–7946.

Llinas, R., Sugimori, M., Lin, J.-W., and Cherskey, B. (1989). *Proc. Natl. Acad. Sci. U.S.A.* **86,** 1689–1693.

Lopez, G. A., Jan, Y. N., and Jan, L. Y. (1991). *Neuron* **7,** 327–336.

McCleskey, E. W., and Almers, W. (1985). *Proc. Natl. Acad. Sci. U.S.A.* **82,** 7149–7153.

McCleskey, E. W., and Schroeder, J. E. (1992). *Curr. Top. Membr. Transp.* **39,** 295–326.

Mikami, A., Imoto, K., Tanabe, T., Niidome, T., Mori, Y., Takeshima, H., Narumiya, S., and Numa, S. (1989). *Nature (London)* **340,** 230–233.

Mori, Y., Friedrich, T., Kim, M.-S., Mikami, A., Nakai, J., Ruth, P., Bosse, E., Hofmann, F., Flockerzi, V., Furuichi, T., Mikoshiba, K., Imoto, K., Tanabe, T., and Numa, S. (1991). *Nature (London)* **350,** 398–402.

Noda, M., Ikeda, T., Suzuki, H., Takeshima, H., Takahashi, T., Kuno, M., and Numa, S. (1986). *Nature (London)* **322,** 826–828.

Olivera, B. M., Gray, W. R., Zeikus, R., McIntosh, J. M., Varga, J., Rivier, J., Desantos, V., and Cruz, L. J. (1985). *Science* **230,** 1338–1343.

Papazian, D. M., Timpe, L. C., Jan, Y. N., and Jan, L. Y. (1991). *Nature (London)* **349,** 305–310.

Perez-Reyes, E., Kim, H. S., Lacerda, A. E., Horne, W., Wei, X., Rampe, D., Campbell, K. P., Brown, A. M., and Birnbaumer, L. (1989). *Nature (London)* **340,** 233–236.

Perez-Reyes, E., Wei, X., Castellano, A., and Birnbaumer, L. (1990). *J. Biol. Chem.* **265,** 20430–20436.

Reuter, H., Stevens, C. F., Tsien, R. W., and Yellen, G. (1982). *Nature (London)* **297,** 501–504.

Rieger, F., Bournaud, R., Shimahara, T., Garcia, L., Pincon-Raymond, M., Romey, G., and Lazdunski, M. (1987). *Nature (London)* **330,** 563–566.

Rios, E., and Brum, G. (1987). *Nature (London)* **325,** 717–720.

Ruth, P., Rohrkasten, A., Biel, M., Bosse, E., Regulla, S., Meyer, H. E., Flockerzi, V., and Hofmann, F. (1989). *Science* **245,** 1115–1118.

Schwartz, L. M., McCleskey, E. W., and Almers, W. A. (1985). *Nature (London)* **314,** 747–751.

Sieber, M., Nastainczyk, W., Zubor, V., Wernet, W., and Hofmann, F. (1987). *Eur. J. Biochem.* **167,** 117–122.

Snutch, T. P., Tomlinson, W. J., Leonard, J. P., and Gilbert, M. M. (1991). *Neuron* **7,** 45–57.

Stuhmer, W., Conti, F., Suzuki, H., Wang, X., Noda, M., Yahagi, N., Kubo, H., and Numa, S. (1989). *Nature (London)* **339,** 597–603.

Takahashi, M., Seagar, M. J., Jones, J. F., Reber, B. F. X., and Catterall, W. A. (1987). *Proc. Natl. Acad. Sci. U.S.A.* **84,** 5478–5482.

Talvenheimo, J. A., Worley, J. F., III, and Nelson, M. T. (1987). *Biophys. J.* **52,** 891–899.

Tanabe, T., Takeshima, H., Mikami, A., Flockerzi, V., Takahashi, H., Kangawa, K., Kojima, M., Matsuo, H., Hirose, T., and Numa, S. (1987). *Nature (London)* **328,** 313–318.

Tanabe, T., Beam, K. G., Powell, J. A., and Numa, S. (1988). *Nature (London)* **336,** 134–139.

Tanabe, T., Mikami, A., Numa, S., and Beam, K. G. (1990a). *Nature (London)* **344,** 451–453.

Tanabe, T., Beam, K. G., Adams, B. A., Niidome, T., and Numa, S. (1990b). *Nature (London)* **346,** 567–569.

Tanabe, T., Adams, B. A., Numa, S., and Beam, K. G. (1992). *Nature (London)* **352,** 800–803.

Triggle, D. J., and Janis, R. A. (1987). *Annu. Rev. Pharmacol. Toxicol.* **27,** 347–369.

Tsien, R. W., Hess, P., McCleskey, E. W., and Rosenberg, R. L. (1987). *Annu. Rev. Biophys. Biophys. Chem.* **16,** 265–290.

Varadi, G., Lory, P., Schultz, D., Varadi, M., and Schwartz, A. (1991). *Nature (London)* **352,** 159–162.

Zagotta, W. N., Hoshi, T., and Aldrich, R. W. (1990). *Science* **250,** 568–571.

Voltage-Dependent Sodium Channels

Sidney A. Cohen* and Robert L. Barchi†
Mahoney Institute of Neurological Sciences and the Departments of Medicine,* Neurology,† and Neuroscience,† University of Pennsylvania School of Medicine, Philadelphia, Pennsylvania 19104

I. Introduction

The past four decades have witnessed tremendous advances in our understanding of the structure and function of voltage-dependent sodium channels. Since the original electrophysiologic description of sodium currents during an action potential some 40 years ago (Hodgkin and Huxley, 1952), the sodium channel protein has been purified, reconstituted, sequenced, functionally expressed, and selectively mutated. Through these manipulations, a detailed understanding of the molecular workings of this complex protein is beginning to emerge. Three-dimensional structural models have been proposed and modified based on experimental results. Functional domains are being defined. The details of structure that distinguish sodium channel isoforms and the factors that govern isoform expression are beginning to emerge.

This review will focus on recent studies of sodium channel structure, function, and expression, focusing on how the interplay of electrophysiology, biochemistry, immunology, and molecular biology have contributed to our current molecular conceptualization of this protein. The vast earlier literature covering electrophysiologic and biochemical studies of the sodium channel will be only briefly summarized; the reader is referred to several excellent reviews for more in-depth coverage (Armstrong, 1981; Hille, 1984; Horn, 1984; Yamamoto, 1985; Catterall, 1986, 1988; Barchi, 1988).

II. Early Studies of Structure and Function

A. Sodium Currents and Channel Kinetics

The central role of sodium currents in the generation of action potentials was first described in the pioneering work of Hodgkin and Huxley (1952).

These studies demonstrated that the rapid upstroke of the action potential is the result of a voltage-dependent increase in membrane sodium conductance, which produces a shift in membrane potential from a point near the potassium equilibrium potential (V_k = ~ −90 mV) through zero toward the sodium equilibrium potential (V_{Na} = ~ +55 mV). This conductance increase is transient, lasting only milliseconds, and then spontaneously inactivates, allowing the membrane to repolarize as potassium permeability once again predominates. In many excitable membranes, repolarization is augmented by the delayed activation of a second channel specific for potassium ions.

The extension of the voltage clamp technique to a variety of cells confirmed and refined the initial kinetic observations made in the squid giant axon. Voltage clamp currents increase and then decrease in a continuous fashion after a rapid depolarization, leading to the concept of sodium channels as aqueous pathways through the membrane that can open progressively as a function of time. The subsequent application of patch clamping, with its ability to resolve current flow through individual ion channels, dramatically changed this view (Horn, 1984). It is now clear that depolarization causes sodium channels to undergo abrupt transitions from the closed state to an open state having a characteristic conductance of 10–30 pS (Fig. 1). After remaining open for variable lengths of time, sodium channels close rapidly to an inactivated state and typically remain inactivated until the membrane is again repolarized. The delay observed between depolarization and channel opening is thought to reflect kinetic delays inherent in passing first through several nonconducting closed states. Any single channel remains open only briefly relative to the overall duration of the macroscopic current. At most potentials, the time course of inactivation observed in voltage-clamp currents reflects the variability in latency to first opening of the channels rather than variability in the duration of channel opening (Aldrich, 1986).

The typical sodium channel conductance of ~20 pS implies an ion transport rate of greater the 10^7 ions/sec/channel, within an order of magnitude of the diffusion-controlled rate of ion movement in free solution (Sigworth and Neher, 1980; Meves and Nagy, 1989). This can be compared with rates of 10^4 ions/sec for small antibiotic ion carriers such as valinomycin and 5×10^5/sec for enzyme reactions such as carbonic anhydrase. Thus, ion transport mediated by the sodium channel is among the most rapid of protein-mediated processes, supporting the concept that ion movement takes place by diffusion through an aqueous pore rather than by a carrier or pump mechanism.

Sodium channel ion transport is highly selective, typically allowing preferential influx of Na^+ over K^+ or Rb^+ in a ratio of 1:0.14:0.03 (Hille, 1972). Based on relative conductance measurements under bi-ionic

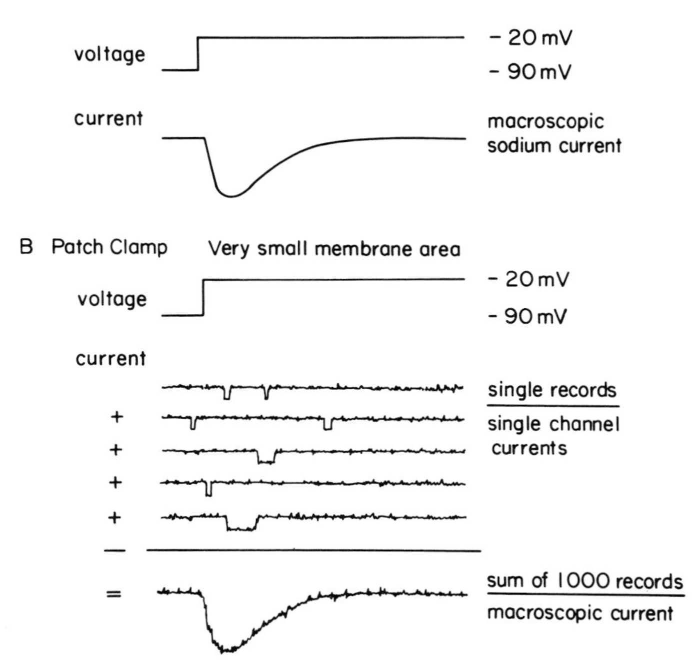

FIG. 1 Voltage clamp vs patch clamp analysis of sodium currents. When a large area of membrane containing many sodium channels is rapidly depolarized under voltage clamp conditions, an inward sodium current develops, peaks, and smoothly inactivates (A). With the patch clamp (B), where current through single ion channels can be resolved, individual sodium channels are seen to open rapidly to a characteristic conductance and then abruptly close to an inactivated state. The smooth macroscopic currents are the statistical summation of stochastic single channel events.

conditions, Hille modeled the selectivity filter as a 3.1 × 5.1 Å rectangular orifice lined by oxygen atoms that function as hydrogen bond acceptors (Hille, 1971). The presence of a carboxyl group in the orifice is suggested by the finding that ion movement is blocked by protonation of a site with a pK_a of ~5.2 (Hille, 1968). Interaction of partially hydrated metal cations with a site in this region is thought to provide the basis for ion selectivity.

Conformational changes associated with channel activation that reflect the action of the membrane potential on charged or dipolar protein segments within the electric field should produce a measurable current in the absence of ion movement. Analysis of sodium channel kinetics predicts that activation of each sodium channel requires the movement of six positive charges from the intracellular to the extracellular side of the membrane (Hodgkin and Huxley, 1952). The existence of this charge movement, or gating current, was first experimentally confirmed in the squid giant axon by Armstrong and Bezanilla (Armstrong and Bezanilla, 1973, 1974). Studies to date have not identified a component of gating current associated with channel inactivation. This has led to the concept that inactivation, although dependent on the prior conformational changes leading to activation, does not simply represent a reversal of this conformational change.

The inactivation process can be modified independently of activation. Internal perfusion of the squid giant axon with pronase, alkaline protease b, or trypsin can remove channel inactivation (Rojas and Armstrong, 1971; Armstrong *et al.*, 1973; Rojas and Rudy, 1976). These latter two enzymes cleave on the carboxyl side of lysine and arginine residues. In a similar manner, intracellularly applied arginine-specific reagents such as glyoxal, phenylglyoxal, and 2,3-butanedione also irreversibly block sodium channel inactivation (Eaton *et al.*, 1978). *N*-Bromoacetamide and the tyrosine-specific reagents *N*-acetylimidazole, tetranitromethane, and iodide plus lactoperoxidase also block sodium channel inactivation when perfused internally (Brodwick and Eaton, 1978). Reagents that modify other amino acids have no effect (Oxford *et al.*, 1978). These studies imply that an intracellular sequence containing lysine, arginine, and tyrosine is involved in sodium channel inactivation and is accessible to these enzymes from the cytoplasm. The actions of pronase and *N*-bromoacetamide are slightly voltage dependent, indicating that at least some susceptible residues are less available when the sodium channel is in an inactivated conformation (Salgado *et al.*, 1985).

B. Sodium Channel Pharmacology

Six major classes of toxin binding sites on the sodium channel have been identified on the basis of electrophysiologic and biochemical measurements (Table I) (see Catterall, 1980, for review). The first site to be characterized (site 1) binds the small, polar, heterocyclic guanidines tetrodotoxin (TTX) and saxitoxin (STX); μ-Conotoxin, a small polypeptide toxin from *Conus geographicus*, binds at an overlapping site (Cruz *et al.*, 1985). Site 1 toxins bind near the outer entrance to the channel and block ion flow either by physically occluding the channel or by inducing a conformational change in the protein.

Several lines of evidence point to the presence of a carboxyl group in the STX/TTX receptor site. Saxitoxin and TTX binding are blocked by protonation of a group with a pK_a of approximately 5.4 (Henderson *et al.*, 1973, 1974; Balerna *et al.*, 1975; Reed and Raftery, 1976; Barchi and

TABLE I

Drug and Toxin Receptor Sites on the Excitable Membrane Sodium Channel

Binding site	Drug/toxin	Electrophysiologic effect
1	Tetrodotoxin Saxitoxin μ-Conotoxin (GIIIa and GIIIb)	Block ionic conductance by either occluding channel or causing conformational change
2	Veratridine Batrachotoxin Aconitine Grayanotoxin	Shift the voltage dependence of activation to more negative potentials causing persistent activation
3	α-Scorpion toxins Sea anemone toxins	Shift the voltage dependence of inactivation to more positive potentials thereby inhibiting inactivation; work synergistically with site 2 toxins to cause persistent channel activation
4	β-Scorpion toxins	Shift the voltage dependence of activation to more negative potentials, causing enhanced activation
5	Brevetoxins Ciguatoxins	Shift the voltage dependence of activation to more negative potentials and block inactivation, causing repetitive firing of nerve cells
6	*Goniopora* toxin *Conus striatus* toxin	Inhibit channel inactivation
Binding sites not characterized	Local anesthetics, antiarrhythmics, and anticonvulsants	Frequency- and voltage-dependent inhibition of ionic conductance

Weigele, 1979). Toxin binding is also blocked by treatment of excitable membranes with carboxyl-modifying reagents (Shrager and Profera, 1973; Baker and Rubinson, 1975, 1976; Reed and Raftery, 1976). That this carboxyl group is distinct from the one responsible for ion selectivity is indicated by the ability of derivatized channels that no longer bind toxin to gate current with normal ion selectivity, although maximal sodium sodium conductance is reduced.

A number of lipid-soluble alkaloid toxins, including batrachotoxin, veratridine, aconitine, and grayanotoxin, bind to a second site (site 2) (Garber and Miller, 1987; Brown, 1988; Behrens *et al.*, 1989). These toxins shift the voltage dependence of channel activation in a hyperpolarizing direction and prevent channel inactivation, resulting in persistent channel activation even at normal membrane potentials. Because of their lipid solubility and the observation that their action is independent of the side of the membrane to which they are applied, these toxins are thought to approach the channel from the membrane lipid phase and interact with a hydrophobic portion of the channel within the membrane interior.

α-scorpion and sea anemone toxins bind externally to site 3 (Meves *et al.*, 1985; Warashina *et al.*, 1988). These polypeptide toxins slow or block inactivation and interact allosterically with the binding of toxins at site 2 to maintain channels in the active state. α- scorpion toxin binding is voltage dependent; the binding site undergoes a conformational change with depolarization that results in decreased affinity for the toxin.

Site 4 binds β-scorpion toxins such as Tityus γ toxin; binding of these toxins, again to an extracellular site, alters the voltage dependence of channel activation (Kirsch *et al.*, 1989). Unlike binding of toxins to site 3, binding of β-scorpion toxins does not allosterically affect the binding of toxins at site 2.

Binding sites 5 and 6 are defined largely on the basis of single toxins whose properties do not fit into the preceding four categories. Brevetoxin binds externally to site 5, markedly enhancing the effects of toxins that bind at site 2 while not competing for binding with the polypeptide toxins which bind at site 3 (Atchison *et al.*, 1986; Poli *et al.*, 1986). Toxins from the coral *Goniopora* bind to site 6 and slow channel inactivation without competing for binding at site 3 (Gonoi *et al.*, 1986).

Local anesthetics and type I antiarrhythmic agents interact with a receptor site(s) that can be approached from either the aqueous or the lipid phase of the bilayer. Both classes of drugs exhibit "use-dependent" effects, inhibiting sodium currents more rapidly and with greater affinity when membranes are repetitively stimulated (Hille, 1977; Woosley, 1991). This effect may be due to voltage-dependent conformational changes in the binding site(s) for these drugs that occurs on channel activation. In addition, the "modulated receptor" and "guarded gate" hypotheses of drug binding invoke the need for channel activation to allow drugs to enter the channel's aqueous pore and gain access to drug binding sites (Hondeghem and Katzung, 1987) .

Both local anesthetics/antiarrhythmic agents and the lipid-soluble toxins aconitine and batrachotoxin modify sodium channel activity with greater speed when sodium channels are stimulated in a repetitive fashion, suggesting that activation produces widespread conformational changes in sodium channel structure. Events at either surface of the channel protein can affect, and are affected by, processes that are distant in the channel's tertiary structure.

C. Sodium Channel Biochemistry

All sodium channels that have been isolated and biochemically characterized contain a large glycoprotein α subunit of 240,000–280,000 Da. In some species the α subunit is associated with one or two smaller β subunits of 30,000–40,000 Da. The single β subunit of the rat muscle channel and the β-1 subunit of rat brain are noncovalently associated with the α subunit, whereas the β-2 subunit found in rat brain is disulfide linked to the α subunit (see Barchi, 1988; Catterall, 1986, 1988, for reviews).

The general topology of α, β-1, and β-2 subunits in the membrane has been inferred from biochemical experiments. All sodium channel subunits are heavily glycosylated with up to 30% carbohydrate by mass, suggesting that both α and β subunits are at least partially exposed on the extracellular side of the membrane (Elmer *et al.*, 1985; Messner and Catterall, 1985; Roberts and Barchi, 1987; Thornhill and Levinson, 1987). α subunits from eel (Costa and Catterall, 1984a), rat brain (Costa *et al.*, 1982), rat heart (Cohen, 1991), and rat skeletal muscle (Yang and Barchi, 1990) are phosphorylated by cAMP-dependent protein kinase both *in vivo* and *in vitro*, indicating that this subunit is also exposed to the intracellular environment. In the rat brain channel, both the α and the β-1 subunits are covalently labeled by photoaffinity derivatives of scorpion toxins, which are known to bind to the external surface of the channel (Beneski and Catterall, 1980; Sharkey *et al.*, 1984; Jover *et al.*, 1988). β-1 and β-2 subunits are preferentially extracted into hydrophobic detergent phases and are labeled heavily by hydrophobic photoaffinity probes, suggesting that they contain substantial hydrophobic domains (Reber and Catterall, 1987). β subunits have not been shown to extend to the cytoplasmic surface of the membrane.

Reconstitution of purified sodium channel preparations from eel electroplax (Duch and Levinson, 1987), rat brain (Hartshorne and Catterall, 1984; Feller *et al.*, 1985; Hartshorne *et al.*, 1985), and rat skeletal muscle (Weigele and Barchi, 1982; Tanaka *et al.*, 1983; Kraner *et al.*, 1985) into defined phospholipid vesicles demonstrated that the protein components of these isolated channels are sufficient to mediate ion fluxes with the voltage dependence, ion selectivity, and pharmacologic properties expected of native sodium channels. The eel electroplax sodium channel, which contains only an α subunit, was functional in the purified state. The skeletal muscle sodium channel could be successfully reconstituted only as a complex of α and β-1 subunits. In the rat brain channel, selective removal of the β-2 subunit after disulfide reduction had no effect on neurotoxin binding, ion flux, single-channel conductance, or the voltage-dependence of purified channels, whereas removal of β-1 subunits by treatment with high ionic strength caused loss of all of these functional characteristics (Messner *et al.*, 1986). Tetrodotoxin binding to the sodium channel prevented the dissociation of the β-1 subunit and associated loss of function whereas carbodiimide-induced formation of intramolecular isopeptide bonds allowed preservation of function in the absence of β-1 subunits (Messner and Catterall, 1986; Tejedor *et al.*, 1988). Although one of the roles of the β subunit may be to stabilize the structure of the α subunit, these small subunits do not seem to be needed for fully functional channels. This conclusion has been supported by recent heterologous expression studies with α-subunit cRNA in oocytes (see Section V).

Sodium channels are extensively modified post-translationally by glycosylation, sulfation, and fatty acylation. Twenty to 30% of α- and β-subunit molecular weight in the eel, rat brain, and rat skeletal muscle channels is contributed by complex carbohydrate chains containing predominantly sialic acid and *n*-acetyl hexosamines (Elmer *et al.*, 1985; Messner and Catterall, 1985; Roberts and Barchi, 1987; Thornhill and Levinson, 1987). Mammalian cardiac sodium channels appear to differ in possessing less complex carbohydrate (Cohen, 1991; Gordon *et al.*, 1988).

Because of the high degree of glycosylation and the resultant negative charge contributed by sialic acid, several groups have examined the possible role of these carbohydrate groups in sodium channel function. Inhibiting N-linked glycosylation with tunicamycin prevents the synthesis and membrane insertion of normal amounts of mature sodium channels in both nerve and muscle cells (Bar-Sagi and Prives, 1983; Waechter *et al.*, 1983; Schmidt *et al.*, 1985; Sherman *et al.*, 1985; Schmidt and Catterall, 1986, 1987). Cultured neurons grown in the presence of tunicamycin synthesize sodium channel α subunits in normal amounts but the channel protein is rapidly degraded before being acylated, sulfated, assembled with β-2 subunits, or released from the endoplasmic reticulum. If core glycosylation is allowed but subsequent processing and sialylation are inhibited with castanospermine or swainsonine, sodium channel protein is synthesized, otherwise post-translationally modified, assembled with β subunits, and inserted into the surface membrane. Although these sodium channels were shown to bind STX and TTX normally, functional studies were not carried out. Similar findings were obtained with sodium channels expressed in squid giant axons (Gilly *et al.*, 1990). These studies imply that core glycosylation is required for the normal synthesis, processing, and insertion of sodium channels into the surface membrane, but it is not clear whether channels with only core carbohydrate function normally.

Removal of terminal sialic acid groups by neuraminidase had no effect on STX and α-scorpion toxin binding but produced channels in which three subconductance states (5, 8, and 14 pS) in addition to the normal conductance state (18 pS) were observed (Scheuer *et al.*, 1988; Recio-Pinto *et al.*, 1990). Neuraminidase treatment also resulted in a large depolarizing shift in the average potential required for channel activation. These studies suggest that sialic acid groups may play a role both in determining the magnitude of the electric field affecting sodium channel activation and in stabilizing the normal conducting conformation of the ionic pore. In the absence of sialic acid, this conducting state may oscillate among several closely related conformations each of which has a different mean conductance value.

Post-translational processing of α subunits includes the addition of palmityl residues to cysteines in thioester linkage and sulfate residues to oligosaccharides (Schmidt and Catterall, 1987). Palmitate incorporation, like core glycosylation, is inhibited by tunicamycin, whereas sulfation, like sialylation, is inhibited by castanospermine. Based on these observations, a model of post-translational processing has been proposed in which core glycosylation occurs in the rough endoplasmic reticulum, allowing proper folding of the sodium channel α subunit and possibly providing a signal for transport to the Golgi. Once transported to the Golgi, palmitation, terminal glycosylation, and sulfation occur but apparently do not play an essential role in subunit assembly and transport of functional sodium channels to the cell surface.

Each sodium channel studied to date is capable of being rapidly and specifically phosphorylated by cAMP-dependent protein kinase (PKA). All sodium channels have high levels of basal phosphorylation (see Table II); activation of endogenous PKa induces phosphorylation of the same residues that are phosphorylated *in vitro*, implying a physiologic role of phosphorylation at these sites (Rossie *et al.*, 1987; Rossie and Catterall, 1989). Neurotoxin-activated $^{22}Na^{+}$ influx into synaptosomes is reduced in the presence of activators of PKA (Seelig and Kendig, 1982), whereas whole-cell voltage clamp of neurons in culture demonstrates that approximately 20% of sodium channels inactivate at more negative membrane potentials in the presence of PKA activators (Coombs *et al.*, 1988). Furthermore, steady-state inactivation of cardiac sodium channels is shifted to more negative membrane potentials in the presence of β-adrenergic agonists, activated G_s, or other agents that increase intracellular cAMP (Schubert *et al.*, 1989). Activators of intracellular cAMP down-regulate the surface expression of sodium channels in cardiac myocytes, possibly

TABLE II
Sodium Channel Phosphorylation

Characteristic	Eel	Rat brain	Rat muscle
% Serine	>77%	<100%	100%
Molar incorporation	1.2–2.3	3–4	1
No. of labeled tryptic fragments	7	7	1
Endogenous phosphorylation	>70%	~50%	~50%
Location	3 NH_3; 1 COOH	ID 1–2	ID 1–2

by phosphorylating sodium channels directly (Taouis *et al.*, 1991). In addition, cell surface receptors for cAMP appear to shift the voltage dependence of channel activation to more negative potentials (Sorbera and Morad, 1991). The results strongly suggest that both PKA and cAMP are involved in mediating channel inactivation and down-regulation; further studies are needed to confirm these hypotheses and demonstrate the detailed molecular mechanisms involved.

Two groups have obtained somewhat different results in studying the modulation of sodium channel activity by activators of protein kinase C (PKC). Using both cultured neurons and rat brain IIa sodium channels heterologously expressed in CHO cells, peak sodium currents were decreased by up to 80% and a marked slowing of channel inactivation was obtained by activators of PKC. Effects on peak current were felt due either to a decrease in the number of active sodium channels or to a decrease in the probability of single-channel opening. The marked slowing of channel inactivation was shown to be due to both an increased lifetime of single-channel opening and an increased probability of single-channel reopening during prolonged depolarizations (Numann *et al.*, 1991). Phosphorylation by PKC of a single residue in the conserved ID 3–4 region of the rat brain IIa sodium channel was found to be responsible for these effects (West *et al.*, 1991).

Similar studies performed with the rat brain IIa channel heterologously expressed in oocytes demonstrated similar effects on peak sodium current but showed alteration of channel activation rather than channel inactivation by activators of PKC (Dascal and Lotan, 1991; Schreibmayer *et al.*, 1991). The explanation for the differences obtained in these two studies includes the use of different heterologous expression systems, different ionic and PKC-activator concentrations, and possible differences in sodium channel α-subunit or PKC isoforms. In support of the first study is the known abnormal inactivation kinetics of rat brain channels expressed in oocytes (see Section V,1) and the fact that identical results were obtained using cultured neurons.

III. Sodium Channel Primary Structure

The elucidation of the complete primary sequences of sodium channels from various tissues and species represents a major advance in the structural characterization of this channel protein. Complementary DNAs that correspond to mRNA transcripts ranging in size between 6.8 and 9.5 kb have been cloned. These cDNAs contain open reading frames of 1820 to 2072 amino acids, corresponding to core protein molecular masses be-

TABLE III
Cloned Mammalian Sodium Channels

Channel	cDNA (bp)	mRNA (kb)	Protein length (AA)	Predicted core protein MW
Eel electroplax	7230	8.0	1820	208,321
Rat				
I	8147	9–9.5	2009	228,758
II	8343	9–9.5	2005	227,840
III	6822	9.0	1951	221,375
Rat skeletal muscle	6957	8.5	1840	208,847
Rat heart	7076	8.5	2018	227,339
Human skeletal muscle	7823	8.5–9.0	1836	208,107
Human heart	8491	9.0	2016	227,159

tween 208 and 228 kDa (see Table III). Sequences for sodium channels from eel (Noda *et al.*, 1984), *Drosophila* (Salkoff *et al.*, 1987), rat brain (Auld *et al.*, 1985, 1988; Noda *et al.*, 1986a; Kayano *et al.*, 1988), and rat and human skeletal (George *et al.*, 1992; Trimmer *et al.*, 1989; Kallen *et al.*, 1990) and cardiac (Rogart *et al.*, 1989; Gellens *et al.*, 1992) muscle have now been reported.

All known sodium channel primary sequences contain four large regions, each encompassing 231 to 327 amino acids, that show strong internal homology at the amino acid level. About half of the residues are either identical or conservatively substituted among these domains in each channel. The corresponding domains of sodium channels from different tissues and species are more highly related (80–90%) than are the different domains within a single sodium channel (45–60%), implying that functional distinctions between the different domains evolved prior to species divergence (Trimmer *et al.*, 1989; Kallen *et al.*, 1990). Each domain (designated D1 through D4) contains multiple hydrophobic segments that are thought to form transmembrane α helices.

In general, inter- and intraspecies sequence homology is highest within the repeat domains, whereas regions linking the domains (interdomain or ID regions) exhibit lower homology and contain large stretches that are present in one species but not in another. This is especially true in the region between D1 and D2 (ID 1–2) where the brain and heart sequences contain a large insertion (~200 amino acids) when compared to the same region of the skeletal muscle or eel electroplax sequences.

The ID 3–4 region is an exception to the variability seen in the other interdomain regions and has been linked to channel inactivation. This segment, the shortest of the three ID regions, is highly conserved among known sodium channels and contains an unusually large number of lysines and arginines, providing multiple potential pronase, trypsin, glyoxal, phenylglyoxal, and 2,3-butanedione cleavage sites. Two vicinal tyrosines are also present, completing the sequence requirements for the proteases and oxidizing agents that affect channel inactivation. The region is also rich in proline residues, which may provide a means of pivoting this short charged segment on the cytoplasmic surface of the channel protein. These observations, coupled with functional experiments that will be discussed below, have led a number of authors to propose that this region between D3 and D4 plays a role in inactivation gating by serving as the "swinging door" or "ball and chain" that blocks the inner mouth of the activated channel (Noda *et al.*, 1986a).

Analysis of the sequence in each repeat domain for relative hydrophobicity identifies six regions capable of forming membrane spanning helices. Two of these regions are hydrophobic segments that contain no charged residues (S5 and S6), three are hydrophobic segments that have several charged residues (S1, S2, and S3), and one segment (S4) is unique in containing a positively charged residue at every third position. When the S4 segment is folded into an α helix, the predicted optimal conformation for the local sequence, the resultant amphipathic helix has a spiral band of positive charge along an otherwise neutral surface (see Fig. 2). The repeating tripeptide motif ((R or K)–X–X) occurs four times in D1/S4, five times in D2/S4, five times in D3/S4, and eight times in D4/S4. All 22 positive charges are conserved in the S4 segments of all sodium channels sequenced to date. Most of the hydrophobic residues in these segments are highly conserved as well.

Homologous S4 helices with the same repeating positive charge motif are found in the putative membrane spanning domains of other voltage-sensing proteins, including the α subunit of voltage-dependent calcium channels (Tanabe *et al.*, 1987) and all of the various voltage-dependent potassium channels related to the *Drosophila* potassium channel (e.g., Shaker, Shal, Shab) (Baumann *et al.*, 1987; Kamb *et al.*, 1987, 1988; Papazian *et al.*, 1987; Pongs *et al.*, 1988; Schwarz *et al.*, 1988). Several authors have independently proposed that this sequence serves as the voltage sensor for these channels and that gating current is the result of movement of these positively charged amino acids within the membrane electric field.

Sodium channel proteins undergo extensive post-translational modifications. All sodium channels that have been fully sequenced have between 15 (rat brain III) and 20 (SkM2) potential N-linked glycosylation sites, at least 75% of which are predicted to be extracellular in current models. Most of these sites are grouped in the first and third domains and are found

S4 Helix

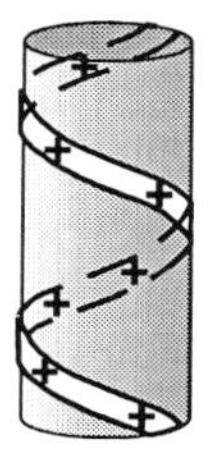

D-4 / S4 Sequences

Sequence	Channel
FRVIRLARIGRILRLIRGAKGIRTL	HH1
FRVIRLARIGRILRLIRGAKGIRTL	SkM2
FRVIRLARIGRVLRLIRGAKGIRTL	SkM1
FRVIRLARIGRILRLIKGAKGIRTL	RB-I
FRVIRLARIGRILRLIKGAKGIRTL	RB-II
FRVIRLARIGRILRLIKGAKGIRTL	RB-III
FRVIRLARIARVLRLIRAAKGIRTL	EEL
LRVVRLARIGRILRLIKGAKGIRTL	FLY

FIG. 2 Proposed structure of S4 helix. The S4 segments of each domain in all cloned sodium channels contain an identical motif of a positively charged residue at every third position. When organized as an α helix, a spiral band of positive charges around an otherwise neutral surface is obtained. These positive charges are thought to pair with fixed negatively charged residues in other helices in each domain. The movement of the positive residues in the membrane electric field is thought to give rise to gating current. The lower panel in this figure demonstrates the marked conservation of both positive and neutral residues in the S4 helices of a variety of sodium channels.

nearly exclusively in the loops joining the proposed S5 and S6 helices. Five N-glycosylation sites are conserved in all sodium channels sequenced to date, adding evidence to the hypothesis that carbohydrate plays an essential role in channel structure and/or function.

All sodium channels have potential sites for phosphorylation by cAMP-dependent protein kinase; while none of these are universally conserved, they do tend to be grouped in the amino terminus, the ID 1–2 region, or the carboxy terminus. Although the rat brain sodium channel has been shown to be a substrate for phosphorylation by protein kinase C (Costa and Catterall, 1984b; Numann *et al.,* 1991), phosphorylation by tyrosine kinase has not been reported.

Rat brain and rat skeletal muscle sodium channels are associated with either one or two β subunits. Due to difficulties in manipulating β subunits biochemically, only the β-1 subunit of the rat brain sodium channel has been sequenced (Isom *et al.,* 1992). Using PCR and library screening techniques, two 1,400 bp clones containing a 656 bp open reading frame encoding a 218 amino acid protein (molecular mass 22,851 Da) were identified. Hydropathy analysis, combined with the presence of a signal peptide in the β-1 subunit primary sequence, leads to a molecular model in which a single transmembrane domain (spanning residues 142–163) separates the extracellular amino-terminus from the intracellular carboxy-terminus. Four of six potential *N*-linked glycosylation sites are extracellular by this analysis; assuming ~3 kDa of carbohydrate per glycosylation site, each β-1 subunit should contain ~12 kDa of carbohydrate.

Each of these predictions is consistent with biochemical studies of rat brain sodium channel β-1 subunits. The deglycosylated β-1 subunit has a apparent molecular mass of 23 kDa while the glycosylated β-1 subunit has an apparent molecular mass of 36 kDa on SDS-PAGE (Messner and Catterall, 1985). In addition, β-1 subunits are preferentially extracted into hydrophobic detergent phases and are labeled by hydrophobic photoaffinity probes, suggesting that they contain at least one hydrophobic domain (Reber and Catterall, 1987).

IV. Sodium Channel Tertiary Structure

A. Structural Models

A major goal of studies of voltage gated sodium channels is to understand how channel function derives from channel structure. Although electrophysiologic studies have provided considerable insight into the processes of channel gating and ion selectivity, they are not sufficient to describe the molecular mechanisms by which channels work. Accomplishing this requires detailed knowledge of protein structure and of how that structure is affected by membrane potential. Although X-ray diffraction of proteins is the most rigorous approach to protein structure currently available, crystals of sodium channel protein suitable for diffraction have not yet been obtained.

An alternate approach to this problem is to develop models of sodium channel tertiary structure based on the known primary amino acid sequences and then experimentally test predictions derived from these models using a variety of methods. This approach is limited by the difficulty of accurately predicting tertiary structure even of moderately sized soluble proteins. Nonetheless, the power of this method lies in its ability to generate predictions that can be tested experimentally.

Each of the models to be discussed below was generated with analytical methods that predict local secondary structure for short segments of the primary sequence. More recently, as sequence information has become available for multiple sodium channel subtypes as well as for other related voltage-gated ion channels, these models have been further refined to reflect areas of either conserved or variable sequence among these channels.

Certain features are shared by most of the models (see Fig. 3) (Noda *et al.*, 1984; Greenblatt *et al.*, 1985; Guy and Seetharamulu, 1986; Noda *et al.*, 1986b; Guy and Conti, 1990). All except one (Kosower, 1985) explicitly recognize the presence of the four internally homologous repeat domains in the primary sequence and suggest that these domains are organized in a similar fashion in the plane of the membrane. Since all sodium channels lack an amino-terminal signal peptide sequence, all models place the amino terminus on the cytoplasmic side of the membrane. Each domain is predicted to have an even number of transmembrane crossings, leading to the placement of the interdomain sequences and carboxy terminus on the

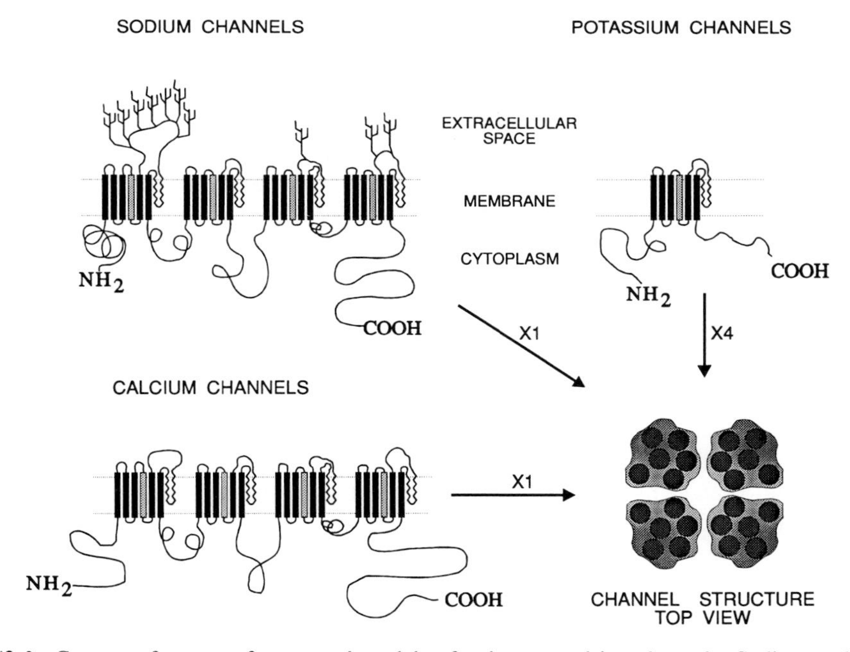

FIG. 3 Common features of structural models of voltage-gated ion channels. Sodium and calcium channel α subunits are modeled to contain four internally homologous repeat domains, whereas potassium channels are thought to be composed of four subunits each of which corresponds to an homologous repeat domain. All models place the amino and carboxy termini and the interdomain linking sequences on the cytoplasmic surface and are organized to optimize the number of potential extracellular N-linked glycosylation sites and intracellular PKA phosphorylation sites. The highly conserved S4 helix is shaded in each homologous domain. The homologous domains are thought to form a donut-shaped structure in which the circularly arranged domains surround a central aqueous pore.

cytoplasmic surface as well. In addition, each of the models is constructed to optimize the number of potential cAMP-dependent protein kinase sites on the intracellular surface of the protein and place the maximum number of consensus N-glycosylation sites on the extracellular surface. The models differ primarily in the details of the organization of the repeat domains and the structure of the ionic pore itself.

Noda *et al.* (1984) initially proposed that the four repeat domains were organized in a pseudosymmetric fashion within the plane of the membrane and that the weakly amphipathic S1 and S2 helices of each domain contributed to the formation of the central aqueous pore. The amphipathic S4 sequence was modeled to be a 3^{10} helix with a vertical stripe of positive charge on a cylinder otherwise covered with nonpolar residues; because of the positive charge, the authors assigned both S4 and the weakly amphipathic S3 helix to a cytoplasmic location.

Guy and Seetharamulu (1986) subsequently proposed that the ionic pore was lined by the amphipathic S4 α helix of each domain whose positive charges were counterbalanced by negative charges in a region designated "S7," placed between helices S5 and S6 of each domain. In this model, the S4 helix was proposed as the voltage sensor and was placed within the potential gradient of the membrane. Channel activation gating was envisioned as a screw-like motion of the S4 helices in one direction and S7 region in the other direction such that each positive charge is neutralized by a negative side-group one residue below on the adjacent helix. A 60° rotation in each helix could result in a 4.5 Å translocation of the S4 helices perpendicular to the plane of the membrane, thus producing a net charge movement that would be detected as a gating current.

A model proposed by Greenblatt *et al.* (1985) is also similar to that described by Noda *et al.* (1984) with several exceptions. The amphipathic S3 helices are assigned to the lining of the aqueous channel pore and the positively charged amphipathic S4 helices are transmembrane in orientation and placed in the interior of each of the homologous domains, with the positive charges in S4 interacting with negative charges in S3 and S7. Two additional weakly hydrophobic sequences containing a number of charged residues (denoted SS1 and SS2 below) were identified between the S5 and the S6 helices of each domain and modeled to form additional membrane-spanning structures. This model also proposed that gating involved movement of the S4 helix in response to changes in the transmembrane potential.

Two of these models have been updated based on sequence comparisons of sodium channels that had been subsequently cloned. The rationale used was that sequences conserved between different sodium channel subtypes are likely to represent regions responsible for common channel function whereas variable regions might correspond to sequences that are either unimportant in channel function or that produce the unique characteristics of a particular channel isoform. Noda *et al.* (1986b) revised their original model by placing all six helical segments in each domain within the membrane. The sequences lining the aqueous channel pore were now derived from the S2 helix, which contain glutamic acid and lysine residues that are highly conserved in sodium channels from different tissues. In addition, the S4 helix was now described as the voltage sensor and was placed within the interior of each repeat domain.

Recently, Guy and Conti (1990) proposed a detailed model of the sodium channel that used information gleaned from sequence comparisons among cloned sodium channels as well as experimental studies of channel topography. These authors further subdivide the S4 and S6 helices into amino-terminal regions (S4n and S6n), carboxy-terminal regions (S4c and S6c), and, for the S4 helix, an α-helical mid-portion region (S4h). The SS1 and SS2 regions first described by Greenblatt *et al.* (1985) and referred to as the "S7" region by Guy and Seetharamulu (1986) are also preserved with minor modifications (Fig. 4). Finally, the transmembrane region of the channel protein is divided into three concentric cylinders perpendicular

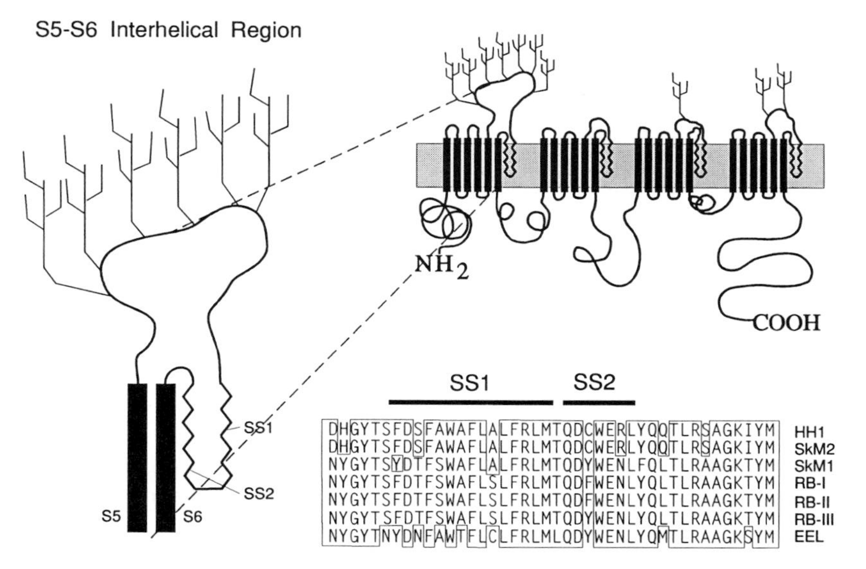

FIG. 4 Sequence conservation and proposed structure of the interhelical S5-S6/SS1 and SS2 regions. Several authors have suggested that a portion of the interhelical S5–S6 sequence, which was originally proposed to be extracellular, dips back into the membrane to form part of the central aqueous pore. This region is highly conserved (including conservative amino acid substitutions) and may form at least part of the binding sites for toxins that bind extracellularly.

to the membrane (interior, middle, and exterior) and into three cross-sectional layers parallel to the membrane (inner, middle, and outer). The exterior cylinder comprises the S1, S2, S3, and S5 helices of previous models and spans all three cross-sectional layers. The interior cylinder is composed of SS2 β-strands and S4c in either a β-strand or α-helical conformation and is present only in the middle layer of the three cross sections, thus forming the narrow portion of an "hour-glass" shaped ionic pore. The middle cylinder of the middle layer comprises buried hydrophobic SS1 and S6n segments in an unspecified orientation. The large portals of the outer and inner layers (i.e., the entrance to and exit from the channel) are lined with segments that contain hydrophilic side chains: the extracellular approach is lined by S4n, S4h (in the open conformation), and segments between S5 and SS1 and between SS2 and S6, whereas the cytoplasmic exit from the channel is lined by the region linking S4 to S5 and by S6a. Charged groups near the ends of S1, S2, S3, and S5 are also modeled to be at least partially exposed to the intracellular lumen in the inner cross-sectional layer.

In this model, voltage-dependent transitions between the open and the "deactivated" states involve conformational changes of only the S4 and SS2 segments and the sequences that link these segments to adjacent membrane-embedded sequences. As opposed to the "helical screw" or "sliding helix" model of activation gating, Guy and Conti suggest a "propagating helix" model in which S4 has an extended or β-stranded structure at its amino and carboxy termini and assumes an α-helical conformation only in its mid-portion. Activation would be associated with an amino-terminal movement of the α-helical structure such that the amino terminus of S4 would go from an extended to an α-helical structure whereas the formerly α-helical mid-portion would revert to an extended structure, thus propagating the α helix from middle cross-sectional region toward the outer cross-sectional region (see Fig. 2).

B. Experimental Studies

Common features of all current models of sodium channel structure include the presence of four membrane-embedded homologous domains, intracellular amino and carboxy termini, and intracellular sequences joining the homologous domains. The models vary in the way in which the sequences within each repeat domain are organized and in the choice of sequences that line the aqueous pore of the channel. Although many elements of these models, such as the overall structure of the protein, the role of the 3–4 interdomain region in channel inactivation, and the role of the S4 helices as voltage sensors, are attractive and intuitively compelling, the true value of these models is that they generate predictions that can be tested and used to constrain the features of future models.

Some of the most direct constraints on the overall organization of the sodium channel protein have been provided by the localization of sites of post-translational modification and by the identification of sequences comprising toxin binding sites. Sodium channels from a variety of tissues are rapidly phosphorylated by cAMP-dependent protein kinase A (and by protein kinase C in the case of the rat brain sodium channel). Since phosphorylation is an intracellular process, identifying the sequences phosphorylated by these kinases *in vitro* constrains their location to the intracellular surface of the protein.

Studies with the eel electroplax, rat brain, and rat skeletal muscle sodium channel proteins have shown that all three channels have a high level of basal phosphorylation (>50%) (see Table II). All are phosphorylated predominantly on serine residues. Tryptic digestion of the eel electroplax (Emerick and Agnew, 1989) and rat brain (Rossie *et al.*, 1987; Rossie and Catterall, 1989) channels yielded seven labeled fragments, whereas similar treatment of the skeletal muscle channel yielded only one prominently labeled fragment (Yang and Barchi, 1990). The phosphorylation sites in the brain and skeletal muscle channels are located in the interdomain 1–2 region whereas those for the eel electroplax channel are found in the amino and carboxy termini. In addition, the site of endogenous phosphorylation of the rat brain sodium channel by protein kinase C has been localized to the ID 3–4 region (West *et al.*, 1991). These studies constrain models of channel tertiary structure to ones in which these four regions are exposed to the intracellular milieu.

Additional constraints on the structure of the first homologous domain have been place by the localization of the sites of attachment of α-scorpion toxin photoaffinity derivatives to the rat brain sodium channel. α-scorpion toxin is a small polypeptide that binds to the extracellular surface of the sodium channel at site 3 (see Table I) in a voltage-dependent manner and modifies channel activity by slowing inactivation. When an azidonitrobenzoyl derivative of toxin V, the principal α-scorpion toxin from *Leiurus quinquestriatus* (LqTx), was used to photoaffinity label sodium channels in either synaptosomes or intact neuronal cells in culture, or after purification and reconstitution into lipid vesicles, both the α and the β-1 subunits of the rat brain sodium channel were specifically labeled (Beneski and Catterall, 1980; Sharkey *et al.*, 1984). Using a combination of proteolysis and immunoprecipitation, the site of attachment of the photoactivatable group in the α subunit was initially localized within domain 1. (Tejedor and Catterall, 1988). Further proteolysis yielded a labeled but nonglycosylated 14-kDa cyanogen bromide fragment derived from within amino acids 313–426, further constraining the labeled site to beyond the five potential glycosylation sites in the proposed extracellular loop joining S5 and S6 in D1. This region overlaps with the SS1 and SS2 sequences that are postulated to form at least part of the transmembrane pore (see Fig. 4).

Since this site of attachment could be adjacent to, rather than form part of, the toxin receptor site, a series of monoclonal and polyclonal antibodies were used in competition studies to further map the toxin binding site (Thomsen and Catterall, 1989). Using this approach, residues 371–400 were identified as comprising a portion of the α-scorpion toxin binding site. A survey of antibodies against other proposed extracellular regions demonstrated that an additional segment encompassing residues 1686–1705, which is located in the region between the S5 and S6 helices of domain 4, is also involved in toxin binding. The kinetics of toxin inhibition is consistent with direct competition between antibody and toxin for binding; decreased binding due to conformational changes in the protein was effectively ruled out.

Since scorpion toxins have a small binding surface consisting of only about 13 amino acids (Fontecilla-Camps *et al.*, 1988), the inhibition of toxin binding by antibodies to these sequences from opposite ends of the molecule suggests that they are closely opposed *in vivo* (see lower right panel of Fig. 3). Since at least one of the β subunits was also labeled by photoreactive toxin derivatives in previous studies, these results provide evidence for a compact toxin binding site composed of at least two sequences in the α subunit and, possibly, a region of the β-1 subunit. Thus, the S5–S6 loops of the first and fourth domains are accessible from the extracellular surface of the membrane. Since toxin binding interferes with the coupling of activation to inactivation (Thomson and Catterall, 1989), some investigators have proposed that the toxin interacts with the S4 helices of both D1 and D4, interfering with the coupling of activation gating charge movements in domain 1 to the conformational changes in domain 4 that initiate closing of the inactivation gating segment.

Limited proteolytic cleavage has also been used to probe the tertiary structure of the skeletal muscle sodium channel protein. Using sequence-specific antisera, the channel protein was shown to be sensitive to endogenous proteases in regions predicted to link the four repeat domains, especially the ID 2–3 region, whereas the repeat domains themselves were resistant to proteolysis (Kraner *et al.*, 1989). Furthermore, when proteolyzed channel protein was probed with iodinated lectin, most of the carbohydrate was located between 22 and 90 kDa from the amino terminus of the channel, consistent with the observed clustering of consensus N-glycosylation sites in the first domain of all channels sequenced to date.

These studies were extended using limited proteolysis with trypsin, chymotrypsin, and V-8 protease in conjunction with a more extensive panel of sequence-specific antisera (Zwerling *et al.*, 1991). When the channel was in its native conformation either in membrane fragments or in mixed detergent-lipid micelles, cleavage occurred at a limited number of sites and in discrete, reproducible steps (Fig. 5). Cleavage at these sites took place in a fixed temporal sequence, suggesting a hierarchy of relative accessibility to the soluble enzymes. Furthermore, the sequence of appearance of the fragments was similar for all three enzymes, suggesting that the observed pattern was determined by the accessibility of selected sites in the tertiary structure rather than by the presence of sites within the primary sequence. The carboxy terminus of the protein was rapidly cleaved at multiple sites by all proteases, whereas the amino terminus proved remarkably protease-resistant. Of the remaining sites, ID 2–3 was the most readily cleaved, whereas ID 1–2 and especially 3–4 were kinetically less accessible. Although D1 and D4 appeared to remain intact throughout proteolysis, limit fragments for epitopes associated with D2 and D3 suggest that cleavage eventually occurred at sites between the putative S5 and S6 helices in these domains. Portions of sequences in these large loops appear to be exposed in the tertiary structure.

Based on the known sequence of the sodium channel and the location of the antibody epitopes and assuming that carbohydrate is distributed evenly at each glycosylation site, the carbohydrate content of each domain was estimated. The weight of the limit fragment arising from D1 suggests that most channel carbohydrate is covalently associated with this domain,

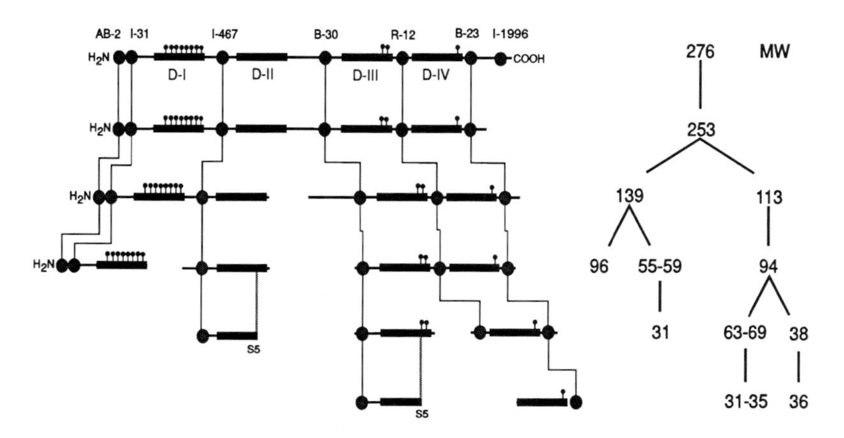

FIG. 5 The location of protease-sensitive regions in the sodium channel primary sequence. Sodium channel protein in native membranes and in mixed micelles was cleaved with trypsin, α-chymotrypsin, and V-8 protease. With all three enzymes, proteolysis occurred first in the carboxy terminus, next in the ID 2–3 region, then in the ID 1–2 region, and then in the ID 3–4 region. Prolonged protease treatment eventually cleaved in the S5–S6 interhelical loops of the second and third domains. The molecular weights corresponding to each of the fragments generated are given on the right. (From Zwerling *et al.*, 1991, with permission.)

supporting previous qualitative results with lectin binding to proteolytic fragments. The weights of the D2, D3, and D4 limit fragments were consistent with none, two, and one complex carbohydrate chains, respectively. Thus, the overall distribution of carbohydrate mass appears to conform to the distribution of potential N-glycosylation sites predicted from structural models.

Results from these proteolytic cleavage studies provide experimental support for the organization of the repeat domains as compactly folded intramembranous units. The relative accessibilities of the sequences linking these domains and at the amino and carboxy termini allow a hierarchical order of tertiary structure to be proposed. The carboxy terminus distal to D4 appears to be in the most extended configuration followed by ID 2–3. Interdomain 1–2 and especially ID 3–4 appear to be more compactly folded into the interior of the cytoplasmic domain of the protein. The amino terminus is most resistant to proteolysis and may assume a more compact globular structure in the native channel. Finally, the sequence between S5 and S6, at least in domains 1 and 4, is sufficiently exposed to allow cleavage by these relatively large protease molecules.

Sequence-specific antibodies can also provide information on the topological location of regions of the channel sequence when used to determine the orientation of their respective epitopes either in intact tissue or in vesicles containing oriented channels. Gordon *et al.* synthesized a 13-amino acid oligopeptide corresponding to residues 1781–1794 at the carboxy terminus and a 12-amino acid oligopeptide corresponding to residues 927–938 in the ID 2–3 of the eel channel primary sequence (Gordon *et al.*, 1987, 1988). Antiserum and immunoaffinity-purified IgG against both of these peptides reacted only with a 260-kDa band on immunoblots of eel membrane protein and specifically labeled the innervated face of the eel electroplax at the light microscopic level. Secondary antibody labeled with colloidal gold was used to detect bound antibody at the electron microscopic level; both antibodies were found exclusively on the cytoplasmic side of the innervated membrane bilayer. These studies provide experimental evidence localizing the amino terminus and the ID 2–3 region to the cytoplasmic surface of the molecule. The results also imply that, if the four repeat domains are homologous in organization, then the protein must cross the membrane an even number of times in each domain.

In an extension of this type of study, monoclonal antibodies were generated against purified rat skeletal muscle sodium channel protein and mapped for location on the primary sequence and for binding interaction in the tertiary structure. Using this approach, two groups of interacting epitopes were identified. The largest group included the binding sites for 16 antibodies that were linked together through mutually competitive interactions (Kraner *et al.*, 1989). Subsequent localization of these epitopes using proteolyzed sodium channel demonstrated that several of the antibodies that were linked by competition studies were located in different halves of the channel primary sequence (Cohen and Barchi, submitted for publication, 1993). Several other epitopes for monoclonal antibodies previously thought to distinguish between sodium channel subunits in skeletal muscle were found to reside close together at the amino terminus of the same channel sequence, suggesting that the appearance of independent subtypes in tissue immunocytochemistry reflected the relative accessibility of these two epitopes, perhaps due to differential interactions of this structurally constrained region with β subunits or with other membrane proteins (Cohen and Barchi, 1992).

Application of a polyclonal antibody raised to a short intracellular segment between D3 and D4 was shown to produce a gradual slowing of sodium channel inactivation in whole-cell recordings after intracellular application (Vassilev *et al.*, 1988). Antibody-induced slowing of inactivation was greater during test depolarizations to more positive membrane potentials or when more negative holding potentials were used prior to the test pulse. At the single-channel level, the antibody did not alter single-channel current amplitude but channel openings were now observed throughout the depolarizing pulse rather than being restricted to the onset of the pulse, consistent with the conclusion that this antibody blocked channel inactivation (Vassilev *et al.*, 1989). Again, these effects were more rapid in onset at more negative holding potentials and at more positive test potentials. These findings indicate that the accessibility of the ID 3–4 segment is sensitive to membrane potential and that the antibody acts by impeding the movement of this segment. An alternate explanation involves antibody-induced favoring of an alternate gating mode characterized by prolonged open states and repetitive channel opening (see V,2). This work both confirms the localization of the ID 3–4 segment to the intracellular side of the membrane and its role in sodium channel inactivation.

V. *In Vitro* Expression and Mutagenesis

The *Xenopus* oocyte expression system has provided electrophysiologists with a means of studying the functional protein products of both mRNA and cRNA for different sodium channel subtypes under controlled conditions. Using either voltage clamp or patch clamp configurations, the unique characteristics of both native channels and channel mutants can be examined. Identification of sequences involved in a particular channel function is possible either by introducing specific mutations or through the formation and expression of channel chimeras. Both approaches have proven useful in associating specific functions with specific regions of the primary structure.

A. Functional Expression

Poly(A)$^+$ mRNA prepared from mammalian brain, heart, or skeletal muscle is capable of directing the synthesis of functional sodium channels when injected into *Xenopus* oocytes (Gunderson *et al.*, 1984; Hirono *et al.*, 1985; Dascal *et al.*, 1986; Sigel, 1987; Krafte *et al.*, 1991). In several studies, size fractionation of crude mRNA preparations using sucrose gradient sedimentation has yielded a single peak of large RNA transcripts that promote the expression of sodium channels in oocytes, implying that only the large α subunit was needed for channel activity (Sumikawa *et al.*, 1984).

Functional sodium channels have subsequently been expressed in oocytes using cRNA encoding brain (Noda *et al.*, 1986b; Stuhmer *et al.*, 1987; Joho *et al.*, 1988; Suzuki *et al.*, 1988), cardiac (Cribbs *et al.*, 1990; Gellens *et al.*, 1992), and skeletal muscle (Trimmer *et al.*, 1989; Kallen *et al.*, 1990; White *et al.*, 1991) sodium channel α subunits. These studies further support the notion that the α subunits of sodium channels contain all the essential elements of fully functional channels when expressed in oocytes. This conclusion, however, assumes that *Xenopus* oocytes do not produce endogenous β-type subunits that are capable of associating with the exogenous α-subunit protein.

Although both high-molecular-weight mRNA and cRNA derived from α-subunit cDNA can direct the synthesis of functional sodium channels, the properties of these channels are not always normal. For many isoforms, the time course of inactivation is much slower for α subunits expressed in oocytes from cRNA than when expressed from total mRNA or studied *in vivo*.

Both rat brain and adult rat skeletal muscle (SkM1) channels show this abnormally slow inactivation in oocytes (Noda *et al.*, 1986b; Trimmer *et al.*, 1989; Joho *et al.*, 1990; Krafte *et al.*, 1990). Coinjection of low-molecular-weight mRNA from rat brain (derived from nondenaturing sucrose-gradient fractionation of poly(A)$^+$ RNA) with the α-subunit cRNA reduced the time course of channel inactivation to values comparable to those seen *in vivo* and increased the cell surface expression of functional sodium channels almost fourfold (Auld *et al.*, 1988). Thus, proteins encoded by low-molecular-weight mRNA can modulate both the functional properties and the cell surface expression of α subunits.

The slow inactivation kinetics seen with the SkM1 muscle isoform in oocytes has recently been examined at the single-channel level and shown to be due to the transition of individual sodium channels between a normal kinetic mode with fast inactivation and an abnormal mode characterized by altered inactivation with multiple late openings or channel reopenings (Zhou *et al.*, 1992). Similar events have been observed in skeletal (Patlak and Ortiz, 1986, 1989) and cardiac muscle (Nilius, 1988) sodium channels *in vivo*, but at a much lower probability. Repetitive or late channel openings have also been implicated in the abnormal behavior of sodium channels in several neuromuscular diseases (see below).

Coinjection of a low-molecular-weight mRNA fraction, or of cRNA prepared from a subgroup of cDNA's that does not itself encode for a sodium channel, can shift the equilibrium between modes to favor normal inactivation (Auld *et al.*, 1988; Zhou *et al.*, 1992). It is not clear whether the low-molecular-weight proteins introduced with mRNA or with cRNA from the cDNA subfraction represent sodium channel β subunits, enzymes involved in post-translational modification of sodium channel α subunits, or other proteins such as G proteins or cytoskeletal elements that interact with the sodium channel α subunit. The recent observation that PKC-dependent phosphorylation slows channel inactivation suggests that protein phosphorylation/dephosphorylation may also be responsible for these different gating modes (Numann *et al.*, 1991).

Coexpression of β-1 cDNA with rat brain IIa cRNA in oocytes produces functional effects that match those obtained when rat brain IIa α subunit cRNA is coexpressed with low molecular weight RNA from rat brain: The abnormally slow inactivation kinetics of rat brain IIa cRNA expressed in oocytes is corrected to normal levels; the voltage-dependence of inactivation is shifted to more negative potentials, and the peak sodium current is increased (Isom *et al.*, 1992). This finding suggests that it is the β-1 mRNA component of the low molecular weight RNA which is responsible for producing these effects.

Complementary DNA for the TTX-insensitive (TTX-I) form of the rat skeletal muscle sodium channel (SkM2) and that for its homologous human heart sodium channel (hH1) both generate cRNA transcripts that encode sodium channels with normal inactivation kinetics in oocytes (Fig. 6) (Gellens *et al.*, 1992; White *et al.*, 1991b). In contrast to TTX-sensitive (TTX-S) adult skeletal muscle channels (SkM1), SkM2 and hH1 channels do not show late channel openings at the single-channel level during depolarizing pulses. These findings suggest that, at least for these muscle channels, other protein elements either are not necessary for normal channel kinetics or are uniquely supplied to these particular channels by endogenous oocyte proteins.

Other electrophysiologic characteristics of these TTX-I and TTX-S channels appear to be maintained when expressed in oocytes. Ion selectivity and drug or toxin sensitivity again resemble those measured *in vivo*. Although similar in most regards, there are several distinct functional

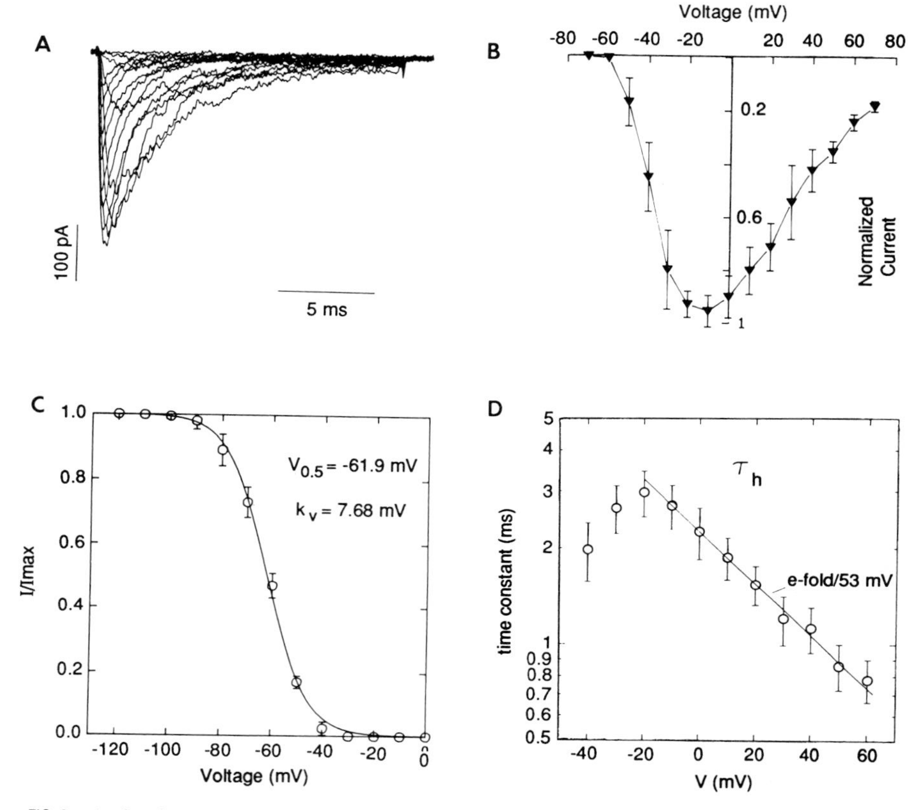

FIG. 6 Activation and inactivation of hH1 sodium currents expressed in oocytes. Full-length cRNA encoding the rH1 channel was injected into oocytes and recordings were made from macropatches 2–3 days later. The basic kinetic properties of these human cardiac sodium channels are shown. (A) A family of voltage clamp currents recorded from a large patch by depolarization from -120 mV to test voltages between -70 and +70 mV in 10-mV increments. (B) Normalized (± SD) peak I–V relationship for six patches. (C) The steady-state inactivation curve obtained from seven outside-out patches. (D) The voltage-dependence of τ_h. (From Gellens *et al.*, 1992, with permission.)

differences between rat and human cardiac transcripts. First, the human channel is less sensitive to block by tetrodotoxin than is the rat channel by a factor of about 3. Second, the hH1 I–V curve from single-channel measurements shows inward rectification with a conductance near 0 mV of ~22 pS, whereas the SkM2 channel has a linear I–V curve with a slope of ~10 pS. This difference could be due to increased negative charge near the extracellular opening of the aqueous pore in the human channel. Because of the higher probability of calcium residing in a blocking site in the pore, calcium ions could produce a curvature in the I–V relationship at negative potentials. The actual role of these additional negative charges in altering toxin binding and sodium flux will need to be determined by further experimentation.

B. Mutagenesis

The availability of both cloned segments and full length transcripts for various sodium channel isoforms has allowed deletions, insertions, mutations, and interchange of segments between isoforms to be performed in an attempt to define functional regions in the sodium channel primary structure. Stuhmer and colleagues constructed mutants of the rat brain II sodium channel that contained amino acid substitutions, deletions, or discontinuities in the channel sequence (Stuhmer *et al.*, 1989). Several predictions were tested in these studies. First, point mutations were introduced into the S4 regions of D1 and D2 that resulted in the replacement of positively charged arginine or lysine residues by either neutral or negatively charged residues in order to evaluate the role of these helices in voltage-dependent activation.

Although no expression of functional sodium channels was observed for any mutation that involved neutralization of more than three positive charges, all single, double, and triple mutations produced functional channels. Analysis of these mutants indicated that a reduction in net positive charge in the S4 segment of D1 produced a decrease in the steepness of the voltage dependence of channel activation that was directly proportional to the number of charges altered.

A similar correlation between positive charge in the homologous S4 helix and voltage dependence of activation has been shown for mutated potassium channels (Lopez *et al.*, 1991; Papazian *et al.*, 1991). Taken together, these results provide experimental support for the direct involvement of S4 positive charges in the voltage-sensing mechanism for channel activation. In addition to the changes in steepness of voltage dependence, some mutants produced significant shifts in the voltage dependence of activation. Similar voltage shifts in the activation curve have been found for mutations of uncharged residues in the S4 helix.

An additional piece of evidence supporting the role of the S4 helix in channel activation involves the electrophysiologic characterization of a rat brain IIa cDNA clone that differed from the native rat brain II sequence at seven amino acids (Auld *et al.*, 1990). When expressed in *Xenopus* oocytes, this transcript produced channels with activation current–voltage relationship shifted 20–25 mV in the depolarizing direction compared to rat brain II; these channels also demonstrated slow inactivation in whole-cell current recording. Coinjection of low- molecular-weight rat brain RNA increased the rate of macroscopic inactivation to levels comparable to channels formed from injection of total rat brain poly(A)$^+$ RNA but had no effect on the altered current–voltage relationship. Each of the seven amino acid differences in the rat brain IIa sequence was then mutated to the corresponding residue in rat brain II; a single amino acid substitution of phenylalanine for leucine at residue 860 in the S4 helix of domain 2 was found to be responsible for the altered current–voltage relationship. Polymerase chain reaction amplification and genomic sequencing confirmed that genomic DNA codes for leucine and not phenylalanine at this position, suggesting a cloning artifact in the rat brain IIa clone. This fortuitous error provides evidence that the alteration of neutral as well as charged residues in the S4 helix can affect the current–voltage relationship of channel activation, suggesting that factors other than side-chain charge influence the voltage-dependent conformational transitions associated with channel activation.

Functional potassium channels appear to be formed by the association of four independently translated subunits, each resembling a single sodium channel repeat domain (Zagotta and Aldrich, 1990; Stuhmer *et al.*, 1991). In order to test whether individual sodium channel domains or several contiguous domains can interact to produce functional sodium channels in a manner analogous to these K^+ channel subunits, cRNA's encoding the relevant regions of the rat brain sodium channel were constructed and coinjected into oocytes (Stuhmer *et al.*, 1989). Neither single domains nor transcripts that together contained less than four contiguous domains were capable of forming functional channels. Coinjection of cRNA's encoding D1 and D2/3/4 produced small sodium currents whereas the combinations of D1/2 with D3/4, D1/2/3 with D4, and a construct of D1/2/3/4 with deletions in the amino and carboxy termini produced functional channels, although the total oocyte sodium currents were somewhat reduced compared to wild-type cRNA. These results suggest that all four domains are required to form functional sodium channels although these domains can associate to form functional channels even when the linkage between domains is cleaved.

Sodium channels expressed from constructs containing deletions in the amino or carboxy termini as well as channels formed from coinjection D1/2 and D3/4 constructs demonstrated activation and inactivation kinetics comparable to channels expressed form the wild-type cRNA. Channels produced by coinjection of the D4 and the D1/2/3 construct, however, demonstrated a dramatic loss of channel inactivation. This was confirmed by measurements at the single-channel level, which demonstrated channel reopenings throughout the voltage step. These findings are similar to the effects reported for native sodium channels treated with intracellularly applied proteolytic enzymes (see Section II,1) or with sodium channels expressed in oocytes that were subsequently exposed to trypsin in outside-out patches, and with rat brain and rat skeletal muscle cDNA's expressed in oocytes.

Additional support for assigning the role of channel inactivation to the ID 3–4 region comes from a related study on potassium channels in which Aldrich and co-workers combined site-directed mutagenesis with the deletion of selected regions of the amino terminus (Hoshi *et al.*, 1990). Using the "ball and chain" model of channel inactivation first proposed by Armstrong in 1977 (Armstrong and Bezanilla, 1977), these investigators demonstrated that the distal amino terminus acted like a ball tethered to the membrane-embedded domain by a chain whose length could, within certain limits, be altered without affecting channel inactivation. Alteration of the size or charge of the distal amino-terminal "ball" resulted in the loss of inactivation, which could be reconstituted by the addition of millimolar amounts of a soluble synthetic peptide corresponding to the first 20 amino acids of the amino terminus (Zagotta *et al.*, 1990). Thus, fast inactivation is influenced by the amino terminus of potassium channels in a manner that may be analogous to the proposed function of the sodium channel ID 3–4 region.

The ball and chain model of channel inactivation first proposed by Armstrong in 1977 has gained support from the above studies and from the observation that the 11 lysines and 1 arginine in the ID 3–4 region can serve as targets for endopeptidases that remove channel inactivation; positively charged residues present on the ball are thought to interact with negatively charged residues on the interior of the channel opening. Thus, removing positive charges should delay channel inactivation without affecting channel activation. To test this prediction, groups of two or three continuous lysines were neutralized or a Glu was substituted for an Arg at position 1461 in the ID 3–4 region of the rat brain III channel and the resulting channel constructs were expressed in oocytes (Moormann *et al.*, 1990). Contrary to the above prediction, inactivation occurred more rapidly rather than more slowly and the Arg-to-Glu mutation at position 1461 resulted in a channel with delayed activation. It is possible that the charge alterations in these experiments altered the electrostatic interactions of the ID 3–4 linker with other parts of the channel protein or that the ID 3–4 region serves as the binding site for another blocking region of the molecule. A shift to a gating mode with faster inactivation is an alternate explaination. Thus, the ID 3–4 region has more complex effects on sodium channel gating than previously thought.

C. Mutations Affecting Toxin Binding

Locating the TTX and STX binding site is of considerable interest not only because of the key role that these toxins have played in studies of sodium channels but also because identification of their binding site will provide further information concerning the structure of the extracellular entrance to the transmembrane pore. Biochemical evidence supporting the presence of a carboxyl group in the binding site for these toxins has already been discussed (see Section II,C). When sodium channel primary sequence information became available for TTX-S sodium channels from several tissues and organisms, conserved negative charges were identified. Excluding sequences modeled to be present either intracellularly or in membrane-embedded helices, two such changes were identified in the SS2 region of the first domain, two in the SS2 region of the second domain, and three in the SS2 region of the fourth domain, making these possible candidates for sites of binding for these toxins. Unexpectedly, the sequences of the TTX-I sodium channels from rat and human heart and rat denervated muscle share all of these conserved residues except for aspartate^{1545} in the SS2 region of domain 4, which is converted to tyrosine^{1728} in the TTX-I channel sequences.

In an effort to localize the residues involved in tetrodotoxin binding and on the basis of the conserved negative charges in the S5–S6 loops of tetrodotoxin-sensitive channels, Noda and colleagues mutated residue 387 located in the SS2 region of D–1/S5–S6 in the rat brain II sodium channel primary sequence from glutamate to glutamine and expressed the altered cRNA in *Xenopus* oocytes (Noda *et al.*, 1989). Although most of the electrophysiologic characteristics of expressed channels matched those of the wild-type controls, two important differences were noted. The first was a virtual loss of sensitivity to TTX and STX, rather than the reduced sensitivity seen with native TTX-I channels: 1 m*M* TTX applied externally had no effect on the sodium current conducted through these channels, whereas 18 n*M* blocked wild-type channels and micromolar concentrations blocked TTX-I sodium channel currents. The second was a reduction in inward sodium current, a finding identical to that obtained with carboxyl-modifying reagents. This reduction in inward current might be due to a decreased concentration of sodium ions near the ion pore due to the localized decrease in negative charge. Although conformational changes propagated through the molecule are another possible explaination for these findings, it seems more likely that this glutamic acid residue is in close proximity to the mouth of the channel and is involved either directly or indirectly in toxin binding.

These initial studies were extended in a recent study in which mutations of charged residues were systematically introduced into the S5–S6 regions of each of the four repeat domains (Terlau *et al.*, 1991). These mutations were located at homologous locations in the carboxy-terminal segment of the S5–S6 loop in each domain of the rat brain II channel. A number of these mutations produced changes in TTX affinity. This work suggests that portions of all four S5–S6 loops may contribute to the formation of the TTX/STX binding site. Furthermore, several of these mutations produced drastic reductions in channel ionic conductance without affecting gating currents, supporting a role for this portion of the primary sequence in the formation of the channel's ion pore.

Another approach to localizing the residues connected with toxin binding involves the construction of channel chimeras. Channel chimeras have been manufactured by swapping the first domains of the TTX-S adult rat skeletal muscle channel and the TTX-I rat channel characteristic of heart and denervated and immature muscle (R. G. Kallen *et al.*, personal communication). Channel chimeras composed of $(D1)_{SkM1}(D2{:}D3{:}D4)_{SkM2}$ and $(D1)_{SkM2}(D2{:}D3{:}D4)_{SkM1}$ were constructed and expressed in oocytes. The TTX phenotype of the chimeric channel directly followed the lineage of the first domain. Comparison of the SS2 regions of the TTX-S SkM1 and TTX-I SkM2/RH1 channels demonstrates differences at only two residues. Three groups have demonstrated that mutating one of these residues (Tyr^{401} in the SkM1 channel and Cys^{374} in the SkM2/RH1 channel) to that present in the other isoform produces TTX-sensitivity which follows with the lineage of the resulting residue (Satin *et al.*, 1992; Backx *et al.*, 1992; Chen *et al.*, 1992). Thus, a Tyr 401 Cys substitution resulted in a TTX-I channel while a Cys 374 Tyr substitution resulted in a TTX-S channel. Thus, the residue which imparts TTX-sensitivity is located in the SS2 region of the S5-S6 extracellular loop in the first domain, consistent with models which place this region back into the membrane as part of the central aqueous pore.

VI. Regulation of Gene Expression *in Vivo*

Excitable membranes contain a number of sodium channel isoforms including four distinct transcripts in rat brain, two in rat skeletal muscle, and five in rat heart. The mechanisms that regulate the induction and tissue-specific expression of these different isoforms are at best poorly understood. *In vitro* studies have identified multiple extra- and intracellular signals that affect sodium channel subtype expression, including electrical activity, thyroid hormone, cytosolic sodium concentration, and a variety of channel active drugs and toxins. Serum factors, growth factors, and cellular or viral oncogenes have also been shown to alter sodium channel expression. Recent advances in molecular biology have provided powerful tools that will lead to a better understanding of the detailed mechanisms responsible for enhancing, repressing, and promoting the expression of each sodium channel subtype in a time- and tissue-specific fashion. In this section, we will review several of the more physiologically oriented studies and then describe recent work directed at understanding the molecular mechanisms responsible for the regulation of sodium channel subtype expression.

A. Brain Sodium Channels

Sodium channel ontogenesis in the central nervous system has been examined using action potential recordings, neurotoxin-binding assays, sodium flux measurements, and, more recently, subtype-specific antibodies. Limited developmental information was available until the subtype-specific probes derived from cloning studies were developed and used to quantitate rat brain subtype mRNA levels during development (Beckh *et al.*, 1989; Beckh, 1990). The three predominant rat brain sodium channels exhibit different temporal and regional patterns of expression in the developing rat central nervous system. Rat brain I mRNA is expressed predominantly at late postnatal stages in the caudal regions of the brain and in the spinal cord, rising to adult levels following the second and third postnatal weeks. Rat brain II mRNA reaches adult levels after the first or second postnatal weeks and is preferentially expressed in rostral rather than caudal regions of the brain and spinal cord; levels in the cerebellum only rise to significant levels in the adult rat. Rat brain II therefore appears to be expressed throughout development and with considerable regional variability. Rat brain III mRNA is expressed diffusely in fetal and early postnatal stages but drops shortly after this period to variable but low adult levels. Quantitation of mRNA levels indicates that rat brain type II is the most highly expressed subtype, with type I present at approximately 5–10% of the level of type II. Type III mRNA is expressed at moderately high levels in embryonic brain, but is barely detectable in the adult.

A variety of factors have been shown to alter the surface expression of sodium channels in nervous tissue. Studies of the biosynthesis and processing of sodium channels in developing neuronal tissues have shown that most newly synthesized sodium channel α subunits are not disulfide-linked to β-2 subunits but exist in a metabolically stable intracellular pool, readily available for rapid membrane insertion (Schmidt *et al.*, 1985). Thus one limitation of the following studies concerns the site of metabolic control; it is not clear whether transcriptional events, translational mechanisms, or transport to the cell surface is the limiting step that is under control by a specific factor or factors. Thus, the time course and sensitivity of altered expression to agents that affect each step in the biosynthetic process need to be defined.

Large increases in surface sodium channel protein expression during neuronal development were initially measured by neurotoxin binding (Berwald-Netter *et al.*, 1981; Jover *et al.*, 1988) and $^{22}Na^{+}$ influx assays (Couraud *et al.*, 1986). These sodium channels were found to be predominantly located on neurites (Boudier *et al.*, 1985). Among the variety of factors involved in the selective induction of brain sodium channels, nerve growth factor is perhaps the best studied.

PC12 is a pheochromocytoma cell line that can be induced to differentiate into sympathetic-like neurons by nerve growth factor (NGF). In the absence of NGF, PC12 cells are electrically inexcitable, whereas after several weeks of NGF treatment both morphologic differentiation and the development of rapid-upstroke action potentials are observed. It was recognized at an early stage that it was the expression of sodium channels in these cells (as apposed to other voltage-gated channels) that best accounted for the ability of differentiated PC12 cells to generate action potentials. Using subtype-specific mRNA probes, selective induction of the rat brain II channel was observed following treatment with NGF; no effect on rat brain I expression was observed (Mandel *et al.*, 1988). $^{22}Na^{+}$ measurements in the presence of channel activators produced similar results with NGF-induced increases in TTX-S channels (presumably rat brain II) but also demonstrated the appearance of TTX-I channels (Rudy *et al.*, 1987). In an extension of these studies, whole-cell clamp studies were used to compare the effects of several factors on both neurite outgrowth and sodium channel expression (Pollock *et al.*, 1990). Treatment with NGF or basic fibroblast growth factor (FGF) induced differentiation of PC12 cells; FGF induced the same or somewhat larger increases in channel density as NGF but produced less neurite outgrowth. Epidermal growth factor, on the other hand, produced no neurite outgrowth but did induce a small increase in sodium channel density. Raising intracellular cyclic AMP levels did not produce neurite outgrowth but did result in a decrease in functional sodium channels. Finally, dexamethasone was found to inhibit NGF's increase in sodium channel expression but did not inhibit NGF's effect on PC12 differentiation. Thus, sodium channel expression parallels the changes in morphology that lead to neurite outgrowth but does not depend on them; these two aspects of neuronal differentiation appear to be independently regulated by factors in the microenvironment.

Since NGF is thought to affect neurite outgrowth in part through activation of cAMP-dependent PKA and phospholipid-dependent PKC, agents that stimulated each of these systems were examined for effects on sodium channel expression using patch clamp of PC12 cells (Kalman *et al.*, 1990). Agents that activate PKC [including phorbol esters and a *ras* oncogene product (p21)] stimulated neurite outgrowth but produced no effect on channel number. On the other hand, agents that increased intracellular cAMP were as effective as NGF in affecting both neurite outgrowth and channel expression, indicating a role for cAMP-dependent protein kinase in the effects of NGF. These results differ from the results of Pollack *et al.* (1990); although the reasons for this disparity are unclear, the techniques used in these two studies differ (whole-cell clamp vs patch clamp) and it is possible that the experimental conditions were not identical in both studies. Additional evidence for both the role and sites of action of PKA in modulating sodium channel expression are provided by studies involving PC12 cell lines deficient in PKA activity (Ginty *et al.*,1992). The expected increase in functional sodium channels following treatment with either NGF or basic FGF does not occur in two PC12 cell lines deficient in both isozymes of PKA. While whole-cell patch clamp recordings do not demonstrate functional sodium channels, Northern blot analysis and radiolabelled saxitoxin binding assays of intact cells indicate that these growth factors are capable of producing increases in both sodium channel mRNA and protein (expressed on the cell surface). Thus, PKA appears to also play an essential posttranslational role in the expression of functional sodium channels, at least in PC12 cells.

Two additional studies shed light on the role of intracellular transducers in sodium channel expression. AtT-20 cells, derived from a mouse anterior pituitary tumor, express both TTX-S and TTX-I sodium channels. Transfection of activated *ras* into the AtT-20 cell line turned off the expression of TTX-I channels but had no effect on the expression of TTX-S sodium channels (Flamm *et al.,* 1990). In a similar fashion, cell-fusion hybrids of a human bladder carcinoma EJ (which are tumorigenic and contain activated c-Ha-ras) and a human fetal lung fibroblast cell line GM2291 (which express TTX-I sodium channels) express levels of activated c-Ha-ras similar to those of the EJ parental cell line but do not express sodium channels (Estacion, 1990). These observation suggest a role for the *ras* protein p21 in the regulation of expression of at least the TTX-I sodium channel.

Efforts to identify the genetic elements involved in the cell-specific expression of the rat brain II sodium channel gene have been made using transient expression assays (Maue *et al.,* 1990). Chimeric reporter constructs composed of progressively smaller portions of the 1051-bp 5′ rat brain II flanking sequence fused with a sequence encoding the bacterial enzyme chloramphenicol acetyltransferase (CAT) were constructed and expressed in several cell lines. These constructs were expressed in neuronal cells but not in muscle-derived cells, fibroblasts, or HeLa cells, consistent with the presence of sequences within the 5′ region that confer cell-specific expression of the rat brain II gene; 5′ deletion studies indicate that the 134-bp promoter region is not sufficient to confer cell specificity. These authors were able to identify at least three upstream elements that appeared to provide neural tissue specificity. Sequences in the 5′ region did not resemble any of the negative elements that have been described for other genes, including the insulin, immunoglobin heavy chain, α-fetoprotein, and myosin heavy chain genes, indicating that unique transcriptional factors are likely to be involved in the cell-specific expression of the rat brain II gene.

Additional deletion studies identified a 28 bp silencer element, called RE1, in the 5′ flanking region of the rat brain II channel (Kraner, *et al.*, 1992). RE1 confers neural-specific expression of rat brain II-CAT fusion genes in transient expression assays. DNA binding activities for RE1 were present in fibroblasts, in skeletal muscle cells, and in a neuronal cell line that does not normally endogenously express the rat brain II channel. RE1 is highly similar to a silencer called NRSE (neural-restrictive silencer element) identified in the SCG10 gene (Mori *et al.*, 1992), giving rise to the possibility that this silencer is a general DNA element for the control of neuron-specific expression in vertebrates.

B. Skeletal Muscle Sodium Channels

Earlier pharmacologic studies demonstrated that mammalian skeletal muscle expresses at least two forms of sodium channel that differ in their sensitivity to TTX (Barchi and Weigele, 1979; Rogart and Regan, 1985). The sodium channel in adult muscle is TTX-S, whereas the channel synthesized in embryonic and denervated muscle is relatively resistant to TTX and, as discussed above, is identical to the TTX-I sodium channel identified in heart. The development of the adult phenotype in mammalian skeletal muscle requires the interaction of muscle fibers with active motor neurons; denervation removes inhibition of expression of the embryonic subtype (Yang *et al.*, 1991). Tetrodotoxin-insensitive currents appear within 2 days of denervation. High-affinity STX binding sites, however, decrease over the first 2 weeks following denervation to approximately 60% of baseline values (Bambrick and Gordon, 1988). Single-fiber voltage clamp measurements indicate that at least 40% of the channels in denervated muscle are of the TTX-I subtype (Pappone, 1980).

Several factors have been implicated in the regulation of expression of skeletal muscle sodium channels. Channel down-regulation can be induced by the sodium ionophores amphotericin B and monensin (Dargent and Couraud, 1990). Channel internalization rather than decreased channel synthesis was favored as an explaination for these observations because of the rapid time course of this phenomenon. Similar results were obtained with batrachotoxin treatment of cultured chick muscle cells in which a 75% reduction in surface channels was measured with a half-time of 3–

6 hr (Bar-Sagi and Prives, 1985). The reduction in surface channels in this study was also felt to be due to channel internalization and was shown to be reversible, with increased channel expression dependent on protein synthesis.

The role of neuromuscular activity in the cell surface expression of sodium channels in muscle was investigated using radiolabeled toxin binding to neonatal rat muscle (Bambrick and Gordon, 1988). During normal development, channel expression increased exponentially with a time constant of 12 days. The period of most rapid channel incorporation coincided with the period of accelerated muscle growth and neuromuscular activity at 2 weeks of age. Eliminating neuromuscular activity retarded both muscle growth and the normal incorporation of sodium channels. Denervation was more effective than treatment with botulinum toxin during development, whereas both were equally effective in reducing channel number in adult tissue, implying different modes of control of sodium channel expression during development and in the adult.

Insight into the intracellular mechanisms responsible for neuromuscular control of sodium channel expression comes from several studies. Electrical activity, intracellular calcium concentration, cAMP, thyroid hormone, and treatment with verapamil or ethanol have all been demonstrated to modulate channel expression in skeletal muscle cells in culture. Thyroid hormone produces a dose-dependent increase in sodium channel expression, which coincides with dose-dependent increases in action potential frequency and rate of rise of the action potential in cultured skeletal myotubes (Brodie and Sampson, 1989). The effect is rapid, with onset at 12 hr and plateau levels attained by 36–48 hr. The increase is blocked by inhibitors of protein synthesis and reduced by increased extracellular calcium. Lowering intracellular calcium concentrations by treatment with TTX or verapamil produces a marked increase in thyroid-stimulated sodium channel synthesis.

In a related study, the normal age-dependent increase in channel expression in cultured muscle cells was decreased by preventing myoblast fusion (Brodie *et al.*, 1989). Down-regulation of sodium channels occurred following treatment with the calcium ionophore A23187, whereas channel up-regulation was induced by eliminating spontaneous electrical and contractile activity with TTX, elevated external KCl, and verapamil. Again, protein synthesis inhibitors blocked channel up-regulation. The authors felt that electrical and mechanical activity, working through alterations in cytosolic calcium, regulate the *de novo* synthesis of sodium channels.

Further evidence that intracellular calcium may play a role in regulating the expression of skeletal muscle sodium channels comes from detailed studies of the effects of verapamil and ethanol on cells in culture (Offord and Catterall, 1989; Brodie and Sampson, 1990, 1991). Inhibition of spontaneous electrical activity with local anesthetics (bupivicaine) or elevated intracellular cAMP caused a twofold increase in sodium channels measured with radiolabeled toxins, whereas treatment with the calcium ionophore A23187 decreased sodium channel density and overcame the effect of blocking electrical activity. Using a rat brain II cDNA probe, these authors demonstrated coincidence of message and channel expression during development, during treatment with bupivicaine, and during treatment with 8 Br-cAMP, indicating that the regulation of α-subunit mRNA levels is the primary mechanism of feedback regulation of sodium channel density by electrical activity in rat muscle cells.

Subtype-specific probes have been used to examine mRNA levels in adult muscle, denervated muscle, and muscle treated with botulinum toxin (Yang *et al.*, 1991). Distinct mechanisms appear to be involved in the regulation of the expression of mRNAs encoding TTX-S and TTX-I sodium channels. In innervated muscle, TTX-I transcript expression is suppressed and the levels of TTX-S channels vary between fast and slow twitch muscle, with fast twitch fibers containing a considerably higher level of expression than slow twitch fibers. Withdrawal of electrical stimulation, whether through surgical or chemical denervation, has little effect on TTX-S mRNA levels, suggesting that the previously described decline in density of high affinity TTX binding sites in denervated sarcolemma may reflect regulation at the translational or post-translational level rather than at the level of transcription. Tetrodotoxon-insensitive mRNA expression, however, is rapidly induced when electrical activity is eliminated (Fig. 7), and, based on studies with botulinum toxin, its level of expression appears to reflect the nonquantal release of a factor(s) from the motor nerve terminal ending. Quantal acetylcholine release, on the other hand, may play a major role in suppressing TTX-I sodium channel expression in innervated or reinnervated muscle.

The effects of growth factors on developing skeletal muscle cells have also been assessed. Both NGF and FGF caused dose-related increases in sodium channel expression as assessed by both STX binding and measurement of the frequency and rate of rise of spontaneously occurring action potentials (Offord and Catterall, 1989). Although both fused and unfused C2 muscle cells derived from mouse skeletal muscle possess TTX-S sodium channels, treatment of these cells with transforming growth factor β-1, an inhibitor of myogenic differentiation, inhibits the expression of TTX-S channels; mitogen withdrawal is required to allow sodium channel expression (Caffrey *et al.*, 1989). Thus, growth factors appear to have similar effects on both nerves and skeletal muscle cells in culture. Again, the detailed mechanisms and the level at which control is exerted need to be further evaluated.

The factors which provide transcriptional regulation of the TTX-I sodium channel are also being studied (Sheng *et al.*, 1992). DNA transfection studies indicate that transcriptional regulation of the SkM2 gene is highly

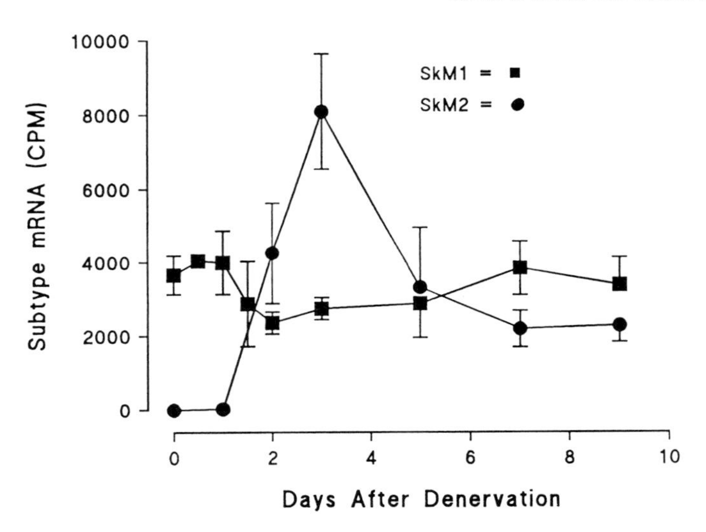

FIG. 7 Time course of changes in SkM1 and SkM2 mRNA expression after axotomy. The sciatic nerve was transected in the mid-thigh region in rats. SkM1 mRNA declined early by a small but significant amount but returned to baseline values by Day 7. SkM2 mRNA, undetectable between 0 and 24 hr, rose rapidly to a peak at 72 hr after axotomy before declining to a steady-state intermediate level. (From Yang *et al.*, 1991, with permission.)

complex. Two nonspecific enhancers, two muscle-specific enhancers, and two nonspecific silencer elements were identified. In addition, four GC-rich elements resembling Sp1 recognition sites, four overlapping C-rich motifs thought important for muscle-specific expression of selected genes, and seven MyoD helix-loop-helix protein binding sites were also identified. Thus, the cell type specific expression of the TTX-I promoter appears to be regulated in a complex manner by multiple interactions of positive and negative cis-acting elements.

VII. Sodium Channels and Disease

A. Chromosomal Localization of Sodium Channel Genes

Using probes specific for the hSkM1 and hH1 gene products, hamster–human chimera cell lines have been screened to determine chromosomal localization of the cognate genes. Probes for the human heart sodium channel (hH1) localize its genomic sequence to chromosome 3 whereas those for the human adult skeletal muscle sodium channel (hSkM1) localize its genomic sequence to chromosome 17 (George *et al.*, 1991). Using a chromosome 17 hybrid panel, the site of the hSkM1 gene, designated SCN4A, was further localized to the region between 17q23.3 and 17q25.1.

Similar techniques were used to localize the gene encoding the human counterpart of the rat brain II sodium channel to chromosome 2. Further analysis with cell hybrid panels and *in situ* hybridization helped to localize this channel to the region between 2q21 and 2q33 (Litt *et al.*, 1989; Han *et al.*, 1991). These data suggested that multiple copies of this gene (designated SCN2A) or multiple related genes might be present within this region.

Mouse brain sodium channel genes may be organized in a similar fashion. Mouse homologs of the rat brain I, II, and III genes have been mapped to a contiguous small segment of the mouse chromosome 2 (Malo *et al.*, 1991). The gene encoding the mouse homolog of the TTX-sensitive skeletal muscle sodium channel (SkM1) is located on chromosome 11 (Ambrose *et al.*, 1992; George *et al.*, 1992) while the cardiac sodium channel gene (SkM2) is on chromosome 9, close to a region that shows homology to human chromosome 3 (George *et al.*, 1992).The human gene encoding the adult skeletal muscle sodium channel on chromosome 17q has been analyzed in detail (George *et al.*, submitted for publication). The complete coding region of this gene, SCN4A, is contained in 32.5 kb of genomic DNA and consists of 24 exons ranging from 54 to 2242 bp in length, and 23 introns varying in size between 97 and 4850 bp. The exon organization of the gene shows no relationship to the predicted functional domains of the channel protein. The position of exon–intron junctions in the human gene, however, correspond exactly to those contained within an 8 kb homologous segment of the rat gene, and 10 of 24 splice junctions in SCN4A are positioned in homologous locations in the putative *Drosophila* sodium channel gene *para*.

B. The Human Periodic Paralyses

The chromosomal localization of the human sodium channel genes has been used in linkage studies of several neurological diseases in which abnormalities in sodium channel function are thought to play a role in disease pathophysiology. The periodic paralyses are a group of rare disorders affecting skeletal muscle characterized by intermittent episodes of weakness or paralysis, often with apparently normal neuromuscular function between attacks. These episodes can be triggered by specific dietary factors such as a large carbohydrate load, by environmental factors such as cold exposure, or by physiological changes such as rest after vigorous exercise. Although there are some recognized acquired forms, most disorders in this category are hereditary, usually transmitted in an autosomal dominant pattern. In some cases, the paralytic phenotype is associated with signs of membrane hyperexcitability in the form of myotonic discharges and stiffness.

The paralytic episodes seen in these unusual disorders are associated with depolarization of the muscle sarcolemma (Rudel *et al.*, 1984; Lehmann-Horn *et al.*, 1987a,b). In a number of cases in which skeletal muscle has been studied *in vitro*, this depolarization is the result of an abnormal increase in membrane permeability to sodium ions. In two disorders, hyperkalemic periodic paralysis (HyPP) and paramyotonia congenita, the abnormal sodium conductance can be blocked by TTX, implicating the voltage-dependent sodium channel as the defective protein (Lehmann-Horn *et al.*, 1987a,b; Ricker *et al.*, 1989). Voltage clamp recordings in HyPP demonstrate a small non-inactivating sodium current that is present only with elevated extracellular potassium and which may be responsible for producing the persistent membrane depolarization that characterizes the paralytic episodes (Lehmann-Horn *et al.*, 1987a).

Single-channel recordings have also been made on myotubes in cultures obtained from a HyPP muscle biopsy (Cannon *et al.*, 1991). These measurements support the hypothesis that a primary sodium channel abnormality underlies the pathophysiology of this disease. In the presence of normal extracellular potassium, the single-channel kinetics of sodium channels in HyPP appeared normal. Increasing $[K^+]_{out}$ to 10 mM had no effect on control channels, but caused a small percentage, of the HyPP sodium channels to exhibit aberrant gating properties. These channels showed an intermittent loss of inactivation, with multiple channel openings during a depolarizing pulse, and bursts of late channel openings. Abnormal inactivation appeared intermittently in epochs, suggesting that the defective channels were able to switch between normal and abnormal kinetic modes over a longer time period. The failure of inactivation seen in HyPP channels with high potassium is consistent with the non-inactivating sodium current previously reported in voltage clamp studies.

C. Linkage Analysis

Involvement of the hSkM1 channel in the pathogenesis of HyPP has been tested by analysis of linkage between the SCN4A gene encoding this channel on chromosome 17 with the phenotypic expression of the disease in several well-characterized families with this disorder. Fontaine *et al.* (1990) used a restriction fragment length polymorphism (RFLP) in the region encoding the ID 2–3 to demonstrate significant linkage between the SCN4A gene locus and a phenotype of HyPP with myotonia, obtaining a

LOD score of 4.01 at 0.00 recombination frequency. Using a different BglII RFLP, Ebers *et al.* (1991) showed comparable linkage in another family expressing Hypp without myotonia. Ptasek *et al.* (1991a) studied a very large family with HyPP with myotonia, and confirmed linkage to the SCN4A sodium channel gene with an exceptionally high LOD score of 10.31. These findings were further confirmed by an analysis by Koch *et al.* (1991) on additional families with HyPP in Germany.

Linkage analyses have also been carried out on families with the paramyotonia congenita phenotype (Ptackek *et al.*, 1991b; Ebers *et al.*, 1991). In two studies, significant linkage was again found with the adult skeletal muscle sodium channel gene, indicating that this disorder is also associated with a defect in the SCN4A gene.

D. Identification of Sodium Channel Mutations

In an attempt to identify specific mutations in the SCN4A gene in HyPP, a known sequence for hSkM1 was first used to identify exon–intron boundaries in the genomic DNA (George *et al.*, submitted for publication). Using this information, Ptacek *et al.* (1991b) generated PCR primers for individual exons and used them to screen genomic DNA from seven unrelated affected individuals for mutations by analysis of single-strand conformational polymorphism. A cytosine-to-thymidine mutation was identified in three of these individuals that resulted in the substitution of a methionine residue for an absolutely conserved threonine at position 704 in S5 of domain 2. This mutation segregated with affected members of two families and appeared as a spontaneous mutation in the third. The defect was not found in either of the parents of the third patient, in unaffected members of the other families, or in any of 109 controls.

Rojas *et al.* (1991) employed an alternative approach in identifying a separate mutation in two other cases of HyPP. In this study, mRNA was prepared from a muscle biopsy of an individual with HyPP. Complementary DNA produced from this mRNA was then analyzed to define an adenine-to-guanine mutation that results in the substitution of a methionine for a highly conserved valine in S6 of domain 4. This mutation was shown to segregate with disease expression in the affected members of this family and to appear as a new mutation in another affected family.

Two separate mutations affecting the same codon were subsequently identified by Ptacek *et al.* (1992) in three unrelated families with paramyotonia congenita. These mutations produced substitutions of either histidine or cysteine for the same arginine residue at position 1448 in the S4 helix of domain 4 (Fig. 8). These mutations result in the neutralization of conserved

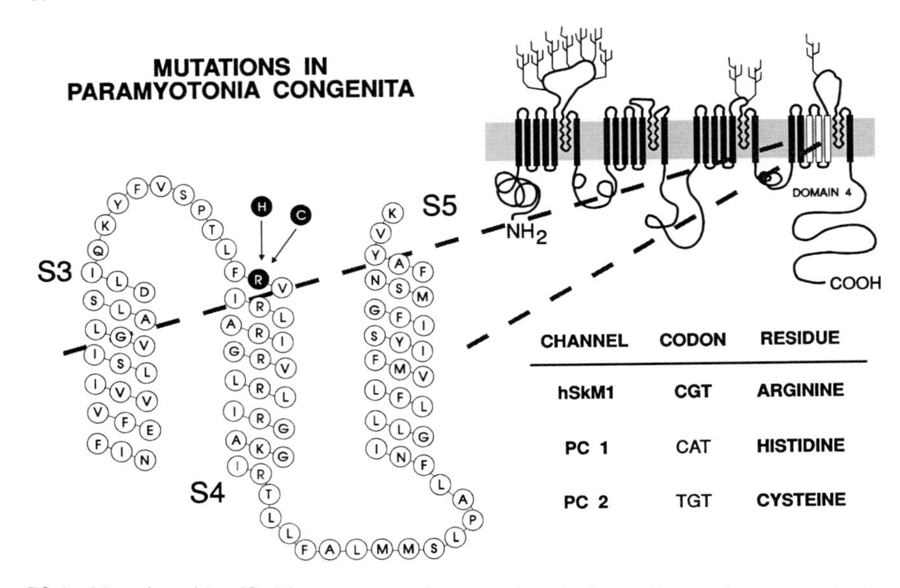

FIG. 8 Mutations identified in paramyotonia congenita. Amino acid mutations occurring in three unrelated families with paramyotonia congenita have been identified. Single nucleotide substitutions result in the replacement of the positively charged arginine at position 1448 in the hSkM1 sequence (located at the origin of the S4 helix in the fourth domain) with either a histidine or a cysteine. The effect of these mutations on channel function has yet to be determined.

positive charges in the region hypothesized to be responsible for voltage-dependent channel activation.

Two other mutations have also been identified in paramyotonia congenita (McClatchey *et al.*, 1992). These mutations affect amino acids near the N-terminal end of the cytoplasmic loop joining domains three and four. This region has been implicated in channel inactivation. In one family, a valine replaces an absolutely conserved glycine (Gly 1306 Val) while in the second mutation, a conserved threonine is replaced by methionine (Thr 1313 Met).

It appears that a variety of point mutations in the sodium channel gene are capable of producing diseases with related phenotypes. A formal analysis of each will be needed to determine the mechanisms through which the mutations act, but these experiments of nature will undoubtedly prove valuable in further analyzing channel structure and function.

VIII. Conclusion

In the short span of 4 decades, the sodium channel has been purified, reconstituted, sequenced, functionally expressed, and selectively mutated. Our understanding of the molecular mechanisms responsible for the complex behavior of this large protein are being revealed using a combination of molecular, immunologic, and protein chemical techniques. Sequences involved in channel activation, inactivation, gating, voltage sensitivity, and toxin binding have been identified. Several axioms which underlie sodium channel structure-function relationships are evolving. It is likely that continued application of these approaches and soon to be available methods of imaging large complex membrane proteins will provide an even more detailed and complex model of sodium channel tertiary structure and function. These findings, when coupled with a better understanding of how sodium channels contribute to the pathophysiology of disease processes in nerve, skeletal muscle, and cardiac muscle, will lead to new approaches to diagnosis and therapy.

References

Aldrich, R. W. (1986). *Trends Neurosci.* **9,** 82–86.

Ambrose, C., Cheng, S., Fontaine, B., Nadeau, J. H., MacDonald, M., and Gusella, J. F. (1992). *Mammalian Genome* **3,** 151–155.

Armstrong, C. M. (1981). *Physiol. Rev.* **61,** 644–683.

Armstrong, C. M., and Bezanilla, F. (1973). *Nature* (*London*) **242,** 459–461.

Armstrong, C. M., and Bezanilla, F. (1974). *J. Gen. Physiol.* **63,** 533–552.

Armstrong, C. M., and Bezanilla, F. (1977). *J. Gen. Physiol.* **70,** 567–590.

Armstrong, C. M., Bezanilla, F., and Rojas, E. (1973). *J. Gen. Physiol.***62,** 375–391.

Atchison, W. D., Luke, V. S., Narahashi, T., and Vogel, S. M. (1986). *Br. J. Pharmacol.* **89,** 731–738.

Auld, V. J., Marshall, J., Goldin, A., Dowsett, A., Catterall, W. A., and Davidson, N. (1985). *J. Gen. Physiol.* **86,** 10a.

Auld, V. J., Goldin, A. L., Krafte, D. S., Marshall, J., Dunn, J. M., Catterall, W. A., Lester, H. A., Davidson, N., and Dunn, R. J. (1988). *Neuron* **1,** 449–461.

Auld, V. J., Goldin, A. L., Krafte, D. S., Catterall, W. A., Lester, H. A., Davidson, N., and Dunn, R. J. (1990). *Proc. Natl. Acad. Sci. U.S.A.* **87,** 323–327.

Backx, P. H., Yue, D. T., Lawrence, J. H., Marban, E., and Tomaselli,G. F. (1992). *Science* **257,** 248–251.

Baker, P. F., and Rubinson, K. A. (1975). *Nature* (*London*) **257,** 442–444.

Baker, P. F., and Rubinson, K. A. (1976). *J. Physiol.* (*London*) **266,** 3P–4P.

Balerna, R., Ritchie, J. M., and Strichartz, G. R. (1975). *Biochemistry* **14,** 5500–5511.

Bambrick, L. L., and Gordon, T. (1988). *J. Physiol.* (*London*) **407,** 263–274.

Barchi, R. L. (1988). *Annu. Rev. Neurosci.* **11,** 455–495.

Barchi, R. L., and Weigele, J. B. (1979). *J. Physiol.* (*London*) **295,** 383–396.

Bar-Sagi, D., and Prives, J. (1983). *J. Cell. Physiol.* **114,** 77–81.

Bar-Sagi, D., and Prives, J. (1985). *J. Biol. Chem.* **260,** 4740–4744.

Baumann, A., Krah, J. I., Mueller, R., Mueller, H. F., Seidel, R., Kecskemethy, N., Casal, J., Ferrus, A., and Pongs, O. (1987). *EMBO J.* **6,** 3419–3430.

Beckh, S. (1990). *FEBS Lett.* **262,** 317–322.

Beckh, S., Noda, M., Luebbert, H., and Numa, S. (1989). *EMBO J.* **8,** 3611–3616.

Behrens, M. I., Oberhauser, A., Bezanilla, F., and Latorre, R. (1989). *J. Gen. Physiol.* **93,** 23–42.

Beneski, C. A., and Catterall, W. A. (1980). *Proc. Natl. Acad. Sci. U.S.A.* **77,** 639–643.

Berwald-Netter, Y., Martin-Moutot, N., Koulakoff, A., and Couraud, F. (1981). *Proc. Natl. Acad. Sci. U.S.A.* **78,** 1245–1249.

Boudier, J. A., Berwald-Netter, Y., Dellmann, H. D., Boudier, J. L., Couraud, F., Koulakoff, A., and Cau, P. (1985). *Dev. Brain Res.* **352,** 137–142.

Brodie, C., and Sampson, S. R. (1989). *Endocrinology (Baltimore)* **125,** 842–849.

Brodie, C., and Sampson, S. R. (1990). *J. Pharmacol. Exp. Ther.* **255,** 1195–1201.

Brodie, C., and Sampson, S. R. (1991). *J. Neurosci. Res.* **28,** 229–235.

Brodie, C., Brody, M., and Sampson, S. R. (1989). *Brain Res.* **488,** 186–194.

Brodwick, M. S., and Eaton, D. C. (1978). *Science* **200,** 1494–1496.

Brown, G. B. (1988). *Int. Rev. Neurobiol.* **29,** 77–116.

Caffrey, J. M., Brown, A. M., and Schneider, M. D. (1989). *J. Neurosci.* **9,** 53.

Cannon, S. C., Brown, R. J., and Corey, D. P. (1991). *Neuron* **6,** 619–626.

Catterall, W. A. (1980). *Annu. Rev. Pharmacol. Toxicol.* **20,** 15–43.

Catterall, W. A. (1986). *Annu. Rev. Biochem.* **55,** 953–985.

Catterall, W. A. (1988). *Science* **242,** 50–61.

Chen, L.-Q., Chahine, M., Kallen, R. G., Barchi, R. L., and Horn, R. (1992). *FEBS Lett.,* **309,** 253–257.

Cohen, S. A. (1991). *Circulation* **84,** II–175.

Cohen, S. A., and Barchi, R. L. (1992). *J. Membr. Biol.* **128,** 219–226.

Cohen, S. A., and Barchi, R. L. (1993). Submitted.

Coombs, J., Scheuer, T., Rossie, S., and Catterall, W. (1988). *Biophys. J.* **53,** 542a.

Costa, M. R. C., and Catterall, W. A. (1984a). *J. Biol. Chem.* **259,** 8210–8218.

Costa, M. R. C., and Catterall, W. A. (1984b). *Cell. Mol. Neurobiol.* **4,** 291–298.

Costa, M. R. C., Casnellie, J. E., and Catterall, W. A. (1982). *J. Biol. Chem.* **257,** 7918–7921.

Couraud, F., Martin-Moutot, N., Koulakoff, A., and Berwald-Netter, Y. (1986). *J. Neurosci.* **6,** 192–198.

Cribbs, L. L., Satin, J., Fozzard, H. A., and Rogart, R. B. (1990). *FEBS Lett.* **275,** 195–200.

Cruz, L. J., Gray, W. R., Olivera, B. M., Zeikus, R. D., Kerr, L., Yoshikami, D., and Moczydlowski, E. (1985). J. Biol. Chem. **260,** 9280–9288.

Dargent, B., and Couraud, F. (1990). *Proc. Natl. Acad. Sci. U.S.A.* **87,** 5907–5911.

Dascal, N., and Lotan, I. (1991). *Neuron* **6,** 165–175.

Dascal, N., Snutch, T. P., Lubbert, H., Davidson, N., and Lester, H. A. (1986). *Science* **231,** 1147–1150.

Duch, S., and Levinson, S. R. (1987). *J. Membr. Biol.* **98,** 43–55.

Eaton, D. C., Brodwick, M. S., Oxford, G. S., and Rudy, B. (1978). *Nature (London)* **271,** 473–476.

Ebers, G. C., George, A. L., Barchi, R. L., Ting-Passador, S. S., Kallen, R. G., Lathrop, G. M., Beckmann, J. S., Hahn, A. F., Brown, W. F., Campbell, R. D., and Hudson, A. J. (1991). *Ann. Neurol.* **30,** 810–816.

Elmer, L. W., Obrien, B. J., Nutter, T. J., and Angelides, K. J. (1985). *Biochemistry* **24,** 8128–8137.

Emerick, M. C., and Agnew, W. S. (1989). *Biochemistry* **28,** 8367–8380.

Estacion, M. (1990). *J. Membr. Biol.* **113,** 169–175.

Feller, D. J., Talvenheimo, J. A., and Catterall, W. A. (1985). *J. Biol. Chem.* **260,** 11542–11547.

Flamm, R. E., Birnberg, N. C., and Kaczmarek, L. K. (1990). *Pfluegers Arch.* **416,** 120–125.

Fontaine, B., Khurana, T. S., Hoffman, E. P., Bruns, G. A., Haines, J. L., Trofatter, J. A., Hanson, M. P., Rich, J., McFarlane, H., Yasek, D. M., Romano, D., Gusella, J., and Brown, R. (1990). *Science* **250,** 1000–1002.

Fontecilla-Camps, J. C., Habersetzer, R. C., and Rochat, H. (1988). *Proc. Natl. Acad. Sci. U.S.A.* **85,** 7443–7447.

Garber, S. S., and Miller, C. (1987). *J. Gen. Physiol.* **89,** 459–480.

Gellens, M. E., George, A. L., Jr., Chen, L., Chahine, M., Horn, R., Barchi, R. L., and Kallen, R. G. (1992). *Proc. Natl. Acad. Sci. U.S.A.* **89,** 554–558.

George, A. L., Ledbetter, D. H., Kallen, R. G., and Barchi, R. L. (1991). *Genomics* **9,** 555–556.

George, A. L., Komisarof, J., Kallen, R. G., and Barchi, R. L. (1992). *Ann. Neurol.* **31,** 131–137.

George, A. L., Iyer, G. S., Kleinfield, R., Kallen, R. G., and Barchi, R. L. Genomic organization of the human skeletal muscle sodium channel gene. Submitted for publication.

Gilly, W. F., Lucero, M. T., and Horrigan, F. T. (1990). *Neuron* **5,** 663–674.

Ginty, D. D., Fanger, G. R., Wagner, J. A., and Maue, R. A. (1992). *J. Cell Biol.* **116,** 1465–1473.

Gonoi, T., Ashida, K., Feller, D., Schmidt, J., Fujiwara, M., and Catterall, W. A. (1986). *Mol. Pharmacol.* **29,** 347–354.

Gordon, D., Merrick, D., Wollner, D. A., and Catterall, W. A. (1988). *Biochemistry,* **27,** 7032–7038.

Gordon, R. D., Fieles, W. E., Schotland, D. L., Hogue, A. R., and Barchi, R. L. (1987). *Proc. Natl. Acad. Sci. U.S.A.* **84,** 308–312.

Gordon, R. D., Li, Y., Fieles, W. E., Schotland, D. L., and Barchi, R. L. (1988). *J. Neurosci.* **8,** 3742–3749.

Greenblatt, R. E., Blatt, Y., and Montal, M. (1985). *FEBS Lett.* **193,** 125–134.

Gunderson, C. B., Miledi, R., and Parker, I. (1984). *Nature (London)* **308,** 421–424.

Guy, H. R., and Conti, F. (1990). *Trends Neurosci.* **13,** 201–206.

Guy, H. R., and Seetharamulu, P. (1986). *Proc. Natl. Acad. Sci. U.S.A.* **83,** 508–512.

Han, J., Lu, C. M., Brown, G. B., and Rado, T. A. (1991). *Proc. Natl. Acad. Sci. U.S.A.* **88,** 335–339.

Hartshorne, R. P., and Catterall, W. A. (1984). *J. Biol. Chem.* **259,** 1667–1675.

Hartshorne, R. P., Keller, B. U., Talvenheimo, J. A., Catterall, W. A., and Montal, M. (1985). *Proc. Natl. Acad. Sci. U.S.A.* **82,** 240–244.

Henderson, R., Ritchie, J. M., and Strichartz, G. R. (1973). *J. Physiol. (London)* **235,** 783–804.

Henderson, R., Ritchie, J. M., and Strichartz, G. R. (1974). *Proc. Natl. Acad. Sci. U.S.A.* **71,** 3936–3940.

Hille, B. (1968). *J. Gen. Physiol.* **51,** 221–236.

Hille, B. (1971). *J. Gen. Physiol.* **58,** 599–619.

Hille, B. (1972). *J. Gen. Physiol.* **59,** 637–658.

Hille, B. (1977). *J. Gen. Physiol.* **69,** 497–515.

Hille, B. (1984). "Ionic Channels of Excitable Membranes." Sinauer, Sunderland, Massachusetts.

Hirono, C., Yanagishi, S., O'Hara, R., Hisanaga, Y., Nakayama, T., and Sugiyama, H. (1985). *Brain Res.* **359,** 57–61.

Hodgkin, A. L., and Huxley, A. F. (1952). *J. Physiol. (London)* **116,** 473–496.

Hondeghem, L. M., and Katzung, B. G. (1987). *Annu. Rev. Pharmacol. Toxicol.* **24,** 387–423.

Horn, R. (1984). "Ion Channels: Molecular and Physiological Aspects," pp. 53–97. Academic Press, New York.

Hoshi, T., Zagotta, W. N., and Aldrich, R. W. (1990). *Science* **250,** 533–538.

Isom, L. L., Dejongh, K. S., Patton, D. E., Reber, B. F. X., Offord, J., Charbonneau, H., Walsh, K., Goldin, A. L., and Catterall, W. A. (1992). *Science* **256,** 839–842.

Joho, R. H., Moorman, J. R., Vandongen, A. M. J., Kirsch, G. E., Silberberg, H., Schuster, G., and Brown, A. M. (1988). *J. Cell Biochem.* **13,** 207.

Joho, R. H., Moorman, J. R., VanDongen, A. M., Kirsch, G. E., Silberberg, H., Schuster, G., and Brown, A. M. (1990). *Brain Res. Mol. Brain Res.* **7,** 105–113.

Jover, E., Massacrier, A., Cau, P., Martin, M. F., and Couraud, F. (1988). *J. Biol. Chem.* **263,** 1542–1548.

Kallen, R. G., Sheng, Z. H., Yang, J., Chen, L. Q., Rogart, R. B., and Barchi, R. L. (1990). *Neuron* **4,** 233–242.

Kalman, D., Wong, B., Horvai, A., Cline, M., and O-Lague, P. (1990). *Neuron* **2,** 355–366.

Kamb, A., Iverson, L. E., and Tanouye, M. A. (1987). *Cell* **50,** 405–413.

Kamb, A., Tseng-Crank, J., and Tanouye, M. A. (1988). *Neuron* **1,** 421.

Kayano, T., Noda, M., Flockerzi, V., Takahashi, H., and Numa, S. (1988). *FEBS Lett.* **228,** 187–194.

Kirsch, G. E., Skattebol, A., Possani, L. D., and Brown, A. M. (1989). *J. Gen. Physiol.* **93,** 67–83.

Koch, M. C., Ricker, K., Otto, M., Grimm, T., Hoffman, E., Rudel, R., Bender, K., Zoll, B., Harper, P., and Lehmann-Horn, F. (1991). *J. Med. Genet.* **28,** 583–586.

Kosower, E. M. (1985). *FEBS Lett.* **182,** 234–242.

Krafte, D. S., Goldin, A. L., Auld, V. J., Dunn, R. J., Davidson, N., and Lester, H. A. (1990). *J. Gen. Physiol.* **96,** 689–706.

Krafte, D. S., Volberg, W. A., Dillon, K., and Ezrin, A. M. (1991). *Proc. Natl. Acad. Sci. U.S.A.* **88,** 4071–4074.

Kraner, S. D., Tanaka, J. C., and Barchi, R. L. (1985). *J. Biol. Chem.* **260,** 6341–6347.

Kraner, S. D., Yang, J., and Barchi, R. L. (1989). *J. Biol. Chem.* **264,** 13273–13280.

Kraner, S. D., Chong, J. A., Tsay, H.-J., and Mandel, G. (1992). *Neuron* **9,** 37–44.

Lehmann-Horn, F., Kuther, G., Ricker, K., Grafe, P., Ballanyi, K., and Rudel, R. (1987a). *Muscle Nerve* **10,** 363–374.

Lehmann-Horn, F., Rudel, R., and Ricker, K. (1987b). *Muscle Nerve* **10,** 633–641.

Litt, M., Luty, J., Kwak, M., Allen, L., Magenis, R. E., and Mandel, G. (1989). *Genomics* **5,** 204–208.

Lopez, G. A., Jan, Y. N., and Jan, L. Y. (1991). *Neuron* **7,** 327–336.

Malo, D., Schurr, E., Dorfman, J., Canfield, V., Levenson, R., and Gros, P. (1991). *Genomics* **10,** 666–672.

Mandel, G., Cooperman, S. S., Maue, R. A., Goodman, R. H., and Brehm, P. (1988). *Proc. Natl. Acad. Sci. U.S.A.* **85,** 8.

Maue, R. A., Kraner, S. D., Goodman, R. H., and Mandel, G. (1990). *Neuron* **4,** 223–231.

Messner, D. J., and Catterall, W. A. (1985). *J. Biol. Chem.* **260,** 10597–10604.

Messner, D. J., and Catterall, W. A. (1986). *J. Biol. Chem.* **261,** 211–215.

Messner, D. J., Feller, D. J., Scheuer, T., and Catterall, W. A. (1986). *J. Biol. Chem.* **261,** 14882–14890.

Meves, H., and Nagy, K. (1989). *Biochim. Biophys. Acta* **988,** 99–106.

Meves, H., Simard, J. M., and Watt, D. D. (1985). *Ann. N.Y. Acad. Sci.* **5,** 897–966.

Moorman, J. R., Kirsch, G. E., Brown, A. M., and Joho, R. H. (1990). *Science* **250,** 688–690.

Mori, N., Schoenherr, C., Vandenbergh, D. J., and Anderson, D. J. (1992). *Neuron* **9,** 45–54.

Nilius, B. (1988). *Biophys. J.* **53,** 857–862.

Noda, M., Shimizu, S., Tanabe, T., Takai, T., Kayano, T., Ikeda, T., Takahashi, H., Nakayama, H., Kanaoka, Y., Minamino, N., Kangawa, K., Matsuo, H., Raftery, M. A., Hirose, T., Inayama, S., Hayashida, H., Miyata, T., and Numa, S. (1984). *Nature (London)* **312,** 121–127.

Noda, M., Ikeda, T., Kayano, T., Suzuki, H., Takeshima, H., Kurasaki, M., Takahashi, H., and Numa, S. (1986a). *Nature (London)* **320,** 188–192.

Noda, M., Ikeda, T., Suzuki, H., Takeshima, H., Takahashi, T., Kuno, M., and Numa, S. (1986b). *Nature (London)* **322,** 826–828.

Noda, M., Suzuki, H., Numa, S., and Stuhmer, W. (1989). *FEBS Lett.* **259,** 213–216.
Numann, R., Catterall, W. A., and Scheuer, T. (1991). *Science* **254,** 115–118.
Offord, J., and Catterall, W. A. (1989). *Neuron* **2,** 1447–1452.
Oxford, G. S., Wu, C. H., and Narahashi, T. (1978). *J. Gen. Physiol.* **71,** 227–247.
Papazian, D. M., Schwarz, T. L., Tempel, B. L., Jan, Y. N., and Jan, L. Y. (1987). *Science* **237,** 749–753.
Papazian, D. M., Timpe, L. C., Jan, Y. N., and Jan, L. Y. (1991). *Nature (London)* **349,** 305–310.
Pappone, P. A. (1980). *J. Physiol. (London)* **306,** 377–410.
Patlak, J. B., and Ortiz, M. (1986). *J. Gen. Physiol.* **87,** 305–326.
Patlak, J. B., and Ortiz, M. (1989). *J. Gen. Physiol.* **94,** 279–302.
Poli, M. A., Mende, T. J., and Baden, D. G. (1986). *Mol. Pharmacol.* **30,** 129–135.
Pollock, J. D., Krempin, M., and Rudy, B. (1990). *J. Neurosci.* **10,** 2626–2637.
Pongs, O., Kecskemethy, N., Muller, R., Krah-Hentgens, I., Baumann, A., Kiltz, H. H., Canal, I., Llamazares, S., and Ferrus, A. (1988). *EMBO J.* **7,** 1087–1096.
Ptacek, L. F., Tyler, F., Trimmer, J. S., Agnew, W. S., and Leppert, M. (1991a). *Am. J. Hum. Genet.* **49,** 378–382.
Ptacek, L. J., George, A. L., Griggs, R. C., Tawil, R., Kallen, R. G., Barchi, R. L., Robertson, M., and Leppert, M. (1991b). *Cell* **67,** 1021–1027.
Ptacek, L. J., George, A. L., Barchi, R. L., Griggs, R. C., Riggs, J. E., Robertson, M., and Leppart, M. F. (1992). *Neuron* **8,** 891–897.
Reber, B. F. X., and Catterall, W. A. (1987). *J. Biol. Chem.* **262,** 11369–11374.
Recio-Pinto, E., Thornhill, W. B., Duch, D. S., Levinson, S. R., and Urban, B. W. (1990). *Neuron* **5,** 675–684.
Reed, J. K., and Raftery, M. A. (1976). *Biochemistry* **15,** 944–953.
Ricker, K., Camacho, L., Grafe, P., Lehmann-Horn, F., and Rudel, R. (1989). *Muscle Nerve* **12,** 883–891.
Roberts, R. H., and Barchi, R. L. (1987). *J. Biol. Chem.* **262,** 2298–2303.
Rogart, R. B., and Regan, L. F. (1985). *Brain Res.* **329,** 314–318.
Rogart, R. B., Cribbs, L. L., Muglia, L. K., Kephart, D. D., and Kaiser, M. W. (1989). *Proc. Natl. Acad. Sci. U.S.A.* **86,** 8170–8174.
Rojas, E., and Armstrong, C. M. (1971). *Nature (London)* **229,** 177–178.
Rojas, E., and Rudy, B. (1976). *J. Physiol. (London)* **262,** 501–531.
Rojas, C., Wang, J., Schwartz, L., Hoffman, E., Powell, B., and Brown, R. (1991). *Nature (London)* **254,** 387–389.
Rossie, S., and Catterall, W. A. (1989). *J. Biol. Chem.* **264,** 14220–14224.
Rossie, S., Gordon, D., and Catterall, W. A. (1987). *J. Biol. Chem.* **262,** 17530–17535.
Rudel, R., Lehmann-Horn, F., Ricker, K., and Kuther, G. (1984). *Muscle Nerve* **7,** 110–120.
Rudy, B., Kirschenbaum, B., Rukenstein, A., and Greene, L. A. (1987). *J. Neurosci.* **7,** 1613–1635.
Salgado, V. L., Yeh, J. Z., and Narahashi, T. (1985). *Biophys. J.* **47,** 567–571.
Salkoff, L., Butler, A., Wei, A., Scavarda, A., Giffen, K., Ifune, C., Goodman, R., and Mandel, G. (1987). *Science* **237,** 744–749.
Satin, J., Kyle, J. W., Chen, M., Bell, P., Cribbs, L. L., Fozzard, H. A., and Rogart, R. B. (1992). *Science* **256,** 1202–1205.
Scheuer, T., Mchugh, L., Tejedor, F., and Catterall, W. (1988). *Biophys. J.* **53,** 541a.
Schmidt, J. W., and Catterall, W. A. (1986). *Cell* **46,** 437–446.
Schmidt, J. W., and Catterall, W. A. (1987). *J. Biol. Chem.* **262,** 23.
Schmidt, J. W., Rossie, S., and Catterall, W. A. (1985). *Proc. Natl. Acad. Sci. U.S.A.* **82,** 4847–4851.
Schreibmayer, W., Dascal, N., Lotan, I., Wallner, M., and Weigl, L. (1991). *FEBS Lett.* **291,** 341–344.

Schubert, B., VanDongen, A. M., Kirsch, G. E., and Brown, A. M. (1989). *Science* **245,** 516–519.

Schwarz, T., Tempel, B., Papazian, D., Jan, Y., and Jan, L. (1988). *Nature (London)* **331,** 137–142.

Seelig, T. L., and Kendig, J. J. (1982). *Brain Res.* **245,** 144–147.

Sharkey, R. G., Beneski, D. A., and Catterall, W. A. (1984). *Biochemistry* **23,** 6078–6086.

Sheng, Z.-H., Barchi, R. L., and Kallen, R. G. (1992). *J. Biol. Chem.,* in press.

Sherman, S., Chrivia, W., and Catterall, W. (1985). *J. Neurosci.* **5,** 1570–1576.

Shrager, P., and Profera, C. (1973). *Biochim. Biophys. Acta* **318,** 141–146.

Sigel, E. (1987). *J. Physiol. (London)* **386,** 73–90.

Sigworth, F. J., and Neher, E. (1980). *Nature (London)* **287,** 447–449.

Sorbera, L. A., and Morad, M. (1991). *Science* **253,** 1286–1289.

Stuhmer, W., Methfessel, C., Sakmann, B., Noda, M., and Numa, S. (1987). *Eur. Biophys. J.* **14,** 131–138.

Stuhmer, W., Conti, F., Suzuki, H., Wang, X., Noda, M., Yahagi, N., Kubp, H., and Numa, S. (1989). *Nature (London)* **339,** 597–603.

Stuhmer, W., Conti, F., Stocker, M., Pongs, O., and Heinemann, S. H. (1991). *Pfluegers Arch.* **418,** 423–429.

Sumikawa, K., Parker, L., and Miledi, R. (1984). *Proc. Natl. Acad. Sci. U.S.A.* **81,** 7994–7998.

Suzuki, H., Beckh, S., Kubo, H., Yahagi, N., Ishida, H., Kayano, T., Noda, M., and Numa, S. (1988). *FEBS Lett.* **228,** 195–200.

Tanabe, T., Takeshima, H., Mikami, A., Flockerzi, V., Takahashi, H., Kangawa, K., Kojima, M., Matsuo, H., Hirose, T., and Numa, S. (1987). *Nature (London)* **328,** 313–318.

Tanaka, J. C., Eccleston, J. F., and Barchi, R. L. (1983). *J. Biol. Chem.* **258,** 7519–7526.

Taouis, M., Sheldon, R. S., Hill, R. J., and Duff, H. J. (1991). *J. Biol. Chem.* **266,** 10300–10304.

Tejedor, F. J., and Catterall, W. A. (1988). *Proc. Natl. Acad. Sci. U.S.A.* **85,** 8742–8746.

Tejedor, F. J., Mchugh, E., and Catterall, W. A. (1988). *Biochemistry* **27,** 2389–2397.

Terlau, J., Heinemann, S. H., Stuhmer, W., Pusch, M., Conti, F., Imoto, K., and Numa, S. (1991). *FEBS Lett.* **293,** 93–96.

Thomsen, W. J., and Catterall, W. A. (1989). *Proc. Natl. Acad. Sci. U.S.A.* **86,** 10161–10165.

Thornhill, W. B., and Levinson, S. R. (1987). *Biochemistry* **26,** 4381–4388.

Trimmer, J. S., Cooperman, S. S., Tomiko, S. A., Zhou, J. Y., Crean, S. M., Boyle, M. B., Kallen, R. G., Sheng, Z. H., Barchi, R. L., and Sigworth, F. J. (1989). *Neuron* **3,** 33–49.

Vassilev, P. M., Scheuer, T., and Catterall, W. A. (1988). *Science* **241,** 1658–1661.

Vassilev, P. M., Scheuer, T., and Catterall, W. A. (1989). *Proc. Natl. Acad. Sci. U.S.A.* **86,** 8147–8151.

Waechter, C. J., Schmidt, J. W., and Catterall, W. A. (1983). *J. Biol. Chem.* **258,** 5117–5123.

Warashina, A., Ogura, T., and Fujita, S. (1988). *Comp. Biochem. Physiol. C* **90,** 351–360.

Weigele, J., and Barchi, R. (1982). *Proc. Natl. Acad. Sci. U.S.A.* **79,** 3651–3655.

West, J. W., Numann, R., Murphy, B. J., Scheuer, T., and Catterall, W. A. (1991). *Science* **254,** 866–868.

White, M. M., Chen, L. Q., Kleinfield, R., Kallen, R. G., and Barchi, R. L. (1991). *Mol. Pharmacol.* **39,** 604–608.

Woosley, R. L. (1991). *Annu. Rev. Pharmacol. Toxicol.* **31,** 427–455.

Yamamoto, D. (1985). *Prog. Neurobiol. (Oxford)* **24,** 257–292.

Yang, J. R., and Barchi, R. L. (1990). *J. Neurochem.* **54,** 954–962.

Yang, J. R., Sladky, J., Kallen, R., and Barchi, R. L. (1991). *Neuron* **7,** 421–427.
Zagotta, W. N., and Aldrich, R. A. (1990). *J. Gen. Physiol.* **95,** 29–60.
Zagotta, W. N., Hoshi, T., and Aldrich, R. W. (1990). *Science* **250,** 568–570.
Zhou, J., Potts, F. F., Trimmer, J. S., Agnew, W. A., and Sigworth, F. J. (1992). *Neuron* **7,** 775–785.
Zwerling, S. J., Cohen, S. A., and Barchi, R. L. (1991). *J. Biol. Chem.* **266,** 4574–4580.

The Vacuolar H^+-ATPases: Versatile Proton Pumps Participating in Constitutive and Specialized Functions of Eukaryotic Cells

Stephen L. Gluck
Departments of Medicine and Cell Biology and Physiology, Washington University School of Medicine, and the Renal Division, Jewish Hospital of St. Louis, St. Louis, Missouri 63110

I. Biochemical Properties of the Vacuolar H^+-ATPases

A. Subunit Structure and Function

Vacuolar H^+-ATPases (of V-ATPases) have been purified from several mammalian, plant, and fungal sources and have remarkably similar structures. They are all large-molecular-weight heteromultimeric proteins with a complicated structure that includes both intrinsic membrane proteins and peripheral membrane proteins. Depending on the species and membrane source, vacuolar H_+-ATPases may consist of nine or more different subunit polypeptides. The subunit structure reported for several H^+-ATPases is summarized in Table I.[1] Because the M_r values for the subunits vary among preparations, a nomenclature suggested by Nelson (1989), and now in general use in the field, will be used for clarity. This designates the subunits of the "catalytic sector" as A–E, the subunits of the "membrane sector" as a–c[2] and designates "accessory polypeptides" by their M_r, which are discussed in more detail below.

[1] For many of the subunits, the reported differences in M_r are within the experimental error of the determination or are due to differences in the gel systems employed. In this discussion, the M_r values for the subunits of the bovine kidney H^+-ATPase will be used, and the corresponding subunit sizes in other systems can be identified in the table.

[2] Anraku and colleagues refer to the yeast subunits using different designations: the M_r 69,000 is their "a" subunit, and the M_r 56,000 is their "b" subunit. However, these will be referred to as the A and B subunits, respectively.

TABLE I

Subunit Structure of Vacuolar H^+-ATPases

Origin of H^+-ATPase	Single-letter designation												Reference
		A	B		C		D	E	a	c			
	(M_r values for polypeptides)												
Bovine kidney microsomes		70	56	45	42	38	32	31	15	14		12	Gluck and Caldwell (1987)
Bovine brain coated vesicle	100	73	58		40	38	34	33		19	17		Arai *et al.* (1989)
Bovine brain coated vesicle	116	70	58				34	33		17			Xie and Stone (1988)
Chromaffin granule	115	72	57		39				20	17			Moriyama and Nelson (1989b)
Lysosome		72	57		41		34	33	18	15			Moriyama and Nelson (1989a)
Golgi		68	58		41					16			Young *et al.*(1988)
Golgi		72	57		41		34	33					Moriyama and Nelson (1989b)
Yeast (*S. cerevisiae*)	100	69	60		42	36	32	27		17			Kane *et al.* (1989)
Oat tonoplast		72	60		41								Randall and Sze (1987)
Beet tonoplast		67	57							15			Manolson *et al.* (1988)
Yeast (*S. cerevisiae*)		67	57							20			Uchida *et al.* (1986)
Insect midgut and Malpighian tubules		70	57		46			29			17		Schweikl *et al.* (1989)

The understanding of the structure of the vacuolar H^+-ATPases was greatly facilitated by the discovery (E. J. Bowman *et al.*, 1989; B. J. Bowman *et al.*, 1989a; Brown *et al.*, 1987; Mandel *et al.*, 1988; Zimniak *et al.*, 1988) that these enzymes had a structure similar to that of the F_0F_1 class of H^+-ATPases (called F-ATPases in more recent nomenclature). F-ATPases, which include the bacterial, chloroplast, and mitochondrial H^+-ATPases, consist of a catalytic sector (the F_1 particle), composed entirely of soluble proteins peripherally associated with the membrane, and a transmembrane sector (the F_0 domain), composed of intrinsic membrane proteins forming a proton channel. The F_1 catalytic head of the F-ATPases is generally composed of five different subunits with a stoichiometry $\alpha_3\beta_3\gamma\delta\varepsilon$. The β and α subunits are assembled as a symmetrical, hexagonal ring of alternating subunits. The β subunit contains the catalytic site for ATP hydrolysis; although 3 replicas of the subunit are present, evidence suggests that only a single subunit is involved in active ATP hydrolysis at any time. The α subunit has a high-affinity ATP binding site and is thought to be a regulatory subunit ( ), although the precise function of the bound nucleotide remains unclear. The ring of α and β subunits is connected to the transmembrane domain by a "stalk" containing one each of a γ,δ, and ε subunit. The assembled F_1 particle retains ATPase activity when detached from the membrane, but rapidly loses activity at 4°C (cold inactivation; Futai and Kanazawa, 1983). The principal structural component of the transmembrane domain is a M_r 8000 proteolipid subunit, subunit c, present in an estimated 12 replicas, that forms the membrane-spanning proton channel. Dicyclohexylcarbodiimide (DCCD), a hydrophobic carboxyl reagent that reacts efficiently with an aspartate residue on the subunit, prevents proton movement through the channel and inhibits ATPase activity of the fully assembled enzyme. The transmembrane domain also contains 2 replicas of a b subunit, and a single c subunit, which participate in binding the F_1 sector and in controlling the permeability of the proton channel (Schneider and Altendorf, 1987).

Like the F-ATPases, the vacuolar H^+-ATPases have a large non-transmembrane catalytic sector (or cytoplasmic domain) containing the catalytic site and a transmembrane domain forming a proton channel through the membrane. Similar to the F-ATPases, the cytoplasmic domain appears to contain two subdomains: a catalytic "head" composed of a hexagonal ring of alternating A and B subunits (Adachi *et al.*, 1990a,b; Puopolo and Forgac, 1990), and a "stalk" domain (Bowman *et al.*, 1989b; Brown *et al.*, 1987; Klein and Zimmerman, 1991a; Taiz and Taiz, 1991), likely consisting of the C, D, and E subunits. (Fig. 1). The A and B subunits of the vacuolar H^+-ATPases share substantial sequence homology with subunits in the catalytic "head" of the F-ATPases.

The cytoplasmic domain of the vacuolar H^+-ATPases may be dissociated from the membrane by several methods. High pH buffers, such as carbonate (Lai *et al.*, 1988), removed the cytoplasmic domain presumably by denaturing the composite subunits, demonstrating that it consists of peripheral membrane proteins. At high concentrations, the chaotropic ions KI, KBr, and KNO_3 dissociated the subunits of the cytoplasmic domain from the membrane (Arai *et al.*, 1989; Kane *et al.*, 1989; Lai *et al.*, 1988; Moriyama and Nelson, 1989b,c; Parry *et al.*, 1989; Puopolo and Forgac, 1990); at lower concentrations, the same anions dissociated an intact cytoplasmic domain when performed at 4°C in the presence of ATP. This property, called "cold inactivation," is also characteristic of the F-ATPases. The dissociated cytoplasmic domain could then reattach to the transmembrane domain restoring ATP-dependent proton transport following removal of the chaotropic ion by dialysis (Puopolo and Forgac, 1990). The dissociated domain consisted of a complex of the A, B, C, D, and E subunits that could be coprecipitated (the V_1 domain), leaving the

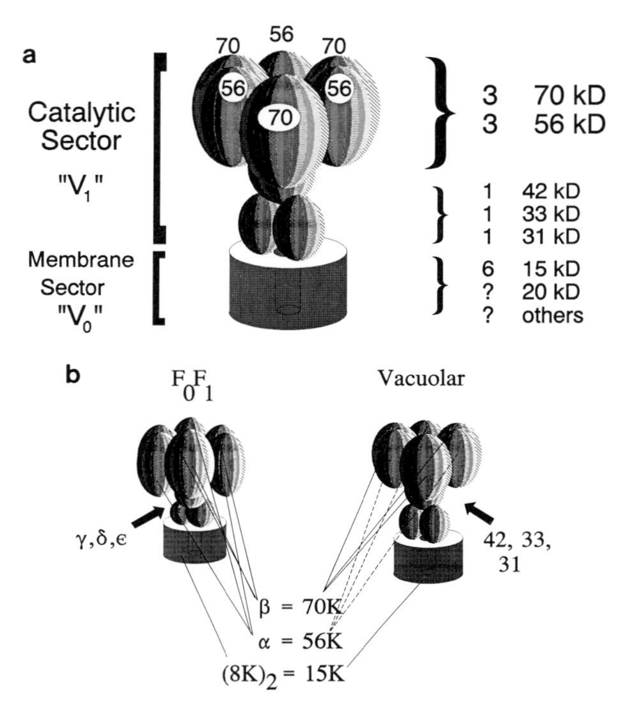

FIG. 1 (a) Diagram depicting structure of the vacuolar H^+-ATPases. (b) Diagram of homologous subunits of F_0F_1 and vacuolar H^+-ATPases.

a and c subunits and the M_r 38,000 and 100,000, and possibly others, remaining in the transmembrane (V_0) domain (Puopolo and Forgac, 1990). Although the F_1 domain dissociated from the F-ATPases retains Mg-ATPase hydrolytic activity (Amzel and Pedersen, 1983; Futai *et al.*, 1989; Schneider and Altendorf, 1987; Senior and Wise, 1983), the V_1 domain did not (Moriyama and Nelson, 1989c; Puopolo and Forgac, 1990). It has been reported that the V_1 domain dissociated from the membrane with urea exhibits Ca^{2+}-activated ATPase activity that is inhibited by low concentrations of Mg^{2+} (Xie and Stone, 1988). In these experiments, the A, B, C, and D subunits were required for Ca^{2+}-activated ATPase activity; a subcomplex of the A and B subunits had no activity. At the present time, it is uncertain whether the Ca^{2+}-activated ATPase activity of this preparation reflects normal catalytic activity of the V_1 preparation, since calcium does not support coated vesicle ATP-dependent proton transport (Puopolo and Forgac, 1990).

The A subunit shares closest homology with the F-ATPase catalytic β subunit. Several observations support the assignment of the A subunit as containing the ATP-binding catalytic site. However, as discussed below, evidence suggests that the B subunit, and possibly other subunits, may modify the catalytic properties of the enzyme. The A subunit in several of the vacuolar H^+-ATPases was labeled by the proton transport and ATPase inhibitors [^{14}C]NBD-Cl or *N*-[^{3}H]ethylmaleimide (NEM) in an ATP-protectable manner (Arai *et al.*, 1987b; Bowman *et al.*, 1983; Mandala and Taiz, 1986; Moriyama and Nelson, 1987b; Percy and Apps, 1986; Randall and Sze, 1987; Uchida *et al.*, 1986). Antibodies to the corn coleoptile H^+-ATPase A subunit inhibited proton transport, but antibodies to the B subunit did not (Rausch *et al.*, 1987). Perhaps the most convincing evidence that the A subunit is the locus of the catalytic nucleotide binding site comes from the deduced sequence of the subunit. cDNA clones encoding the subunit have been isolated from a number of sources (Bowman *et al.*, 1989; Manolson *et al.*, 1988; Puopolo *et al.*, 1991; Pan *et al.*, 1991; Zimniak *et al.*, 1988). The deduced amino acid sequences from fungal, plant, and mammalian sources have an approximate 60–70% identity among themselves and 25% identity with the sequence of the catalytic β subunit of the F-ATPases (Nelson, 1989). Most interestingly, the sequences of the H^+-ATPase of archaebacteria (Denda *et al.*, 1988a; Inatomi *et al.*, 1989) show that they are more closely related to the vacuolar H^+-ATPases than to the F-ATPases of eubacteria. The A subunit contains a P-loop motif thought to participate in catalytic nucleotide binding (Walker *et al.*, 1982), and an ATP synthase motif (Gogarten *et al.*, 1989), shared by the α and β subunits of the F-ATPases and their homologs., and also thought to participate in the ATP binding site. Unlike the F-ATPases, the vacuolar H^+-ATPases are highly sensitive to sulfhydryl reagents. This property is likely attributable to a cysteine residue, C^{254}, highly conserved among species, lying within the P-loop motif. Feng and Forgac (1992) demonstrated that only the cysteine residue responsible for NEM sensitivity was capable of forming a disulfide bond with cysteine by disulfide exchange with cystine (Feng and Forgac, 1992). The remaining free sulfhydryl groups of the cysteine-modified enzyme were allowed to react with NEM and the enzyme was then treated with dithiothreitol to regenerate the cysteine-protected sulfhydryl. This protected sulfhydryl group was subsequently labeled with a fluorescent maleimide and the labeled subunit was digested with V8 protease, revealing a fluorescently tagged peptide. Protein sequence analysis of the labeled peptide was consistent with labeling of C^{254} (Feng and Forgac, 1992). The four other cysteine residues in the A subunit retain a high degree of conservation in their relative positions, but are not located within putative nucleotide binding domains. The mid-portion of the coding region of the A subunit is the most highly conserved and probably contains the catalytic domain. The amino and carboxyl termini show wide variability among species; the function of these domains is unknown, and they may serve in regulation or targeting of the subunit. Currently available evidence suggests that there is only one gene for the A subunit (Puopolo *et al.*, 1991).

The B subunit is homologous to the F-ATPase α subunit, the regulatory subunit with the high-affinity noncatalytic nucleotide binding site (Futai *et al.*, 1989; Senior and Wise, 1983; Senior, 1988). Nucleotide labeling studies support the presence of a nucleotide binding site on the vacuolar H^+-ATPase B subunit. Binding of a photoaffinity analog of ATP, 3-O-(4-benzoyl)benzoyladenosine-5′-triphosphate, has been demonstrated in the vacuolar H^+-ATPase of the beet tonoplast (Manolson *et al.*, 1985). Treatment of the bovine brain coated vesicle H^+-ATPase with TNP-ATP reduced the susceptibility of this subunit to proteolytic digestion (Adachi *et al.*, 1990a). cDNA clones for this subunit have been isolated from several fungal, plant, and animal sources (Bernasconi *et al.*, 1990; Bowman *et al.*, 1989a; Manolson *et al.*, 1988; H. Nelson *et al.*, 1989b; R. Nelson *et al.*, 1992; Puopolo *et al.*, 1992; Südhof *et al.*, 1989) and from the homologous subunit in archaebacteria (Denda *et al.*, 1988b; Inatomi *et al.*, 1989). Like the A subunits, all of the B subunits have an ATP synthase signature (Gogarten *et al.*, 1989) thought to compose part of an ATP-binding domain. However, the vacuolar H^+-ATPase B subunit lacks a P-loop present in the A subunit. There are at least two different genes encoding the B subunit (Bernasconi *et al.*, 1990; Nelson *et al.*, 1992; Puopolo *et al.*, 1992; Südhof *et al.*, 1989). A third isoform was reported (Puopolo *et al.*, 1992), but appears to have been an artifact of the polymerase chain reaction (PCR). A comparison of the deduced sequences of the mammalian subunits shows that the isoforms have a nearly identical mid-coding region. In contrast, for at least two of the genes (Nelson *et al.*, 1992), the amino and carboxyl termini are entirely different, but the differences are highly conserved across species. The function of the B subunit is unresolved, but it may serve to modify the overall enzymatic properties of the vacuolar H^+-ATPases, as discussed more below. The unique amino and carboxyl termini may have other functions such as in regulation or targeting.

The C, D, and E subunits probably reside in a stalk domain of the H^+-ATPase connecting the hexagonal catalytic "head" to the transmembrane domain. Electron microscopic images (Bowman *et al.*, 1989b; Brown *et al.*, 1987; Taiz and Taiz, 1991) and experiments using cross-linking to detect adjacent polypeptides (Adachi *et al.*, 1990b) support this topographical arrangement of the subunits. If this model is correct, these subunits likely serve partly as a proton conduit, but they may have other functions not related to proton transport. cDNA clones have been reported for the C (Beltrán *et al.*, 1992; Nelson *et al.*, 1990a) and E (Foury, 1990; Hirsch *et al.*, 1988) subunits; at this writing, no sequence is yet available for the D subunit. The degree of identity between the bovine and yeast protein sequences were 39 and 34%, respectively, below the 60% conservation for the A and B subunits. Unlike the A and B subunits, the deduced amino acid sequence for these "stalk domain" subunits have no discernible homology with the γ, δ, and ε subunits of the F-ATPases, suggesting that they may have acquired some functions unique to the vacuolar H^+-ATPases. As indicated above, some evidence suggests that they may be required to activate the catalytic center of the H^+-ATPase, even if they do not participate directly in catalysis (Stone *et al.*, 1990; Xie and Stone, 1988), analogous to the function of the γ subunit of the F-ATPases, which activates a complex of the α and β subunits (Senior, 1988).

Far less is known about the structure of the transmembrane (V_0) domain of the vacuolar H^+-ATPases. The dissociation experiments described above suggest that the domain consists of membrane-embedded a and c subunits, M_r 38,000, and 100,000 polypeptides, and likely others. Zhang *et al.* studied the properties of the V_0 sector of the bovine coated vesicle H^+-ATPase, isolated after dissociation of the V_1 sector with KI. The isolated V_0 retained the ability to restore H^+-ATPase activity when reconstituted with the V_1 sector and the isolated V_0 retained the same DCCD labeling and susceptibility to tryptic cleavage as in the intact H^+-ATPase. When reconstituted into a phospholipid liposome, the isolated V_0 did not have a detectable proton conductance, suggesting that the proton channel is closed until the cytosolic domain associates with the transmembrane domain (Zhang *et al.*, 1992).

The c subunit of the V_0 sector is highly hydrophobic and may be extracted from the membrane by organic solvents (Arai *et al.*, 1987a; Randall and Sze, 1986; Sun *et al.*, 1987), a property for which it is classified as a proteolipid. The vacuolar H^+-ATPase proteolipid is strongly reactive with DCCD, which inactivates the H^+-ATPase (Arai *et al.*, 1987a; Randall and Sze, 1986; Sun *et al.*, 1987). This DCCD binding subunit is functionally analogous to the DCCD binding proteolipid of the F-ATPases. cDNA cloning of the vacuolar H^+-ATPase proteolipid from several sources (G. Gillespie *et al.*, 1991; Hanada *et al.*, 1991; Mandel *et al.*, 1988; Nelson and Nelson, 1989a; Umemoto *et al.*, 1990) revealed that it shares significant sequence similarity with the F-ATPase proteolipid. The proteolipid of the F-ATPases has an approximate M_r of 8000, only about half the size of the vacuolar H^+-ATPase proteolipid and has two membrane-spanning α-helices predicted from hydropathy plots of the deduced sequence (Mandel *et al.*, 1988). The hydropathy plot from the deduced amino acid sequence of vacuolar H^+-ATPase subunit predicts four membrane-spanning α-helices (helices I–IV), and both the amino- and the carboxyl-terminal halves of the vacuolar c subunit share sequence homology with the F-ATPase proteolipid (Mandel *et al.*, 1988). Hence, the vacuolar subunit appears to have evolved through gene duplication and modification. The proteolipid of the F-ATPases is the major structural component of the transmembrane proton channel. The isolated F-ATPase proteolipid was shown to form an electrically conductive proton channel in liposomes (Kagawa, 1978); valinomycin and potassium concentration gradients were used to generate an inside negative potential in liposomes, and reconstitution of the F-ATPase proteolipid in the membrane provided the capacity for potential-drive uptake of protons and generation of a pH gradient (Kagawa, 1978). Some evidence suggests that the vacuolar H^+-ATPase proteolipid may also form a proton channel in the absence of other subunits. When the subunit was isolated by toluene extraction of coated vesicle H^+-ATPase and reintroduced into liposomes containing bacteriorhodopsin, it was able to collapse a light-generated pH gradient (Sun *et al.*, 1987). Whether the increase in membrane proton permeability was due to detergent-like effects on the lipid bilayer or was truly mediated by a functional channel is not entirely clear, although the protonophoric effects of the extracted proteolipid were inhibited by DCCD (Sun *et al.*, 1987). In contrast, the isolated c subunit from the F_0 domain of the bacterial F-ATPase did not exhibit proton channel activity (Schneider and Altendorf, 1985).

In *Saccharomyces,* a cDNA clone encoding a second proteolipid was recently isolated, which had a high degree of similarity to the cDNA from bovine chromaffin granules and other species (VMA11; Umemoto *et al.*, 1990). The gene was isolated using a cloning strategy based on assaying transferred genes for their ability to restore a normal phenotype in yeast mutants defective in vacuolar acidification (see below). Most surprisingly, deletion of either isoform of the proteolipid subunit produced a lethal mutation, implying that both proteolipids coexist and are necessary for function within a single H^+-ATPase complex. Several isoforms were also discovered for the oat vacuolar H^+-ATPase proteolipid subunit, although the coding region of these isoforms was identical (Lai *et al.*, 1991). No isoforms of this subunit have yet been described for the mammalian protein.

The sequence of the yeast and mammalian proteolipids did not reveal any putative signal sequences, raising the interesting issue of how, and in what compartment of the cell, it inserts in the membrane. Although spontaneous insertion of membrane proteins occurs in bacteria, the process requires a transmembrane potential. Even if a similar process were to occur for this subunit, how would it distinguish among the many vacuolar compartments that maintain potential differences across the membrane? Answers to these questions may soon be forthcoming in the recently developed systems for analyzing H^+-ATPase assembly in yeast.

Many of the vacuolar H^+-ATPases isolated have an associated subunit of M_r 100–116,000. This has been classified as an "accessory" polypeptide by Nelson (1989) and its actual requirement for H^+-ATPase function remains controversial. The subunit does not appear to be required for ATPase or proton transport activity, as the bovine kidney (Gluck and Caldwell, 1987) and Golgi (Moriyama and Nelson, 1989b; Young *et al.*, 1988) enzymes lack such polypeptides, but retain enzymatic and transport activity. The presence or absence of the polypeptide appears highly dependent on the type of detergent used to solubilize the membrane (J. Gillespie *et al.*, 1991; Xie and Stone, 1988). For example, the kidney H^+-ATPase shows an associated M_r 115,000 subunit when the membrane is solubilized with Triton X-100, but not when solubilized with CHAPS and nonyl glucopyranoside (J. Gillespie *et al.*, 1991; Gluck and Caldwell, 1987), yet the Triton-solubilized enzyme does not retain any activity (J. Gillespie *et al.*, 1991). Removal of the M_r 116,000 polypeptide from the coated vesicle H^+-ATPase with nonyl glucopyranoside reduced, but did not eliminate, proton transport and ATPase activity (Xie and Stone, 1988). Initial preparations of the yeast H^+-ATPase lacked this subunit (Uchida *et al.*, 1985). A M_r 100,000 polypeptide was reported associated with the yeast vacuolar H^+-ATPase in a more recent purification from another laboratory (Kane *et al.*, 1989). This polypeptide was highly susceptible to proteolytic digestion, and often missing or detectable only as a proteolytic fragment on sodium dodecylsulfate (SDS) gels, yet without apparent effect on the ATPase activity (Kane *et al.*, 1992). Unfortunately, none of these observations has demonstrated conclusively whether these polypeptides are essential. As discussed below, experiments using yeast genetics to delete individual genes for the H^+-ATPase have provided a powerful means for resolving such issues. The gene for the yeast M_r 100,000 polypeptide had not been reported at this writing. However, the cDNA sequence of a M_r 116,000 polypeptide from the rat vacuolar H^+-ATPase has been reported (Perin *et al.*, 1991). The deduced amino acid sequence predicts a protein with a hydrophilic amino-terminal domain and at least six putative membrane-spanning domains, but whose function remains unclear. The identification of a homologous gene was recently reported in *Caenorhabditis elegans*, whose sequence has a high degree of identity (Sulston *et al.*, 1992), and additional genes were identified in *C. elegans*, suggesting that this protein might be a member of a gene family. These new findings may provide an additional means for determining the function of the protein using genetic approaches to screen for *C. elegans* mutants.

A second M_r 116,000 polypeptide vanadate-sensitive ATPase has also been found in preparations of the chromaffin granule and coated vesicle enzymes, and differs from the M_r 116,000 polypeptide associated with the H^+-ATPase (Moriyama *et al.*, 1991; Xie *et al.*, 1989). No role in the function of the H^+-ATPase is currently ascribed for this second ATPase.

A cDNA clone for the M_r 38,000 polypeptide has been reported (Wang *et al.*, 1988). The deduced amino acid sequence encoded a 32,000-Da protein with no evident similarity to any of the F-ATPase subunits or to the other vacuolar H^+-ATPase subunits. A hydropathy analysis of the polypeptide revealed one putative transmembrane domain of only 25 residues. No cDNA clone for an equivalent or homologous subunit has yet been identified in yeast.

Most of the information on the necessity and function of the different polypeptides isolated in the preparations of H^+-ATPase has come from partial reconstitution studies cited above. However, the isolation of several of the H^+-ATPase genes in *Saccharomyces* over the last few years has enabled investigators to use yeast genetics to study the effects of deletions and mutations in individual polypeptides on overall vacuolar acidification and subunit localization. A finding crucial for this type of analysis was that disruption of the vacuolar H^+-ATPase genes produces a conditional lethal phenotype (Kane *et al.*, 1992; Nelson and Nelson, 1990b; Yamashiro *et al.*, 1990). Yeasts with defective vacuolar H^+-ATPase activity do not grow in media with a pH value higher than 6.5 and have a narrow pH optimum for growth at 5.5 (Nelson and Nelson, 1990b; Yamashiro *et al.*, 1990). Disruption of the genes encoding the 69-kDa (VMA1, TFP1; Hirata *et al.*, 1990; Kane *et al.*, 1992; Noumi *et al.*, 1991), 60-kDa (VAT2, VMA2; Kane *et al.*, 1992; Nelson and Nelson, 1990b), 42-kDa (VATC;[3] Beltrán *et al.*, 1992), 27-kDa (VMA4; Foury, 1990), and 17-kDa subunits (VMA3; Nelson and Nelson, 1990b; Umemoto *et al.*, 1990) produces a phenotype with no acidification of the vacuole in intact cells and no ATPase activity in the vacuolar membrane, no growth at neutral pH, and slow growth at acid pH. These findings provide strong evidence that these polypeptides are indispensable components required for H^+-ATPase function or assembly. Since most of the gene disruption studies performed also resulted in a failure of the H^+-ATPase to assemble on the vacuolar membrane (Kane *et al.*, 1992; Umemoto *et al.*, 1990, 1991), they did not provide conclusive information on the participation of these subunits in catalytic activity or other functions of the enzyme.

An analysis of the phenotype of the H^+-ATPase-deficient mutants has provided some important new insights into the function of vacuolar acidification. It has been proposed that the acid-medium requirement of these mutants reflects the capacity for vesicles formed by fluid phase endocytosis (Reizman, 1985) to fuse with and acidify vesicles in the *trans* Golgi network, some of which may enter the vacuolar trafficking pathway (Nelson and Nelson, 1990b). Mutants with disrupted 70-kDa (Hirata *et al.*, 1990), 57-kDa, or 17-kDa subunits (Noumi *et al.*, 1991) grew slowly at pH 5.5 and also failed to grow either at low extracellular Ca^{2+} (in medium with EGTA) or in the presence of 100 m*M* CaCl (Hirata *et al.*, 1990; Noumi *et*

[3] This would be equivalent to VMA3 in the nomenclature of Anraku (Hirata *et al.*, 1990) and Stevens (Kane *et al.*, 1992)

al., 1991; Shih *et al.*, 1988). In the yeast, the vacuolar H^+-ATPase is responsible for accumulation of calcium in the vacuole by driving Ca/H exchange through creation of a proton electrochemical gradient across the vacuolar membrane (Ohsumi and Anraku, 1983). This suggests that one principal function of the low pH requirement in the H^+-ATPase mutants may be to drive vacuolar calcium transport. Active vacuolar transport, therefore, appears to be obligatory in low extracellular Ca^{2+} to provide a sufficient driving force for uptake. The basis for the sensitivity to high calcium is more complex. Anraku and colleagues found that calcium-sensitive (cls) yeast mutants (killed by 100 m*M* $CaCl_2$) showing a Pet phenotype (inability to utilize fermentable carbon sources such as glycerol, lactate, and succinate) all had defects in vacuolar H^+-ATPase activity, and two strains proved to have defects in the A (69 kDa) and c (proteolipid) subunits (Ohya *et al.*, 1991). All of the genetic defects described previously yielding the Pet phenotype were in proteins involved in mitochondrial oxidative phosphorylation (Tzagoloff and Dieckmann, 1990). The Pet– cls mutants had resting cytosolic calcium levels of 900 n*M*, compared to 150 n*M* in the wild-type cells. These mutants also required inositol for growth and had defects in phosphatidylethanolamine synthesis and phosphatidylserine metabolism. The basis for the Pet phenotype is not entirely clear, but in high extracellular Ca^{2+}, vacuolar H^+-ATPase mutants appear unable to take up sufficient vacuolar calcium to regulate cytosolic calcium activity; this may result in accumulation in the mitochondria of cytosol, generating defects in mitochondrial function and lipid metabolism. Evidence for the interaction of vacuolar H^+-ATPases and mammalian vacuolar calcium transport has also recently been reported (Ozawa and Schulz, 1991).

The ability to recognize vacuolar H^+-ATPase mutants by phenotypic properties has provided a means for structure–function analysis of individual subunit polypeptides. Noumi *et al.* (1991) analyzed the effect of mutations in the 17-kDa proteolipid by transforming proteolipid-defective mutants with a plasmid carrying the mutated proteolipid genes. Plasmid constructs with the bovine proteolipid protein replacing most of the yeast subunit and those with the last two of the predicted transmembrane helices of the yeast proteolipid replaced in-frame with the gene for the *Escherichia coli* proteolipid (c subunit) both produced inactive proteins. The effects of single amino acid substitutions produced throughout the yeast proteolipid were also studied. In the F-ATPases, DCCD reacts with a single aspartate residue residing in one of the transmembrane helices, producing a nonconducting proton channel (Futai *et al.*, 1989; Kagawa, 1978; Senior and Wise, 1983; Senior, 1988). Alignment of the deduced amino acid sequences of the proteolipids from the F-ATPases and the vacuolar H^+-ATPase suggests a glutamic acid residue, E137, as the putative DCCD binding site

(Nelson and Nelson, 1989a, 1990b). Replacement of this glutamic acid residue with aspartate yielded yeast that grew slowly, indicating partial function; all other substitutions at this site gave inactive enzyme (Noumi *et al.,* 1991). Many of the amino acid substitutions in the vicinity of this residue also produced an inactive protein. Similar studies on the effects of mutations on other subunits of the H^+-ATPase have not yet been reported.

The yeast provides a powerful model for analyzing the structural requirements for several H^+-ATPase functions, but yeast lacks some of the complexity of the mammalian system. For example, only one gene for the 57-kDa subunit has been identified (Nelson *et al.,* 1989), in contrast to the mammalian subunit, and the vacuolar H^+-ATPase appears restricted to the yeast vacuole (Kane *et al.,* 1992), whereas the enzyme appears in a diversity of intracellular compartments and the plasma membrane, at widely differing densities, in insect and mammalian tissues (Brown *et al.,* 1987; Klein and Zimmerman, 1991a; Rodman *et al.,* 1991). However, the diversity of vacuolar compartments in yeast may prove to be more complex, as a pathway for vesicular traffic between the vacuole of parent cells and the bud vacuole was recently discovered (Weisman and Wickner, 1988).

B. Enzymatic Properties of the H^+-ATPases

The vacuolar H^+-ATPases were first identified as electrogenic proton pumps that were resistant to the F-ATPase inhibitors azide and oligomycin, and to the E1E2 ATPase inhibitor vanadate, but were extremely sensitive to several sulfhydryl reagents. The sensitivities of these enzymes to a variety of inhibitors has been reviewed previously (Forgac, 1989; Nelson, 1989).

Recently, the bafilomycins, a class of macrolide antibiotics produced by species of *Streptomyces,* were identified as high-affinity inhibitors of the vacuolar H^+-ATPases (Bowman *et al.,* 1988). All vacuolar H^+-ATPases so far tested have been inhibited by nanomolar concentrations of bafilomycin A_1 (Bowman *et al.,* 1988; Hanada *et al.,* 1990), although considerable variability in the I_{50} and IC_{50} values have been reported. Recently, Hanada *et al.* (1990) demonstrated that the bafilomycins bound stoichiometrically to the chromaffin granule H^+-ATPase (1 mol antibiotic : 1 mol enzyme) and the degree of inhibition was directly related to the amount of enzyme present, independent of concentration. They also demonstrated that the membrane domain of the enzyme contained the binding site (Hanada *et al.,* 1900). In this respect, bafilomycins resemble the oligomycins, high-affinity macrolide inhibitors of the mitochondrial F-ATPase. An increasing number of reports have emerged in which bafilomycin has been used in intact cells as a specific inhibitor of the vacuolar H^+-ATPases. Such studies must be viewed with some caution, as the effects of the bafilomycins on the function of other cellular proteins have not been investigated sufficiently.

II. Biosynthesis, Assembly, and Targeting of the Vacuolar H^+-ATPase

The synthesis, assembly, and targeting of several of the H^+-ATPase subunits have been studied in *Saccharomyces* (Beltrán *et al.*, 1992; Kane *et al.*, 1992; Nelson and Nelson, 1990b; Noumi *et al.*, 1991; Umemoto *et al.*, 1990). Perhaps the most interesting and surprising result is that the yeast A subunit is synthesized as a 119-kDa precursor polypeptide from which a 50-kDa spacer peptide is excised to yield the final 69-kDa subunit (Hirata *et al.*, 1990; Kane *et al.*, 1990). The spacer protein resides in the middle of the polypeptide; synthesis of the precursor and processing to the final form were demonstrated in pulse–chase labeling experiments (Hirata *et al.*, 1990; Kane *et al.*, 1990). As intriguing as this finding is, no other species has been identified with a similar processing mechanism. In the archaebacterial, *Neurospora*, plant, and mammalian enzymes the cDNA encodes the final form of the polypeptide.

The assembly of the vacuolar H^+-ATPase was studied in wild-type yeast and in mutants with defective A, B, or c subunits (Kane *et al.*, 1992). H^+-ATPase staining of the vacuolar membrane was demonstrated in wild-type cells, but not in any of the mutants with the defective subunits, although the subunits were detectable in cell lysates by immunoblotting (Kane *et al.*, 1992). In contrast, the M_r 100,000 subunit appeared to require the presence of an intact c subunit; in the absence of the c subunit, the M_r 100,000 subunit was not detectable in yeast by either immunofluorescence or immunoblotting (Kane *et al.*, 1992). Unfortunately, the M_r 100,000 subunit was unusually susceptible to proteolysis even in the presence of protease inhibitors and in protease-deficient mutants, and its absence in the subunit c-deficient cells may have been a consequence of enhanced exposure to proteases.

Studies of mammalian H^+-ATPase synthesis and degradation have been performed in the pig kidney cell line LLC-PK_1 (J.-Y. Fu and S. Gluck, unpublished observations). Cells were labeled with $[^{35}S]$methionine and H^+-ATPase was immunoprecipitated with a monoclonal antibody that recognizes the intact enzyme. Labeled enzyme was detectable in immunoprecipitates within 10 min, but 24–48 hr were needed to reach steady-state levels. The degradation rate of the enzyme was also assessed in cells prelabeled to steady state. These experiments showed a half-life of over

48 hr for the enzyme. Taken together, the results showed rapid initial labeling of the enzyme but a long half-life suggests that the rapid labeling may occur in a dynamic pool of H^+-ATPase subunits that exchange with preexisting assembled subunits. Immunoprecipitates in cells pulse-labeled for 10 min and chased with cold methionine showed that labeled D and E subunits of the H^+-ATPase were associated with the assembled H^+-ATPase at 10 min, and labeled A and B subunits did not appear in the immunoprecipitates until 20 min. Immunoblots of cytosol spun to remove any membranes showed that A and B subunits were present, but no E subunits were detected. These results suggest that these cells have a cytosolic pool of unassembled A and B subunits, and that the D and E subunits may have a role in controlling and/or initiating the assembly of the A and B subunits on the membrane.

Specific proteins participating in a chaperonin-like role in the biosynthesis of the vacuolar H^+-ATPase have not been identified. However, two yeast vacuolar protein-sorting mutants, vps3 and vps6, have defective vacuolar acidification with decreased amounts of the A and B subunits on the vacuolar membrane, but neither of the genes corresponds to any of the known subunits of the enzyme (Kane *et al.*, 1992); it is possible that these proteins participate in the assembly of the enzyme without remaining associated. Recently, two nuclear-encoded genes (ATP11 and ATP12) that are required for assembly of the F-ATPase α and β subunits into an active oligomer have been identified (Ackerman and Tzagoloff, 1990).

III. Participation in Constitutive Cellular Functions

The vacuolar class of H^+-ATPase is responsible for acidification of the endocytic and exocytic pathways. The distribution and function of compartments participating in intracellular acidification in endocytosis and secretion have been reviewed previously (Forgac, 1989; Mellman *et al.*, 1986). Perhaps the most interesting aspect of these compartments is their ability to maintain different steady-state intravesicular pH values required for normal endocytic processing, sorting, and post-translational processing.

Chloride ion is an important regulator of vacuolar acidification. Evidence from several laboratories suggests that the vacuolar H^+-ATPases may be stimulated directly by chloride (Kaunitz *et al.*, 1985; Randall and Sze, 1986; Stone *et al.*, 1990). In contrast, we (Gluck and Caldwell, 1987, 1988; Wang and Gluck, 1990) and others (Arai *et al.*, 1989; Van Dyke, 1986, 1988) found no evidence for a direct stimulatory effect of chloride on the enzyme; increasing concentrations of chloride added to the ATPase assay buffer decreased ATPase activity slightly (Gluck and Caldwell, 1987; Wang and Gluck, 1990). These findings suggest that chloride functions primarily by dissipating the potential difference across the vacuolar membrane created by electrogenic proton transport (Arai *et al.*, 1989; Forgac, 1989). Even if changes in cytosolic chloride have little effect on proton transport, recent studies suggest that modulation of vacuolar chloride permeability may affect the rate of acidification. Treatment of endosome-containing renal membranes (Bae and Verkman, 1990) or coated vesicles (Mulberg *et al.*, 1991) with cAMP-dependent protein kinase decreased the initial rate of acidification by 30–35%. It is still uncertain, however, whether kinetic defects of this magnitude would affect the steady-state pH of the vacuole, unless the proton leak rate were high.

Other anions, such as nitrate, had profound inhibitory effects on ATPase activity (Arai *et al.*, 1989; Bowman *et al.*, 1989b; Moriyama and Nelson, 1989b,c). Some of these effects are likely a consequence of cold inactivation and dissociation of the cytoplasmic domain (Moriyama and Nelson, 1989c). Nitrate is an inhibitor of the purified kidney H^+-ATPase at concentrations above 5 m*M*. Our results suggest that this effect is due to an uncompetitive-type inhibition on ATPase activity, not due to dissociation of subunits from the set of plasmic domain of the enzyme (Wang and Gluck, 1990). Sulfite, and to a lesser extent sulfate, had similar inhibitory effects on the H^+-ATPase. These observations suggest that the kidney H^+-ATPase may have an oxyanion binding site that regulates activity. We tested the effect of different concentrations of bicarbonate on the enzyme under conditions of constant pH and found little effect. However, under conditions where greater than 0.1 m*M* carbonate ion was present, as estimated from the bicarbonate–carbonate proton K_d, a 20–30% inhibition of ATPase activity was observed (Z.-Q. Wang and S. Gluck, unpublished observations). Hence one function of the anion binding site of the kidney H^+-ATPase may be as a receptor for carbonate ion. Carbonate ion would be present at significant concentrations only in an alkaline cytosol or in the setting of a significant unstirred layer of bicarbonate ion, and it might serve to inhibit further proton extrusion from the cell. However, it seems unlikely that carbonate could function to maintain different steady-state pH values among compartments.

It has been suggested that the Na,K-ATPase might regulate vacuolar pH by controlling the magnitude of the transvesicular membrane potential opposing electrogenic proton transport (Fuchs *et al.*, 1989); more recent results indicate that this mechanism may operate only in some cells (Sipe *et al.*, 1991). As described below, our laboratory has recently identified cytosolic factors, most likely proteins, that activate or inhibit the vacuolar H^+-ATPase directly. The activator acts preferentially on brush-border and microsomal H^+-ATPases and has a lesser effect on lysosomes (Zhang *et al.*, 1992b). This suggests the possibility that H^+-ATPase regulatory proteins may act selectively on subpopulations of the H^+-ATPase to modulate the pH of vacuolar compartments.

IV. Participation in Specialized Cellular Functions

A. Transcellular Transport in Epithelia

To maintain acid–base balance, the kidney must reabsorb all of the filtered bicarbonate and excrete a quantity of acid equal to that produced by metabolic proton generation. Renal hydrogen ion excretion takes place in several nephron segments (Alpern, 1990). The proximal tubule reabsorbs 90% of the filtered bicarbonate and an additional 5% of the filtered bicarbonate is reabsorbed in a thick ascending limb. The collecting duct is responsible for the final reabsorption of bicarbonate in the regulation of net acid excretion by the kidney. In the collecting duct, hydrogen ion excretion is carried out by the intercalated cells, a specialized cell population rich in carbonic anhydrase (Brown, 1989; Madsen and Tisher, 1985). These cells employ a vacuolar H^+-ATPase as a predominant means for hydrogen ion transport (Brown *et al.*, 1988a,b). Properties of the vacuolar proton pump in membranes from both cortex (primarily proximal tubule) and medulla (largely from collecting duct) have been characterized by a number of investigators (Burckhardt and Warnock, 1990; Gluck and Al-Awqati, 1984; Gurich *et al.*, 1991; Jehmlich *et al.*, 1991; Kaunitz *et al.*, 1985; Kinne-Saffran *et al.*, 1982; Sabolic and Burckhardt, 1986, 1988; Sabolic *et al.*, 1985; Simon and Burckhardt, 1990; Turrini *et al.*, 1989) and largely resemble those from the sources discussed previously. The vacuolar H^+-ATPase purified from bovine kidney and medulla also had a subunit composition similar to that for other members of the class (Gluck and Caldwell, 1987, 1988; Wang and Gluck, 1990).

Polyclonal antibodies to the enzyme were prepared and used to localize the distribution of the enzyme in the mammalian kidney (Brown *et al.*, 1988a,b). In the proximal tubule, vacuolar H^+-ATPase staining was abundant on the brush-border microvilli in the initial portion of the proximal tubule, and in the invaginations of the base of the apical microvilli throughout the proximal tubule. Moderate plasma membrane staining for H^+-ATPase was also observed in the thick ascending limb and the distal convoluted tubule. In the connecting tubule and collecting duct, heavy staining was observed in the proton-transporting intercalated cells. Rapid-freeze deep-etch micrographs confirm that the vacuolar H^+-ATPase resides on the luminal membrane of the acid secreting cells (Brown *et al.*, 1987). In these cells, proton pumps are present at a density of over $16{,}000/\mu m^2$. Vacuolar H^+-ATPases are also present in many of the intracellular vacuolar compartments of eukaryotic cells (Yurko and Gluck, 1987), where they serve in the processing of membranes and intraluminal molecules in endocytosis and secretion. However, immunocytochemical studies suggest that the proton pump in these intracellular compartments is present at far lower densities than in the plasma membrane of proton-transporting cells (Rodman *et al.*, 1991).

The H^+-ATPase in the plasma membrane of proton-transporting cells, therefore, displays several distinct features not evident in most other mammalian cells: cell-specific amplification, polarization to one or more plasmalemmal domains, and physiologic regulation to preserve acid–base homeostasis (discussed in more detail below). How is the enzyme able to accomplish this highly specialized function of transcellular proton transport while functioning to acidify a variety of intracellular compartments?

Several lines of evidence suggest that vacuolar ATPases exist as a family of enzymes with differences in structure and function. H^+-ATPase isolated from bovine kidney microsomes by high-pressure liquid chromatography (HPLC) ion exchange contained two partially resolved peaks of ATPase activity that differed in their polypeptide composition (Gluck and Caldwell, 1987). H^+-ATPase isolated from distinct membrane fractions from bovine kidney had different enzymatic and structural properties (Wang and Gluck, 1990, and unpublished observations) (Table II). H^+-ATPase

TABLE II

Properties of H^+-ATPase in Different Membrane Compartments

Compartment:	Microsomes	Brush border	Lysosomes
pH optimum	ATP : GTP selectivity 4 : 1	ATP : GTP selectivity 2 : 1	ATP : GTP selectivity 3 : 1
Multivalent cation effects	Inhibited 100% by Cu^{2+}; no inhibition by Al^{3+}	Inhibited 50% by Cu^{2+}; no inhibition by Al^{3+}	Inhibited 100% by Cu^{2+}; inhibited 80% by Al^{3+}
Anion effects	No Cl reqt.; inhibited by NO_3^-, SO_3^{2-}, SO_4^{2-}, and CO_3^{2-}	No Cl reqt.; inhibited by NO_3^-, SO_3^{2-}, SO_4^{2-}, and CO_3^{2-}	No Cl reqt.; inhibited by NO_3^-, SO_3^{2-}, SO_4^{2-}, and ? CO_3^{2-}
Structure : 31 kDa	1 principal 31-kDa polypeptide	Additional lower-mobility immunoreactive 31-kDa polypeptides	1 principal 31-kDa polypeptide
Structure: 56 kDa	Both 56- and 58-kDa polypeptides; several different pIs for 56 kDa	Both 56- and 58-kDa polypeptides; fewer pIs for 56 kDa than in microsomes	Little or no 58-kDa polypeptide; fewer pIs for 56 kDa than in microsomes

was isolated from a purified brush-border membrane fraction, from a kidney lysosomal fraction, and from kidney microsomes (a mixture of membranes containing plasma membranes from intercalated cells) by immunoaffinity chromatography on identical monoclonal antibody columns. The H^+-ATPase was isolated from these fractions differed in pH optimum, in effects of added lipids on activity, and in sensitivity to divalent and trivalent cations (Wang and Gluck, 1990); similar results were obtained by Burckhardt and colleagues on H^+-ATPase from brush-border reconstituted into liposomes (Simon and Burckhardt, 1990). The H^+-ATPase affinity purified from the different membrane fractions had subtle differences in structure that could be detected by immunoblotting on two-dimensional gels. The brush-border H^+-ATPase had a pH optimum of ~7.3, the approximate intercellular pH of the proximal tubule cell. The proximal tubule functions primarily in bulk bicarbonate reabsorption and is not the site for fine regulation of net hydrogen in excretion. The pH optimum of the brush-border H^+-ATPase is, therefore, poised for the enzyme to function maximally at a normal intracellular pH. In contrast, the pH optimum of the microsomal enzyme is ~6.3, a 1pH-unit difference from the intracellular pH of intercalated cells. H^+-ATPase in these cells may, therefore, respond with an increase in activity under conditions where intracellular pH decreases, such as a decrease in the extracellular fluid bicarbonate concentration. However, the magnitude of the change is only about 30% of the total activity between pH 6.3 and 7.3, and it is, therefore, likely that other regulatory mechanisms control the rate of transport. The pH optimum of the kidney lysosomal H^+-ATPase was ~6.7 (Table II). Cu^{2+} ion inhibited the microsomal enzyme activity by 100%. A concentration of copper up to 5 m*M* inhibited the brush-border H^+-ATPase by approximately 50%. The lysosomal H^+-ATPase is also relatively insensitive to copper, but is highly sensitive to aluminum ion, whereas aluminum had no effect on the brush-border and microsomal H^+-ATPases. These studies serve to demonstrate that differences in enzymatic properties can be detected in H^+-ATPases from different kidney membrane fractions.

As discussed above, several observations indicate that the vacuolar H^+-ATPases have structural microheterogeneity that may underlie the observed differences in enzymatic properties. Both two-dimensional electrophoresis (Wang and Gluck, 1990) and genomic Southern blots (Puopolo *et al.*, 1991) suggest that there is only one structural form of the A subunit, suggesting that the B subunit, which had sequence homology with the regulatory α subunit of the F-ATPases, may be a regulatory subunit that affects the overall enzymatic properties of the vacuolar H^+-ATPases. The structures of the microsomal brush-border lysosomal ATPases were compared on two-dimensional gels, and significant differences were noted in both the E subunit and the B subunit among the different ATPase preparations (Wang and Gluck, 1990). The E subunit had an isoelectric point of approximately 8.1 with a large pH spread. In the microsomal ATPase, the subunit appeared as a single polypeptide at M_r 31,000. The E subunit in the brush-border microvillar H^+-ATPase appeared as a predominant polypeptide at 31,000 with several polypeptides at an identical pI, but of lower mobility (higher relative molecular mass; Wang and Gluck, 1990). Mice were immunized with a 10-residue peptide from the carboxyl terminus of the deduced amino acid sequence of the E subunit and a series of monoclonal antibodies reacting with the native subunit by immunoblot were generated. When these antibodies were examined by immunoblotting on two-dimensional gels and immunocytochemistry, significant differences in reactivity were discovered (Hemken *et al.*, 1992). Antibody E11 yielded a pattern of immunocytochemistry in rat kidney identical to the pattern that we had described previously using rabbit polyclonal antibodies. E11 stained the brush-border microvilli in rat proximal tubules, the invaginations at the base of the microvilli, the apical and basolateral poles of cortical intercalated cells, and the apical poles of medullary intercalated cells. On two-dimensional immunoblots, E11 reacted with a single M_r 31,000 polypeptide on two-dimensional immunoblots of the affinity-purified microsomal H^+-ATPase. However, E11 reacted with a series of polypeptides on two-dimensional immunoblots of the affinity-purified brush-border H^+-ATPase. The polypeptides had the same isoelectric point as the E subunit, but the additional polypeptides had a lower mobility, a pattern identical to that observed on the two-dimensional protein gels of the affinity-purified brush-border H^+-ATPase. A second antibody, H8, also reacted on immunoblots with the native E subunit from kidney microsomes. On immunocytochemical staining of rat kidney, antibody H8 stained the invaginations at the base of the microvilli and the intercalated cells with a pattern identical to that observed with antibody E11. However, H8 showed no staining at all of the brush-border microvilli. On immunoblots, this antibody stained the single M_r 31,000 polypeptide in microsomal ATPase but had nearly absent immunoreactivity against the brush-border enzyme. These results suggest that the structure of the E subunit differs in brush border appearing as several polypeptide forms with a mobility lower than the most prevalent form of the subunit found in the kidney.

The basis for higher M_r forms of the E subunit is unclear. Bovine kidney cDNA libraries were screened at low stringency with the M_r 31,000 cDNA probe in an attempt to isolate isoforms of the subunit, but none was obtained by this method. More recently, we looked for isoforms of the subunit using genomic cloning (Hemken *et al.*, 1992). A Southern blot was performed on human leukocyte DNA, cleaved with different restriction enzymes and probed with a PCR fragment spanning the carboxyl-terminal portion of the coding region of the M_r 31,000 cDNA and a small portion of the 3′ untranslated region. This probe hybridized to several fragments on Southern blots of leukocyte genomic DNA. A full-length human M_r 31,000 cDNA was isolated from a human kidney library and used to screen a human genomic library. Several genomic clones were obtained. Two of these were purified and the λ-DNA was cleaved with the same restriction enzymes used for the genomic Southern blots and analyzed for hybridization to the same PCR fragment. The two clones accounted for all but one of the fragments observed on the Southern blots. Sequence analysis of the two clones showed that one was a true gene with a stretch of nucleotides identical to the human cDNA. The second genomic clone appeared to be a pseudogene, as it contained a stretch with a predicted protein sequence highly similar to human cDNA, but with several frameshift changes (Hemken *et al.*, 1992). The clone was also missing the stop codon found on the human E subunit cDNA, and the carboxyl-terminal 5 amino acids were absent in all three reading frames. Work is in progress to resolve this issue.

The B subunit also showed differences in structure among the different H^+-ATPases isolated by immunoaffinity purification (Wang and Gluck, 1990). In two-dimensional protein gels of both the brush-order and microsomal enzymes are a cluster of polypeptides at M_r 58,000 and 56,000. In both H^+-ATPase preparations, the M_r 58,000 B subunit appeared as a single polypeptide, where the M_r 56,000 appeared as a series of four to six polypeptides. The pattern of spots in the microsomal enzyme was consistently more complex than in the brush border, which may reflect the source of the microsomal ATPase as a more complicated mixture of membrane compartments than the brush-border H^+-ATPase. The lysosomal H^+-ATPase also retained a M_r 56,000 polypeptide, detected as several spots on two-dimensional gels, but the lysosomal H^+-ATPase did not contain the M_r 58,000 polypeptide (Wang and Gluck, 1993). cDNA cloning from our laboratory (Nelson *et al.*, 1992) and others (Bernasconi *et al.*, 1990; Puopolo *et al.*, 1992; Südhof *et al.*, 1989) has shown that at least three isoforms of the M_r 56,000 subunit exist; hence, this subunit appears to comprise a multigene family. Comparison of the sequences of two human and bovine B-subunit isoforms shows that the middle portion of the coding region is highly conserved, but both the amino- and the carboxyl-terminal sequences of the two isoforms are entirely different (Nelson *et al.*, 1992). These differences, in sequence, are probably not due to random mutations because they are highly conserved in the corresponding isoform across species.

Ribonucleic acid blots from different human tissues revealed that the kidney has the greatest levels of expression of the "kidney" (M_r 58,000) isoform (or isoform 1) of any tissue. Moderate levels were detected in the placenta and low levels were detected in the lung; hybridization was undetectable in polyA mRNA from other human tissues. Similar results were found on RNA hybridization studies performed on total RNA from various bovine tissues. In contrast, RNA hybridization blots performed on human poly$(A)^+$ RNA with the "brain" (M_r 56,000 isoform (isoform 2) showed nearly equal levels expressed in all tissues, but with somewhat higher levels of expression in brain (Nelson *et al.*, 1992). Puopolo *et al.* (1992) also found differences in the levels of expression of different isoforms of the B subunit in different tissues and noted that transcript size varied as well.

Nelson *et al.* (Nelson *et al.*, 1992) performed immunoblots on microsomal membranes from various bovine tissues to examine the protein distribution of the kidney isoform of the subunit, which has a M_r of 58,000 on SDS gels. The results were similar to those of the RNA blots: the subunit isoform protein was detectable only in kidney. The cellular distribution of the isoform in kidney was examined by immunocytochemistry on rat kidney sections using a polyclonal antibody to a peptide from the unique carboxyl-terminal sequence of the kidney isoform of the B subunit. As a control, the distribution of H^+-ATPase was examined using a monoclonal antibody to the E subunit of the H^+-ATPase. Antibody to the E subunit stains a number of segments of the nephron (Brown *et al.*, 1988; Bastani *et al.*, 1991). The antibody exhibited heavy staining in the initial part of the proximal tubule and in the invaginations at the base of the microvilli throughout the entire proximal tubule. It showed moderate staining of the apical membrane and subapical vesicles in the thick ascending limb, and moderate staining in the apical region of the distal convoluted tubule. In the collecting duct, the monoclonal antibody to the E subunit gave intense staining of the intercalated cells. In the cortex, there are two functionally distinct types of intercalated cells: an acid-secreting intercalated cell, or A cell, and a bicarbonate-secreting intercalated cell, or B cell. The A cells have staining predominantly in the apical pole. In contrast, the B cells have staining variously in the basolateral membrane, or diffusely throughout the cytoplasm, or, in some cells, in both the apical and the basal poles of the cell. In the medullary collecting duct, only the A-type intercalated cells are present and staining with the E-subunit antibody showed apical staining only (Brown *et al.*, 1988; Bastani *et al.*, 1991).

In the kidney sections stained with the antibody to the kidney isoform (isoform 1) of the B subunit, intense staining was observed only in the plasma membrane of the intercalated cells (Nelson *et al.*, 1992). No staining was observed in any part of the proximal tubule or in the thick ascending limb. Moderate staining was observed in the apical membrane of the distal convoluted tubule and weak staining was found in the apical membrane of the principle cells of the inner medullary collecting duct. In the cortical collecting duct, staining was observed in the apical or basolateral membranes of some intercalated cells, suggesting that the kidney isoform of the subunit may contain structural information important to plasma membrane targeting, but apparently does not contain structural determinants for selective apical or basolateral polarization. The kidney isoform of the B subunit is amplified selectively in the intercalated cells but is expressed in other tissues. Both the kidney and the brain isoforms of the B subunits were detected in three different cell lines, including the LLC-PK_1 and MDBK cell lines. Immunocytochemistry using the antibody to the carboxyl-terminal peptide of the kidney isoform revealed staining of a perinuclear vacuolar membrane compartment that was unidentified (Nelson *et al.*, 1992). Thus, the current evidence supports the premise that the proton-transporting apparatus of the intercalated cell represents an amplification of a vacuolar compartment that is present, in general, in eukaryotic cells and that has been modified to serve in physiologically regulated transcellular hydrogen ion secretion. The specialized compartments of intercalated cells may be analogous to the amplified endoplasmic reticulum-like compartment observed in the UT-1 cell lines (Chin *et al.*, 1982) that overexpresses HMG-coenzyme A (CoA) reductase.

B. Regulation of the Vacuolar H^+-ATPase in Renal Epithelia

Net renal hydrogen ion excretion is regulated with precision, such that the amount of acid excreted by the kidney is equal to the rate of generation of nonvolatile metabolic protons. In principle, regulation could occur by three different mechanisms (*a*) an increase in the kinetic activity of the H^+-ATPase without a change in the quantity or distribution; (*b*) a change in the distribution of the H^+-ATPase, such that the enzyme is recruited to the plasma membrane from an intercellular pool or retrieved from the plasma membrane into an intercellular compartment; and (*c*) an overall increase in the quantity of H^+-ATPase without a change in its kinetics or relative distribution between different membrane compartments. Several lines of investigation from our laboratory have focused recently on these mechanisms for regulation.

Prior studies in kidney (McKinney and Davidson, 1987) and in model proton-transporting epithelia, such as the turtle urinary bladder (Steinmetz, 1986), provided evidence that kinetic changes in the rate of ATPase activity on the rate of proton pump ATPase activity contribute to the overall regulation of transepithelial hydrogen ion transport. Our studies on the properties of the isolated enzyme showed that the ATPase activity is affected somewhat by pH changes in the normal range of intracellular pH (Gluck and Caldwell, 1987; Wang and Gluck, 1990, 1992) and this may account, in part, for kinetic changes. However, the H^+-ATPase is not sufficiently sensitive to pH conditions likely to be encountered in the cytosol for this to be a principal direct regulator of activity in the intact cell. The effects of concentrations of other cations and anions likely to be encountered in the cytosol also did not have a significant effect on activity (Gluck and Caldwell, 1987; Wang and Gluck, 1990). It was found that cytosol might contain additional regulatory factors not isolated with the H^+-ATPase by the immunoaffinity purification procedure that may interact with the ATPase under certain conditions (Zhang *et al.*, 1992a,b).

Bovine kidney cytosol contains both inhibitory and activating factors that directly modified the ATPase activity of the immunoaffinity-purified enzyme (Zhang *et al.*, 1992a,b) (see Table III). H^+-ATPase activity, immunoaffinity-purified from different kidney membrane fractions, was assayed on monoclonal antibody beads, with and without added cytosolic fractions. The difference with and without cytosol was used to determine the degree of inhibition or activation.

The inhibitor was purified from cytosol by a sequential combination of ultracentrifugation, ammonium sulfate fractionation, acid precipitation, cation exchange chromatography, anion exchange chromatography, and HPLC anion exchange chromatography. The precise degree of purification of the inhibitor could not be determined because the starting cytosol

TABLE III
Properties of Inhibitor and Activator of H^+-ATpase

	Inhibitor	Activator
Actions	Inhibits solubilized immunoaffinity-purified H^+-ATPase; inhibits ATP-dependent proton transport	Activates solubilized immunoaffinity-purified H^+-ATPase
Mode of action	Saturable effect on H^+-ATPase; Hill coefficient of 1.46 suggests dimerization required; inhibition is not readily reversible by washing	Saturable effect on H^+-ATPase; Hill coefficient of 1.0; nonenzymatic effect; probably works by binding
pH effects	% Inhibition increases for pH > 7.2	Binding increased at pH ≤ 6.5
Composition	Heat labile protein; M_r 6300 polypeptide on SDS gels; active MW of ~12,000 by gel filtration; probably active as a dimer	Heat-stable protein; active size of ~40,000 by gel filtration
Specificity	Highly specific for vacuolar H^+-ATPases; slight effect on F_0F_1 H^+-ATPase; not selective for subclass of vacuolar H^+-ATPase	Highly specific for vacuolar H^+-ATPases; slight effect on F_0F_1 H^+-ATPase; selective for brush-border > microsomal $\gg$ lysosomal H^+-ATPase

contained activator activity that masked the inhibitor activity. However, starting from the anion exchange column fractionation step, a 118-fold purification was achieved. The isolated inhibitor was a heat-labile protein that appeared on SDS gels as a 6300-MW polypeptide (Zhang *et al.*, 1992a). The active fraction had an approximate relative mass of 12,000 by gel filtration, suggesting that the active form of the inhibitor was a dimer. Inhibition was concentration-dependent and saturable, with a Hill coefficient of 1.46, consistent with a requirement for dimerization. The inhibitor affected the immunoaffinity purified enzyme and inhibited ATP-dependent proton transport in bovine kidney microsomal vesicles. The percentage of ATPase activity inhibited by the inhibitor increased under pH conditions greater than 7.2. Thus, under conditions representing an increase in cytosolic pH in the physiologic range, the inhibitor was more efficient in suppressing H^+-ATPase activity. The inhibitor was highly specific for the vacuolar H^+-ATPase with little or no effect on Na^+,K^+- or Ca^+-ATPases, but it did have a modest inhibitory effect on the mitochondrial F-ATPase. The isolated inhibitor had equal effects on the isolated microsomal, brush-border, and lysosomal vacuolar H^+-ATPases from kidney and, therefore, had the potential for a role in controlling H^+ transport in several cellular compartments (Zhang *et al.*, 1992a).

The H^+-ATPase activator was partially purified from bovine kidney cytosol by a sequential combination of ultracentrifugation, ammonium sulfate precipitation, acid precipitation, quarternary aminoethyl anion exchange chromatography, and aminohexylagarose chromatography, yielding a 75-fold purification with 27% recovery of activity (Zhang *et al.*, 1992b). The activator was heat-stable, retaining 77% of its activity when heated at 75°C for 10 min, but retaining only 4% of its activity when boiled for 10 min; it was sensitive to trypsin and, therefore, is likely to be a protein. The activator inhibited the immunoaffinity-purified H^+-ATPase and, thus, like the inhibitor, exerts a direct effect on the enzyme. Activation was concentration-dependent and saturable, with a Hill coefficient of 1.0 (Zhang *et al.*, 1992b). Seventy to 80% of the activating effect occurred within 1 min and the time course for activation was not affected by the concentration of activator; hence, it probably works by a nonenzymatic mechanism, likely by binding to the H^+-ATPase. The activation was reversible when the beads were washed, but the binding affinity of the activator was pH-dependent. At pH values below 7.5, the activator washed off the H^+-ATPase beads with a slower time course than at pH values at 7.5 or above. This property may be physiologically important, as a drop in cytosolic pH could promote binding of the activator to the H^+-ATPase and stimulate proton extrusion from the cell. ATPase activation was highly specific for the vacuolar H^+-ATPases with no effect on the Na^+,K^+- or

Ca^{2+}-ATPases, and with only a slight effect on the mitochondrial F-ATPase. Unlike the inhibitor, the activator had a much greater effect on the isolated brush-border and microsomal H^+-ATPases than it did on the lysosomal H^+-ATPase. Consequently, it may have a selective function in controlling vacuolar H^+-ATPases residing on the plasma membrane involved in transcellular proton transport (Zhang *et al.*, 1992b).

Changes in the distribution of H^+-ATPase within the intercalated cell also play a role in the physiologic regulation of hydrogen ion secretion. This property was studied in a model of chronic acid administration in rats (Bastani *et al.*, 1991). Rats given ammonium chloride in their drinking water develop an acidic blood pH (metabolic acidosis) within a day and their urine pH drops from 0.5 to 0.7 pH units. Over a period of 2 weeks of chronic acid administration, the kidney shows physiologic adaptive changes, increasing hydrogen ion secretion such that the animals are restored to normal acid–base status. The role of the H^+-ATPase in the nephron was examined in this adaptational response. We first examined whether any changes occurred in the quantity of vacuolar H^+-ATPase protein or in the steady-state levels of mRNA for the enzyme. These studies were conducted using an immunoassay for the E subunit (Bastani *et al.*, 1991) and the cDNA clone for the rat E subunit (Hirsch *et al.*, 1988). Kidneys from the rats were examined at five different time points over the 2-week period of acid administration. No change in either the quantity of immunoreactive H^+-ATPase or the levels of steady-state message for the E subunit was detected over the 2-week period (Hirsch *et al.*, 1988). Since these results suggested that changes in the quantity of the enzyme did not have a role in adaptation, we used anti-H^+-ATPase immunocytochemistry in the kidney to examine whether a change in the distribution of the enzyme occurred with acid loading. As a means for quantitating changes in the distribution of the H^+-ATPase, we counted the percentage of intercalated cells with plasma membrane staining. In control rats or rats subjected to chronic administration of bicarbonate, H^+-ATPase in intercalated cells of the collecting tubule from kidney medulla was distributed predominately in intercellular vesicles in the cytoplasm. In contrast, in the rats subjected to chronic acid administration over 2 weeks, H^+-ATPase in the majority of medullary intercalated cells was detected mostly in the plasma membrane. We found that there was a time-dependent increase in recruitment of H^+-ATPase from cytoplasm to the plasma membrane over the 2-week period (Bastani *et al.*, 1991). Changes in intercalated cells in the cortical collecting duct were somewhat more complex because of the multitude of different morphologic types of H^+-ATPase staining. Nevertheless, an increase in polarization to the apical membrane was also observed in the cortical intercalated cells. These results suggest that the principle means for adaptation in the kidney is not change in the overall quantity of the H^+-ATPase, but rather a redistribution from an intercellular pool of vesicles with H^+-ATPase to the plasma membrane.

C. Transcellular Transport in Other Cell Types

Vacuolar H^+-ATPases have been found in the plasma membrane of several cell types other than renal epithelia. Macrophages have a plasma membrane vacuolar H^+-ATPase (Bidani *et al.*, 1989; Bidani and Brown, 1990; Lukacs *et al.*, 1990a,b; Swallow *et al.*, 1990, 1991a), which functions to preserve cytoplasmic pH and the ability of macrophages to generate a respiratory burst within the relatively anaerobic environment of an abscess or tumor (Swallow *et al.*, 1991a). The acid loading of macrophages induced a Na^+- and HCO_3^--independent pH recovery that was inhibited by bafilomycin and other inhibitors of vacuolar H^+-ATPases; the induction was not prevented by microtubule inhibitors, nor was it associated with detectable exocytosis, and appeared to result from kinetic activation of a H^+-ATPase already resident in the membrane (Swallow *et al.*, 1990); nitric oxide suppressed the H^+-ATPase activity, probably by inducing guanylate cyclase and elevating cGMP (Swallow *et al.*, 1991b). Human neutrophils have a plasma membrane H^+-ATPase that is normally quiescent, but is activated by phorbol esters (Nanda and Grinstein, 1991). During the respiratory burst triggered by activation, neutrophil proton production increases several orders of magnitude. The cytosolic pH of activated neutrophils is normally maintained by a Na^+/H^+ antiporter, but restoration of cytoplasmic pH by a vacuolar H^+-ATPase was demonstrated in phorbol ester treatment cells in which the Na^+/H^+ antiporter was inhibited by removal of external sodium. These observations in the macrophage and neutrophil serve to underscore the importance of kinetic regulation of the H^+-ATPase.

The osteoclast, which is derived from the macrophage hematopoietic lineage, uses a vacuolar H^+-ATPase to generate an acidic enclosed space at its attachment site on bone in order to dissolve bone hydroxyapatite (Baron, 1989; Blair *et al.*, 1989). Like renal intercalated cells, osteoclasts exhibit high levels of H^+-ATPase expression and are capable of polarizing the enzyme to the ruffled border as a means for efficient bone resorption (reviewed in Gluck, 1992); bafilomycin A_1 inhibits osteoclast bone resorption (Sundquist *et al.*, 1990). Regulated polarization may be one mechanism for controlling osteoclast proton secretory activity similar to that in the kidney intercalated cell (Gluck, 1992). A recent study demonstrated that nitric oxide inhibits bone resorption in the osteoclast but, unlike the macrophage, the effect did not appear to be mediated through guanylate cyclase (Macintyre *et al.*, 1991).

Insect tissues, like the renal intercalated cell and osteoclast, exhibit cell-specific amplification of vacuolar H^+-ATPases. Physiologic studies on the mechanism of potassium secretion in insect midgut (Chao *et al.*, 1991; Harvey *et al.*, 1983; Wieczorek *et al.*, 1989, 1990, 1991) and Malpighian tubule (Bertram *et al.*, 1991; Maddrell and Overton, 1988) led to the discovery that a vacuolar H^+-ATPase functions primarily to generate a membrane potential that energizes potassium transport in these tissues. The apical membrane in each tissue has an electrogenic K^+/H^+ antiporter that mediates K^+ secretion. Although the H^+-ATPase is located in the luminal membrane of these epithelia, they do not exhibit net proton secretion. Antithetically, the midgut generates a highly alkaline lumen as a consequence of the operation of the K^+/H^+ antiporter in parallel with the H^+-ATPase. The vacuolar H^+-ATPase from midgut was purified (Wieczorek *et al.*, 1989, 1990) and it had a structure highly similar to that of other vacuolar H^+-ATPases (see Table I). Immunocytochemistry using antibodies to the insect vacuolar H^+-ATPase showed staining in the apical membrane of the midgut and malpighian tubule, in agreement with the physiologic evidence for the enzyme at those sites (Klein *et al.*, 1991b). High levels of the vacuolar H^+-ATPase were also discovered in the plasma membrane of the enveloping cells in insect sensillum, the sensory organ (Klein and Zimmerman, 1991a). The vacuolar H^+-ATPase in these cells probably functions to generate the cell potential needed to drive the signaling currents initiated by activation of the sensory cells.

V. Conclusions

Structural analysis of vacuolar H^+-ATPases from plant, animal, and fungal sources has revealed the remarkable structural similarity to the F-ATPases. Half of the yeast genes have been cloned and powerful methods for selection of mutants on the basis of phenotype are now available; the yeast system will undoutedly provide the most rapid advances in the structure–function analysis of the enzyme.

It is apparent that the structure of the H^+-ATPase in mammalian cells is likely to be far more complex than that in yeast. The mammalian enzyme exists as a family of related enzymes; the B subunit comprises a multigene family with multiple members expressed within cell lines, and in tissues, it appears to be responsible, at least in part, for tissue-specific differences in expression. Heterogeneity of the E subunit also exists, but the molecular basis and significance of this are only beginning to emerge.

Acidification in intracellular compartments is carefully regulated to maintain different steady-state pH values among several compartments. Kinase regulation of the chloride permeability may participate in this regulation. Cytosolic H^+-ATPase regulatory proteins that appear to act selectively on subsets of the H^+-ATPase have also been identified and these may serve in regulating the pH of distinct compartments.

Several types of cells have a vacuolar H^+-ATPase on the plasma membrane; in epithelia and osteoclasts, it functions in transcellular proton secretion, whereas in macrophages it is responsible for maintaining cytosolic pH under acidic conditions. Regulation in the kidney and osteoclast partly controlled polarization, enabling the cell to govern the number of proton pumps on the plasma membrane. Kinetic regulation of the H^+-ATPase participates in regulating activity in all three systems, which, in the macrophage and osteoclast, is greatly influenced by nitric oxide generation.

Acknowledgment

I thank my colleagues Bahar Bastani, Raoul Nelson, Beth Lee, Xiao-Li Guo, Kun Zhang, Zhi-Qiang Wang, Ji-Yi Fu, David Underhill, and Huan Lo for their contributions to the studies discussed in this review. I am grateful to Edna Major and Debbie Windle for secretarial assistance. This work was supported by NIH Grants DK38848, DK09976, and AR32087, by a grant from the Monsanto–Washington University Fund, and by the Washington University George M. O'Brien Kidney and Urological Diseases Center P50-DK45181. S.L.G. is a Sandoz Pharmaceutical Corporation Established Investigator of the American Heart Association.

References

Ackerman, S. H., and Tzagoloff, A. (1990). *Proc. Natl. Acad. Sci. U.S.A.* **87,** 4986–4990.
Adachi, I., Arai, H., Pimental, R., and Forgac, M. (1990a). *J. Biol. Chem.* **265,** 960–966.
Adachi, I., Puopolo, K., Marquez-Sterling, N., Arai, H., and Forgac, M. (1990b). *J. Biol. Chem.* **265,** 967–973.
Alpern, R. J. (1990). *Physiol. Rev.* **70,** 79–114.
Amzel, L. M., and Pedersen, P. L. (1983). *J. Membr. Biol.* **73,** 105–124.
Arai, H., Berne, M., and Forgac, M. (1987a). *J. Biol. Chem.* **262,** 11006–11011.
Arai, H., Berne, M., Terres, G., Terres, H., Puopolo, K., and Forgac, M. (1987b). *Biochemistry* **26,** 6632–6638.
Arai, H., Pink, S., and Forgac, M. (1989). *Biochemistry* **28,** 3075–3082.
Bae, H., and Verkman, A. (1990). *Nature (London)* **348,** 637–639.
Baron, R. (1989). *Anat. Rec.* **224,** 317–324.

Bastani, B., Purcell, H., Hemken, P., Trigg, D., and Gluck, S. (1991). *J. Clin. Invest.* **88,** 126–136.
Beltrán, C., Kopecky, J., Yu-Ching, E. P., Nelson, H., and Nelson, N. (1992). *J. Biol. Chem.* **267,** 774–779.
Bernasconi, P., Rausch, T., Struve, I., Morgan, L., and Taiz, L. (1990). *J. Biol. Chem.* **265,** 17428–17431.
Bertram, G., Schleithoff, L., Zimmerman, P., and Wessing, A. (1991). *J. Insect Physiol.* **37,** 201.
Bidani, A., and Brown, S. E. S. (1990). *Am. J. Physiol.* **259,** C586–C598.
Bidani, A., Brown, S. E. S., Heming, T. A., Gurich, R., and DuBose, T. D. (1989). *Am. J. Physiol.* **257,** C65–C76.
Blair, H. C., Teitelbaum, S. L., Ghiselli, R., and Gluck, S. (1989). *Science* **245,** 855–857.
Bowman, B. J., Allen, R., Wechser, M. A., and Bowman, E. J. (1989a). *J. Biol. Chem.* **263,** 14002–14007.
Bowman, B., Oshida, W., Harris, T., and Bowman, E. J. (1989b). *J. Biol. Chem.* **264,** 15606–15612.
Bowman, E. J., Mandala, S., Taiz, L., and Bowman, B. J. (1983). *J. Biol. Chem.* **258,** 15238–15244.
Bowman, E. J., Siebers, A., and Altendorf, K. (1988). *Proc. Natl. Acad. Sci. U.S.A.* **85,** 7972–7976.
Bowman, E. J., Tenney, K., and Bowman, B. J. (1989). *J. Biol. Chem.* **263,** 13994–14001.
Brown, D. (1989). *Am. J. Physiol.* **256,** F1–F12.
Brown, D., Gluck, S., and Hartwig, J. (1987). *J. Cell Biol.* **105,** 1637–1648.
Brown, D., Hirsch, S., and Gluck, S. (1988a). *Nature (London)* **331,** 622–624.
Brown, D., Hirsch, S., and Gluck, S. (1988b). *J. Clin. Invest.* **82,** 2114–2126.
Burckhardt, G., and Warnock, D. G. (1990). *Semin. Nephrol.* **10,** 93–103.
Chao, A. C., Moffett, D. F., and Koch, A. (1991). *J. Exp. Biol.* **155,** 403–414.
Chin, D. J., Luskey, K., Anderson, R. G. W., Faust, J. R., Goldstein, J., and Brown, M. (1982). *Proc. Natl. Acad. Sci. U.S.A.* **79,** 1185–1189.
Denda, K., Konishi, J., Oshima, T., Date, T., and Yoshida, M. (1988a). *J. Biol. Chem.* **263,** 6012–6015.
Denda, K., Konishi, J., Oshima, T., Date, T., and Yoshida, M. (1988b). *J. Biol. Chem.* **263,** 17251–17254.
Diaz-Diaz, F. D., LaBelle, E. F., Eaton, D. C., and DuBose, T. D. (1986). *Am. J. Physiol.* **251,** F297–F302.
Feng, Y., and Forgac, M. (1992). *J. Biol. Chem.* **267,** 5817–5822.
Forgac, M. (1989). *Physiol. Rev.* **69,** 765–796.
Foury, F. (1990). *J. Biol. Chem.* **265,** 18554–18560.
Fuchs, R., Schmid, S., and Mellman, I. (1989). *Proc. Natl. Acad. Sci. U.S.A.* **85,** 539–543.
Futai, M., and Kanazawa, H. (1983). *Microbiol. Rev.* **47,** 285–312.
Futai, M., Noumi, T., and Maeda, M. (1989). *Annu. Rev. Biochem.* **58,** 111–136.
Gillespie, G. A., Somlo, S., Germino, G. G., Weinstat-Saslow, D., and Reeders, S. T. (1991). *Proc. Natl. Acad. Sci. U.S.A.* **88,** 4289–4293.
Gillespie, J., Ozanne, S., Percy, J., Warren, M., Haywood, J., and Apps, D. K. (1991). *FEBS Lett.* **282,** 69–72.
Gluck, S. (1992). *Am. J. Med. Sci.* **303,** 134–139.
Gluck, S., and Al-Awqati, Q. (1984). *J. Clin. Invest.* **73,** 1704–1710.
Gluck, S., and Caldwell, J. (1987). *J. Biol. Chem.* **262,** 15780–15789.
Gluck, S., and Caldwell, J. (1988). *Am. J. Physiol.* **254,** F71–F79.
Gogarten, J. P., Kibak, H., Dittrich, P., Taiz, L., Bowman, E. J., Bowman, B. J., Manolson, M. F., Poole, R. J., Date, T., Oshima, T., Konishi, J., Denda, K., and Yoshida, M. (1989). *Proc. Natl. Acad. Sci. U.S.A.* **86,** 6661–6665.

Gurich, R. W., and DuBose, T. D. (1989). *Am. J. Physiol.* **257,** F777–F784.
Gurich, R. W., Codina, J., and DuBose, T. D. (1991). *J. Clin. Invest.* **87,** 1547–1552.
Hanada, H., Moriyama, Y., Maeda, M., and Futai, M. (1990). *Biochem. Biophys. Res. Commun.* **170,** 873–878.
Hanada, H., Hasebe, M., Moriyama, Y., Maeda, M., and Futai, M. (1991). *Biochem. Biophys. Res. Commun.* **176,** 1062–1067.
Harvey, W. R., Cioffi, M., Dow, J. A. T., and Wolfersberger, M. G. (1983). *J. Exp. Biol.* **106,** 91–117.
Hemken, P., Guo, X.-L., Wang, Z.-Q., Zhang, K., and Gluck, S. (1992). *J. Biol. Chem.* **267,** 9948–9957.
Hirata, R., Ohsumi, Y., Nakano, A., Kawasaki, H., Suzuki, K., and Anraku, Y. (1990). *J. Biol. Chem.* **265,** 6726–6733.
Hirsch, S., Strauss, A., Masood, K., Lee, S., Sukhatme, V., and Gluck, S. (1988). *Proc. Natl. Acad. Sci. U.S.A.* **85,** 3004–3008.
Inatomi, K. I., Eya, S., Maeda, M., and Futai, M. (1989). *J. Biol. Chem.* **264,** 10954–10959.
Jehmlich, K., Sablotni, J., Simon, B. J., and Burckhardt, G. (1991). *Kidney Int.* **40,** S64–S70.
Kagawa, Y. (1978). *Biochim. Biophys. Acta* **505,** 45–93.
Kane, P. M., Yamashiro, C. T., and Stevens, T. H. (1989). *J. Biol. Chem.* **264,** 19236–19244.
Kane, P. M., Yamashiro, C. T., Wolczyk, D. F., Neff, N., Goebl, M., and Stevens, T. (1990). *Science* **250,** 651–657.
Kane, P. M., Kuehn, M. C., Howald-Stevenson, I., and Stevens, T. H. (1992). *J. Biol. Chem.* **267,** 447–454.
Kaunitz, J. D., Gunther, R. D., and Sachs, G. (1985). *J. Biol. Chem.* **260,** 11567–11573.
Kinne-Saffran, E., Beauwens, R., and Kinne, R. (1982). *J. Membr. Biol.* **64,** 67–76.
Klein, U., and Zimmerman, B. (1991a). *Cell Tissue Res.* **266,** 265–273.
Klein, U., Löffelmann, G., and Weiczorek, H. (1991b). *J. Exp. Biol.* **161,** 61–75.
Lai, S., Randall, S. K., and Sze, H. (1988). *J. Biol. Chem.* **263,** 16731–16737.
Lai, S., Watson, J. C., Hansen, J. N., and Sze, H. (1991). *J. Biol. Chem.* **266,** 16978–16084.
Lukacs, G. L., Rotstein, O. D., and Grinstein, S. (1990a). *J. Biol. Chem.* **265,** 21099–21107.
Lukacs, G. L., Rotstein, O. D., and Grinstein, S. (1990b). *J. Biol. Chem.* **266,** 24540–24548.
Macintyre, I., Zaidi, M., Alam, A. S. M. T., Datta, H. K., Moonga, B. S., Lidbury, P. S., Hecker, M., and Vane, J. R. (1991). *Proc. Natl. Acad. Sci. U.S.A.* **88,** 2936–2940.
Maddrell, S. H. P. and Overton, J. A. (1988). *J. Exp. Biol.* **137,** 265–276.
Madsen, K. M., and Tisher, C. C. (1985). *Fed. Proc.* **44,** 2704–2709.
Mandala, S., and Taiz, L. (1986). *J. Biol. Chem.* **261,** 12850–12855.
Mandel, M., Moriyama, Y., Hulmes, J. D., Pan, Y.-C., Nelson, H., and Nelson, N. (1988). *Proc. Natl. Acad. Sci. U.S.A.* **85,** 5521–5524.
Manolson, M. F., Rea, P. A., and Poole, R. J. (1985). *J. Biol. Chem.* **260,** 12273–12279.
Manolson, M. F., Ouellette, B. F., Filion, M., and Poole, R. J. (1988). *J. Biol. Chem.* **263,** 17987–17994.
McKinney, T. D., and Davidson, K. K. (1987). *Am. J. Physiol.* **253,** F816–F822.
Mellman, I., Fuchs, R., and Helenius, A. (1986). *Annu. Rev. Biochem.* **55,** 663–700.
Moriyama, Y., and Nelson, N. (1987a). *J. Biol. Chem.* **262,** 9175–9180.
Moriyama, Y., and Nelson, N. (1987b). *J. Biol. Chem.* **262,** 14723–14729.
Moriyama, Y., and Nelson, N. (1989a). *Biochim, Biophys. Acta.* **980,** 241–247.
Moriyama, Y., and Nelson, N. (1989b). *J. Biol. Chem.* **264,** 18445–18450.
Moriyama, Y., and Nelson, N. (1989c). *J. Biol. Chem.* **264,** 3577–3582.
Moriyama, Y., Nelson, N., Masatomo, M., and Futai, M. (1991). *Arch. Biochem. Biophys.* **286,** 252–256.
Mulberg, A. E., Tulk, B. M., and Forgac, M. (1991). *J. Biol. Chem.* **266,** 20590–20593.

Nanda, A., and Grinstein, S. (1991). *Proc. Natl. Acad. Sci. U.S.A.* **88,** 10816–10820.
Nelson, H., and Nelson, N. (1989a). *FEBS Lett.* **247,** 147–153.
Nelson, H., and Nelson, N. (1990b). *Proc. Natl. Acad. Sci. U.S.A.* **87,** 3503–3507.
Nelson, H., Mandiyan, S., and Nelson, N. (1989b). *J. Biol. Chem.* **264,** 1775–1778.
Nelson, H., Sreekala, M., Noumi, T., Moriyama, Y., Miedel, M., and Nelson, N. (1990a). *J. Biol. Chem.* **265,** 20390–20393.
Nelson, N. (1989). *J. Bioenerg. Biomembr.* **21,** 553–571.
Nelson, R., Guo, X.-L., Masood, K., Kalkbrenner, M., and Gluck, S. (1992). *Proc. Natl. Acad. Sci. U.S.A.* **89,** 3541–3545.
Noumi, T., Beltrán, C., Nelson, H., and Nelson, N. (1991). *Proc. Natl. Acad. Sci. U.S.A.* **88,** 1938–1942.
Ohsumi, Y., and Anraku, Y. (1983). *J. Biol. Chem.* **258,** 5614–5617.
Ohya, Y., Umemoto, N., Tanida, I., Ohta, A., Iida, H., and Anraku, Y. (1991). *J. Biol. Chem.* **266,** 13971–13977.
Ozawa, T., and Schulz, I. (1991). *Biochem. Biophys. Res. Commun.* **180,** 755–764.
Pan, Y. X., Xu, J., Strasser, J. E., Howell, M., and Dean, G. E. (1991). *FEBS Lett.* **293,** 89–92.
Parry, R. V., Turner, J. C., and Rea, P. A. (1989). *J. Biol. Chem.* **264,** 20025–20032.
Percy, J. M., and Apps, D. K. (1986). *Biochem. J.* **239,** 77–81.
Perin, M. S., Fried, V. A., Stone, D. K., Zie, X. S., and Sudhof, T. C. (1991). *J. Biol. Chem.* **266,** 3877–3881.
Puopolo, K., and Forgac, M. (1990). *J. Biol. Chem.* **265,** 14836–14841.
Puopolo, K., Kumamoto, C., Adachi, I., and Forgac, M. (1991). *J. Biol. Chem.* **266,** 24564–24572.
Puopolo, K., Kumamoto, C., Adachi, I., Magner, R., and Forgac, M. (1992). *J. Biol. Chem.* **267,** 3696–3706.
Randall, S. K., and Sze, H. (1986). *J. Biol. Chem.* **261,** 1364–1371.
Randall, S. K., and Sze, H. (1987). *J. Biol. Chem.* **262,** 7135–7141.
Rausch, T., Butcher, D. N., and Taiz, L. (1987). *Plant Physiol.* **85,** 996–999.
Reizman, H. (1985). *Cell* **40,** 1001–1009.
Rodman, J. S., Stahl, P. D., and Gluck, S. (1991). *Exp. Cell Res.* **192,** 445–452.
Rothman, J. H., Yamashiro, D. T., Raymond, C. K., Kane, P. M., and Stevens, T. H. (1989). *J. Cell Biol.* 93–100.
Sabatini, S., Kurtzman, N. A., and Yang, B.-L. (1990a). *Kidney Int.* **37,** 79–84.
Sabatini, S., Laski, M. E., and Kurtzman, N. A. (1990b). *Am. J. Physiol.* **258,** F297–F304.
Sabolic, I., and Burckhardt, G. (1986). *Am. J. Physiol.* **250,** F817–F826.
Sabolic, I., and Burckhardt, G. (1988). *Biochim. Biophys. Acta* **937,** 398–410.
Sabolic, I., Haase, W., and Burckhardt, G. (1985). *Am. J. Physiol.* **248,** F835–F844.
Satlin, L. M., and Schwartz, G. J. (1989). *J. Cell Biol.* **109,** 1279–1288.
Schneider, E., and Altendorf, K. (1985). *EMBO J.* **4,** 515–518.
Schneider, E., and Altendorf, K. (1987). *Microbiol. Rev.* **51,** 477–497.
Schwartz, G. J., Barasch, J., and Al-Awqati, Q. (1985). *Nature (London)* **318,** 368–371.
Schwartz, G. J., Satlin, L. M., and Bergmann, J. E. (1988). *Am. J. Physiol.* **255,** F1003–F1014.
Schweikl, H., Klein, U., Schindbeck, M., and Wieczorek, H. (1989). *J. Biol. Chem.* **264,** 11136–11142.
Senior, A. E. (1988). *Phys. Rev.* **68,** 177–231.
Senior, A. E., and Wise, J. G. (1983). *J. Membr. Biol.* **73,** 105–124.
Shih, C.-K., Wagner, R., Feinstein, S., Kanik-Ennulat, C., and Neff, N. (1988). *Mol. Cell. Biol.* **8,** 3094–3103.
Simon, B. J., and Burckhardt, G. (1990). *J. Membr. Biol.* **117,** 141–151.

Sipe, D., Jesurum, A., and Murphy, R. F. (1991). *J. Biol. Chem.* **266,** 3469–3474.
Steinmetz, P. R. (1986). *Am. J. Physiol.* **251,** F173–F187.
Stetson, D. L., and Steinmetz, P. R. (1986). *Pfluegers Arch.* **407,** Suppl. 2, S80–S84.
Stone, D. K., Crider, B. P., and Xie, X.-S. (1990). *Kidney Int.* **38,** 649–653.
Südhof, T. C., Fried, V. A., Stone, D. K., Johnston, P. A., and Xie, X.-S. (1989). *Proc. Natl. Acad. Sci. U.S.A.* **86,** 6067–6071.
Sulston, J., Du, Z., Thomas, K., Wilson, R., Hillier, L., Staden, R., Halloran, N., Green, P., Thierry-Mieg, J., Qiu, L., Dear, S., Coulson, A., Craxton, M., Durbin, R., Berks, M., Metzstein, M., Hawkins, T., Ainscough, R., and Waterston, R. (1992). *Nature (London)* **356,** 37–41.
Sun, S.-Z., Xie, X.-S., and Stone, D. K. (1987). *J. Biol. Chem.* **262,** 14790–14794.
Sundquist, K., Lakkakorpi, P., Wallmark, B., and Väänänen, K. (1990). *Bichem. Biophys. Res. Commun.* **168,** 309–313.
Swallow, C., Grinstein, S., and Rotstein, O. D. (1990). *J. Biol. Chem.* **265,** 7645–7654.
Swallow, C., Grinstein, S., Sudsbury, R. A., and Rotstein, O. D. (1991a). *Clin. Invest. Med.* **14,** 367–378.
Swallow, C. J., Grinstein, S., Sudsbury, R. A., and Rotstein, O. D. (1991b). *J. Exp. Med.* **174,** 1009–1020.
Taiz, S. L., and Taiz, L. (1991). *Bot. Acta* **104,** 117–121.
Turrini, F., Sabolic, I., Zimolo, Z., Moewes, B., and Burckhardt, G. (1989). *J. Membr. Biol.* **107,** 1–12.
Tzagoloff, A., and Dieckmann, C. L. (1990). *Microbiol. Rev.* **54,** 211–225.
Uchida, E., Ohsumi, Y., and Anraku, Y. (1985). *J. Biol. Chem.* **260,** 1090–1095.
Uchida, E., Ohsumi, Y., and Anraku, Y. (1986). *J. Biol. Chem.* **261,** 45–51.
Umemoto, N., Yoshihisa, T., Hirata, R., and Anraku, Y. (1990). *J. Biol. Chem.* **265,** 18447–18453.
Umemoto, N., Yoshikazu, O., and Anraku, Y. (1991). *J. Biol. Chem.* **266,** 24526–24532.
Van Adelsberg, J., and Al-Awqati, Q. (1986). *J. Cell Biol.* **102,** 1638–1645.
Van Dyke, R. W. (1986). *J. Biol. Chem.* **261,** 15941–15948.
Van Dyke, R. W. (1988). *J. Biol. Chem.* **263,** 2603–2611.
Verlander, J. W., Madsen, K. M., Larsson, L., Cannon, J. K., and Tisher, C. C. (1989). *Am. J. Physiol.* **256,** F454–F462.
Verlander, J. W., Madsen, K. M., and Tisher, C. C. (1991). *Semin. Nephrol.* **11,** 465–477.
Walker, J. E., Saraste, M., Runswick, W. J., and Gay, N. J. (1982). *EMBO J.* **1,** 945–951.
Wang, S.-Y., Yoshinori, M., Mandel, M., Hulmes, J. D., Pan, Y.-C., Danho, W., Nelson, H., and Nelson, N. (1988). *J. Biol. Chem.* **263,** 17638–17642.
Wang, Z.-Q., and Gluck, S. (1990). *J. Biol. Chem.* **265,** 21957–21965.
Wang, Z.-Q., and Gluck, S. (1993). In preparation.
Weisman, L. S., and Wickner, W. (1988). *Science* **241,** 589–591.
Wieczorek, H., Weerth, S., Schindlbeck, M., and Klein, U. (1989). *J. Biol. Chem.* **264,** 11143–11148.
Wieczorek, H., Cioffi, M., Klein, U., Harvey, W. R., Schweikl, H., and Wolfersberger, M. G. (1990). *Methods Enzymol.* **192,** 608–616.
Wieczorek, H., Putzenlechner, M., Zeiske, W., and Klein, U. (1991). *J. Biol. Chem.* **266,** 15340–15347.
Xie, X.-S., and Stone, D. K. (1988). *J. Biol. Chem.* **263,** 9859–9867.
Xie, X.-S., Stone, D. K., and Racker, E. (1989). *J. Biol. Chem.* **264,** 1710–1714.
Yamashiro, C. T., Kane, P. M., Wolczyk, D. F., Preston, R. A., and Stevens, T. H. (1990). *Mol. Cell. Biol.* **10,** 3737–3749.
Young, G., Qiao, J.-Z., and Al-Awqati, Q. (1988). *Proc. Natl. Acad. Sci. U.S.A.* **85,** 9590–9594.

Yurko, M., and Gluck, S. (1987). *J. Biol. Chem.* **262,** 15770–15779.
Zhang, J., Myers, M., and Forgac, M. (1992). *J. Biol. Chem.* **267,** 9773–9778.
Zhang, K., Wang, Z.-Q., and Gluck, S. (1992a). **267,** 14,539–14,542.
Zhang, K., Wang, Z.-Q., and Gluck, S. (1992b). *J. Biol. Chem.* **267,** 9701–9705.
Zimniak, L., Dittrich, P., Gogarten, J. P., Kibak, H., and Taiz, L. (1988). *J. Biol. Chem.* **263,** 9102–9112.

Structure of the Na,K-ATPase

Robert W. Mercer
Departments of Cell Biology and Physiology,
Washington University School of Medicine,
St. Louis, Missouri 63110

I. Introduction

The Na,K-ATPase is a membrane-associated enzyme responsible for maintaining the high internal K concentration and low internal Na concentration characteristic of most animal cells. This enzyme, also known as the sodium pump or sodium–potassium pump, couples the hydrolysis of ATP to the transport of Na and K across the plasma membrane against their respective electrochemical gradients. The cation gradients created by the Na,K-ATPase are fundamental to such diverse cellular functions as the regulation of cell volume, nutrient uptake, and membrane excitability. The importance of the Na,K-ATPase is revealed in the fact that up to one-third of an animal cell's energy requirement is consumed in fueling the Na,K-ATPase. In electrically active tissues or tissues involved in salt transport, up to 70% of the cell's total energy requirement may be used by the pump.

The Na,K-ATPase belongs to a widely distributed class of E_1E_2- or P-type cation-transporting proteins that includes the Ca-ATPase of the sarcoplasm reticulum and plasma membrane, the H,K-ATPase from gastric mucosa, the H-ATPases of plants and fungi, and the bacterial Cd- and K-ATPases. These enzymes all use the hydrolysis of ATP to drive the transport of cations against an electrochemical potential. The E_1E_2 designation refers to the two general enzymatic conformations, E_1 and E_2, which are formed during the catalytic cycle. However, to distinguish these ATPases from other transport ATPases that also undergo conformational changes, these enzymes are usually termed P-type ATPases (Pedersen and Carofoli, 1987). This nomenclature refers to the unique characteristic of these enzymes in forming a transient, phosphorylated aspartyl residue during the reaction cycle. Thus, besides sharing a common reaction mechanism, all P-type ATPases have a similar structure. The region around the phosphorylation aspartate and several regions considered to be involved in ATP binding exhibit a high degree of amino acid homology. Besides implying a common evolutionary ancestor, the similarities in the structure and the reaction mechanism of these widely divergent proteins often make conclusions drawn from one transporter relevant to the others.

The Na,K-ATPase has at least two noncovalently linked polypeptides, a catalytic α-subunit with a molecular weight of about 110,000 and a smaller glycosylated β-subunit with a molecular weight of approximately 55,000 (Glynn, 1985; Jorgensen, 1982). A small peptide with a molecular weight of approximately, 10,000, termed the γ-subunit, has also been identified in purified preparations of the enzyme (Hokin *et al.*, 1973; Forbush *et al.*, 1978). This polypeptide was considered a contaminant of purification until it was shown that the γ-polypeptide, with the α-subunit, could be covalently labeled from the extracellular surface by a photoaffinity-labeled derivative of the specific Na,K-ATPase inhibitor ouabain (Forbush *et al.*, 1978; Rogers and Lazdunski, 1979a,b; Collins *et al.*, 1982). The specific labeling of the γ-subunit by photolabeled ouabain strongly suggests that it is a structural component of the Na,K-ATPase.

Most functions of the Na,K-ATPase have been localized to the α-subunit. The α-subunit contains the binding sites for ATP and ouabain; it is phosphorylated by ATP and undergoes ligand-dependent conformational changes accompanying the binding, occlusion, and translocation of ions (Glynn, 1985). Although the exact functions of the β- and γ-subunits remain unknown, recent evidence suggests that the β-subunit has an important role in Na,K-ATPase function. In ouabain-resistant HeLa cells, both α- and β-subunits are amplified, suggesting that both are required for the acquisition of ouabain resistance (Schneider *et al.*, 1985; Mercer *et al.*, 1986). Moreover, both the α- and the β-subunits are required for the functional expression of the Na,K-ATPase in *Xenopus* oocytes (Noguchi *et al.*, 1987), yeast (Horowitz *et al.*, 1991), and virally infected insect cells (De Tomaso *et al.*, 1991). Although both the α- and the β-subunits are required for the functional expression of enzymatic activity, it appears the γ-subunit is not essential for either ATP hydrolysis or transport (De Tomaso *et al.*, 1991; Hardwicke and Freytag, 1981). However, it has not been determined whether the γ-subunit influences development, regulation, or polarized sorting of the Na,K-ATPase.

Since Skou's first description in 1957 of a Na- and K-dependent ATPase in crab nerve (Skou, 1957) and the subsequent studies proving its identity with the Na,K pump, much information about the structure and function of the Na,K-ATPase has been acquired (reviewed by Glynn, 1985; Jorgensen, 1982). However, an important advance in the understanding of the structure of the Na,K-ATPase was obtained when the complete amino acid sequences for the Na,K-ATPase subunits were determined from their complementary DNA (cDNA) (Shull *et al.*, 1985, 1986b; Kawakami *et al.*, 1985, 1986a; Noguchi *et al.*, 1986; Mercer *et al.*, 1991). These results, with earlier biochemical studies, have helped to elucidate some aspects of the structure of the Na,K-ATPase subunits.

II. Structure of the α-Subunit

The complete amino acid sequences of the α-subunit from brine shrimp (Baxter-Lowe *et al.*, 1989), *Drosophila* (Lebovitz *et al.*, 1989), *Xenopus* (Verrey *et al.*, 1989), *Torpedo* (Kawakami *et al.*, 1985), chicken (Takeyasu *et al.*, 1990), and several mammals (Shull *et al.*, 1985, 1986a; Kawakami *et al.*, 1986b; Ovchinnikov *et al.*, 1986) have been determined. The alignment of the amino acid sequences from these species is shown in Fig. 1. The deduced primary structure of the α-subunits reveals a polypeptide consisting of approximately 1000 amino acids with a molecular weight of about 113,000. In the mammalian α-subunits the initiation methionine if 5 amino acids before the amino terminus of the mature polypeptide. It appears that an amino-terminal pentapeptide is cleaved from the primary translation product during processing. The amino terminus exhibits the highest degree of variability; the brine shrimp α-subunit is slightly shorter, lacking 23 amino terminal residues present in the mammalian and *Torpedo* subunits, whereas the *Drosophila* has an extra 18 residues.

Overall, all Na,K-ATPase α-subunits have a highly conserved amino acid sequence. The α-subunits from human and brine shrimp are nearly 60% identical, whereas the homology between the human and the other mammalian proteins is approximately 98%. As mentioned, the Na,K-ATPase also has a high degree of homology with several other P-type cation-transporting ATPases. As shown in Fig. 2, there are several highly conserved regions in the Cd- and K-ATPases from bacteria (Hesse *et al.*, 1984; Silver *et al.*, 1989), the H-ATPase from *Neurospora* and yeast (Hager *et al.*, 1986; Serrano *et al.*, 1986), and the mammalian Ca- and H,K-ATPases (MacLennan *et al.*, 1985; Shull and Lingrel, 1986). These regions include the phosphorylation site, and several regions thought to be involved in ATP binding (see below). Consequently, regions of the ATPases that are conserved may be involved in the basic energy transduction process common to these proteins, whereas the nonhomologous regions may be involved in the different cation selectivites of the enzymes.

The amino acid sequence, site-specific labeling (Forbush *et al.*, 1978; Farley *et al.*, 1984; Kirley *et al.*, 1984; Ohta *et al.*, 1986a; Le, 1986), immunological (Collins *et al.*, 1983; Farley *et al.*, 1986; McDonough *et al.*, 1982), and proteolytic digestion studies (Jorgensen, 1975, 1977; Castro

```
ATNA$TORCA  MG-KG---AAS-EKYQPAAT--------SENAK-----NSKKSKSKTTDLDELKKEVSIDDHKLNLDELHQKYGTDLTQGLTPARAKEILARDGPNALTPPPTTPEWIKFCRQLFGGFSILLWIGAILCFLAYGIQVATVDNPANDNLYL  132
ATN1$PIG    MG-KG---VGR-DKYEPAAV--------SEHGD-----K--KKAKKERDMDELKKEVSMDDHKLSLDELHRKYGTDLSRGLTPARAAEILARDGPNALTPPPTTPEWVKFCRQLFGGFSMLLWIGAILCFLAYGIQAATEEEPQNDNLYL  130
ATN1$SHEEP  MG-KG---VGR-DKYEPAAV--------SEHGD-----K--KKAKKERDMDELKKEVSMDDHKLSLDELHRKYGTDLNRGLTTARAAEILARDGPNALTPPPTTPEWVKFCRQLFGGFSMLLWIGAVLCFLAYGIQAATEEEPQNDNLYL  130
ATN1$HUMAN  MG-KG---VGR-DKYEPAAV--------SEQGD-----KKGKKGKKDRDMDELKKEVSMDDHKLSLDELHRKYGTDLSRGLTSARAAEILARDGPNALTPPPTTPEWIKFCRQLFGGFSMLLWIGAILCFLAYSIQAATEEEPQNDNLYL  132
ATN1$RAT    MG-KG---VGR-DKYEPAAV--------SEHGD-----KKSKKAKKERDMDELKKEVSMDDHKLSLDELHRKYGTDLSRGLTPARAAEILARDGPNALTPPPTTPEWVKFCRQLFGGFSMLLWIGAILCFLAYGIRSATEEEPPNDDLYL  132
ATN1$CHICK  MG-KG---AGR-DKYEPTAT--------SEHGT-----KKKK--AKERDMDELKKEISMDDHKLSLDELHRKYGTDLSRGLTTARAAEILARDGPNALTPPPTTPEWVKFCRQLFGGFSILLWIGSILCFLAYGITSVMEGEPNSDNLYL  130
ATNA$DROME  MALRSDYEHGRADSYRVATVIATDDDNRTADGQYKSRRKMPAKVNKKENLDDLKQELDIDFHKISPEEMYQRFQTHPENGLSHAKAKENLERDGPN-LTPPKQTPEWVKFCEDLFG-VAMLLWIGAILCFVAYSIQASTSEEPADDNLYL  148
ATNA$ARTSA  MG----------------------------------------K-----KQGKQLSDLKKELEIDDHKIPLEELCRRLGTNTETGLTSSQAKSHLEKYGPNALTPPRTTPEWIKFCKQLFGGFQMLLWIGSILCFIAYTMEKYKNPDVLGDNLYL  109

ATNA$TORCA  GVVLSTVVIITGCFSYYQEAKSSKIMDSFKNMVPQQALVIRDGEKSSINAEQVVVGDLVEVKGGDRIPADIRIISACSCKVDNSSLTGESEPQSRSPEYSSENPLETKNIAFFSTNCVEGTARGIVINIGDHTVMGRIATLASGLEVGQT  282
ATN1$PIG    GVVLSAVVIITGCFSYYQEAKSSKIMESFKNMVPQQALVIRNGEKMSINAEEVVVGDLVEVKGGDRIPADLRIISANGCKVDNSSLTGESEPQTRSPDFTNENPLETRNIAFFSTNCVEGTARGIVVYTGDRTVMGRIATLASGLEGGQT  280
ATN1$SHEEP  GVVLSAVVIITGCFSYYQEAKSSKIMESFKNMVPQQALVIRNGEKMSINAEEVVVGDLVEVKGGDRIPADLRIISANGCKVDNSSLTGESEPQTRSPDFTNENPLETRNIAFFSTNCVEGTARGIVVYTGDRTVMGRIATLASGLEGGQT  280
ATN1$HUMAN  GVVLSAVVIITGCFSYYQEAKSSKIMESFKNMVPQQALVIRNGEKMSINAEEVVVGDLVEVKGGDRIPADLRIISANGCKVDNSSLTGESEPQTRSPDFTNENPLETRNIAFFSTNCVEGTARGIVVYTGDRTVMGRIATLASGLEGGQT  282
ATN1$RAT    GVVLSAVVIITGCFSYYQEAKSSKIMESFKNMVPQQALVIRNGEKMSINAEDVVVGDLVEVKGGDRIPADLRIISANGCKVDNSSLTGESEPQTRSPDFTNENPLETRNIAFFSTNCVEGTARGIVVYTGDRTVMGRIATLASGLEGGQT  282
ATN1$CHICK  GVVLAAVVIITGCFSYYQEAKSSKIMESFKNMVPQQALVIRNGEKMSINAEGVVVGDLVEVKGGDRIPADLRIISAHGCKVDNSSLTGESEPQTRSPDFSNENPLETRNIAFFSTNCVEGTAVGIVISTGDRTVMGRIASLASGLEGGKT  280
ATNA$DROME  GIVLSAVVIMTGMFSYYQESKNSKIMESFKNMVPQFAIVIREGEKPSLRAEDLVIGMLVELEFGDLIPLVYRIIEARDFKVDNSSLTGESEPQSRGAEFTHENPLETKNLAFFSTNAVEALPKGVVISCGDHTVMGRIANLASGLDTGHT  297
ATNA$ARTSA  GLALLFVVIMTGCFRYYQDHNASKIMDSFKNLMPQFARVIRDGKKIQLKAEEVTVGDLVEVKFGDRIPADIRITSCQSMKVDNSSLTGESEPQSRSTECTNDNPLETKNLAFFFTNTLEGTGRGIVINVGDDSVMGRIACLASSLDSGKT  259

ATNA$TORCA  PIAAEIEHFIHIITGVAVFLGVSFFILSLILGYTWLEAVIFLIGIIVANVPEGLLATVTVCLTLTAKRMARKNCLVKNLEAVETLGSTSTICSDKTGTLTQNRMTVAHMWFDNQIHEADTTENQSGISFDKTSLSWNALSRIAALCNRAV  432
ATN1$PIG    PIAAEIEHFIHIITGVAVFLGVSFFILSLILEYTWLEAVIFLIGIIVANVPEGLLATVTVCLTLTAKRMARKNCLVKNLEAVETLGSTSTICSDKTGTLTQNRMTVAHMWFDNQIHEADTTENQSGVSFDKTSATWLALSRIAGLCNRAV  430
ATN1$SHEEP  PIAAEIEHFIHIITGVAVFLGVSFFILSLILEYTWLEAVIFLIGIIVANVPEGLLATVTVCLTLTAKRMARKNCLVKNLEAVETLGSTSTICSDKTGTLTQNRMTVAHMWFDNQIHEADTTENQSGVSFDKTSATWLALSRIAGLCNRAV  430
ATN1$HUMAN  PIAAEIEHFIHIITGVAVFLGVSFFILSLILEYTWLEAVIFLIGIIVANVPEGLLATVTVCLTLTAKRMARKNCLVKNLEAVETLGSTSTICSDKTGTLTQNRMTVAHMWFDNQIHEADTTENQSGVSFDKTSATWLALSRIAGLCNRAV  432
ATN1$RAT    PIAAEIEHFIHIITGVAVFLGVSFFILSLILEYTWLEAVIFLIGIIVANVPEGLLATVTVCLTLTAKRMARKNCLVKNLEAVETLGSTSTICSDKTGTLTQNRMTVAHMWFDNQIHEADTTENQSGVSFDKTSATWFALSRIAGLCNRAV  432
ATN1$CHICK  PIAMEIEHFIHIITGVAVFLGVSFFILSLILGYTWLEAVIFLIGIIVANVPEGLLATVTVCLTLTAKRMARKNCLVKNLEAVETLGSTSTICSDKTGTLTQNRMTVAHMWFDNQIHEADTTENQSGASFDKSSATWLALSRIAGLCNRAV  430
ATNA$DROME  PIAREIEHFIHIITGVAVFLGVTFFVIAFILGYHWLDAVIFLIGIIVANVPEGLLATVTVCLTLTAKRMASKNCLVKNLEAVETLGSTSTICSDKTGTLTQNRMTVAHMWFDNQIIEADTTEDQSGVQYDRTSPGFKALSRIATLCNRAE  447
ATNA$ARTSA  PIAKEIEHFIHIITAMAVFLAAVFAVISFLYGYTWLEAMIFMIGIIVAKVPEGLLATVTVCLTLTAKRMASKNCLVKNLEAVETLGSTSTICSDKTGTLTQNRMTVAHMWFDDKIVTADTTENQSGNQLYRGSKGFPELIRVAALCSRAE  409

ATNA$TORCA  FQAGQDSVPILKRSVAGDASESALLKCIELCCGSVSQMRDRNPKIVEIPFNSTNKYQLSIHENDKA-DSRYLLVMKGAPERILDRCSTILLNGEDKPLNEEMKEAFQNAYLELGGLGERVLGFCHLKLSTSKFPEGYPFDVEEPNFPITD  581
ATN1$PIG    FQANQENLPILKRAVAGDASESALLKCIELCCGSVKEMRERYTKIVEIPFNSTNKYQLSIHKNPNTAEPRHILVMKGAPERILDRCSSILIHGKEQPLDEELKDAFQNAYLELGGLGERVLGFCHLFLPDEQFPEGFQFDTDDVNFPLDN  580
ATN1$SHEEP  FQANQDNLPILKRAVAGDASESALLKCIEVCCGSVKEMRERYAKIVEIPFNSTNKYQLSIHKNANAGEPRHLLVMKGAPERILDRCSSILLHGKEQPLDEELKDAFQNAYLELGGLGERVLGFCHLMLPDEQFPEGFQFDTDDVNFPVDN  580
ATN1$HUMAN  FQANQENLPILKRAVAGDASESALLKCIELCCGSVKEMRERYAKIVEIPFNSTNKYQLSIHKNPNTSEPQHLLVMKGAPERILDRCSSILLHGKEQPLDEELKDAFQNAYLELGGLGERVLGFCHLFLPDEQFPEGFQFDTDDVNFPIDN  582
ATN1$RAT    FQANQENLPILKRAVAGDASESALLKCIEVCCGSVMEMREKYTKIVEIPFNSTNKYQLSIHKNPNASEPKHLLVMKGAPERILDRCSSILLHGKEQPLDEELKDAFQNAYLELGGLGERVLGFCHLLLPDEQFPEGFQFDTDEVNFPVDN  582
ATN1$CHICK  FQANQENVPILKRAVAGDASESALLKCIELCCGSVKEMRERYPKVVEIPFNSTNKYQLSIHKNANAGESRHLLVMKGAPERILDRCDSILIHGKVQPLDEEIKDAFQNAYLELGGLGERVLGFCHLALPDDQFPEGFQFDTDEVNFPVEK  580
ATNA$DROME  FKGGQDGVPILKKEVSGDASEAALLKCMELALGDVMNIRKRNKKIAEVPFNSTNKYQVSIHETEDTNDPRYLLVMKGAPERILERCSTIFINGKEKVLDEEMKEAFNNAYMELGGLGERVLGFCDFMLPSDKYPNGFKFNTDDINFPIDN  597
ATNA$ARTSA  FKTEHAHIPVLKRDVAGDASEAALLKFAEMSTGDVMNIRSKQKKVSEVPFNSAKKYQVSVHEREDKSG--YFLVMKGAPERILERCSTILIDGTEIPLDNHMKECFNNAYMELGGMGERVLGFCDFELPSDQYPRGYVFDADEPNFPISG  557

ATNA$TORCA  LLFVGLMSMIDPPRAAVPDAVGKCRSAGIKVIMVTGDHPITAKAIAKGVGIISEGNETVEDIAARLNIPVNQVNPRDAKACVIHGTDLKDLSHENLDDILHYHTEIVFARTSPQQKLIIVEGCQRQGAIVAVTGDGVNDSPALKKADIGV  731
ATN1$PIG    LCFVGLISMIDPPRAAVPDAVGKCRSAGIKVIMVTGDHPITAKAIAKGVGIISEGNETVEDIAARLNIPVSQVNPRDAKACVVHGSDLKDMTSEQLDDILKYHTEIVFARTSPQQKLIIVEGCQRQGAIVAVTGDGVNDSPALKKADIGV  730
ATN1$SHEEP  LCFVGLISMIDPPRAAVPDAVGKCRSAGIKVIMVTGDHPITAKAIAKGVGIISEGNETVEDIAARLNIPVSQVNPRDAKACVVHGSDLKDMTPEQLDDILKYHTEIVFARTSPQQKLIIVEGCQRQGAIVAVTGDGVNDSPALKKADIGV  730
ATN1$HUMAN  LCFVGLISMIDPPRAAVPDAVGKCRSAGIKVIMVTGDHPITAKAIAKGVGIISEGNETVEDIAARLNIPVSQVNPRDAKACVVHGSDLKDMTSEQLDDILKYHTEIVFARTSPQQKLIIVEGCQRQGAIVAVTGDGVNDSPALKKADIGV  732
ATN1$RAT    LCFVGLISMIDPPRAAVPDAVGKCRSAGIKVIMVTGDHPITAKAIAKGVGIISEGNETVEDIAARLNIPVNQVNPRDAKACVVHGSDLKDMTSEELDDILRYHTEIVFARTSPQQKLIIVEGCQRQGAIVAVTGDGVNDSPALKKADIGV  732
ATN1$CHICK  LLFVGLMSMIDPPRAAVPDAVGKCRSAGIKVIMVTGDHPITAKAIAKGVGIISDGNETVEDIAARLNIPVSQVNPRDAKACVIHGSDLKDMTSEQLDDILLHHTEIVFARTSPQQKLIIVEGCQRQGAIVAVTGDGVNDSPALKKADIGV  730
ATNA$DROME  LIFVGLMSMIDPPRAAVPDAVAKCRSAGIKVIMVTGDHPITAKAIAKSVGIISEGNETVEDIAQRLNIPVSEVNPREAKAAVVHGAELRDVSSDQLDEILRYHTEIVFARTSPQQKLIIVEGCQRMGAIVAVTGDGVNDSPALKKADIGV  747
ATNA$ARTSA  LLFVGLMSMIDPPRAAVPDAVSKCRSAGIKVIMVTGDHPITAKAIARQVGIISEGHETVDDIAARLNIPVSEVNPRSAQAAVIHGNDLKDMNSDQLDDILRHYREIVFARTSPQQKLIIVEGVQRQGEFVAVTGDGVNDSPALKKADIGV  707

ATNA$TORCA  AMGIAGSDVSKQAADMILLDDNFASIVTGVEEGRLIFDNLKKSIAYTLTSNIPEITPFLVFIIANVPLPLGTVTILCIDLGTDMVPAISLAYERAESDIMKRQPRN-PK--TDKLVNERLISMAYGQIGMIQALGGFFSYFVILAENGFL  878
ATN1$PIG    AMGIAGSDVSKQAADMILLDDNFASIVTGVEEGRLIFDNLKKSIAYTLTSNIPEITPFLIFIIANIPLPLGTVTILCIDLGTDMVPAISLAYEQAESDIMKRQ-PRNPK--TDKLVNEQLISMAYGQIGMIQALGGFFTYFVILAENGFL  877
ATN1$SHEEP  AMGIAGSDVSKQAADMILLDDNFASIVTGVEEGRLIFDNLKKSIAYTLTSNIPEITPFLIFIIANIPLPLGTVTILCIDLGTDMVPAISLAYEQAESDIMKRQ-PRNPQ--TDKLVNERLISMAYGQIGMIQALGGFFTYFVILAENGFL  877
ATN1$HUMAN  AMGIAGSDVSKQAADMILLDDNFASIVTGVEEGRLIFDNLKKSIAYTLTSNIPEITPFLIFIIANIPLPLGTVTILCIDLGTDMVPAISLAYEQAESDIMKRQ-PRNPK--TDKLVNERLISMAYGQIGMIQALGGFFTYFVILAENGFL  879
ATN1$RAT    AMGIVGSDVSKQAADMILLDDNFASIVTGVEEGRLIFDNLKKSIAYTLTSNIPEITPFLIFIIANIPLPLGTVTILCIDLGTDMVPAISLAYEQAESDIMKRQ-PRNPK--TDKLVNERLISMAYGQIGMIQALGGFFTYFVILAENGFL  879
ATN1$CHICK  AMGIAGSDVSKQAADMILLDDNFASIVTGVEEGRLIFDNLKKSIAYTLTSNIPEITPFLIFIIANIPLPLGTVTILCIDLGTDMVPAISLAYEQAESDIMKRQ-PRNPK--TDKLVNERLISMAYGQIGMIQALGGFFTYFVILAENGFL  877
ATNA$DROME  AMGIAGSDVSKQAADMILLDDNFASIVTGVEEGRLIFDNLKKSIAYTLTSNIPEISPFLAFILCDIPLPLGTVTILCIDLGTDMVPAISLAYDHAEADIMKRP-PRDPF--NDKLVNSRLISMAYGQIGMIQAAAGFFVYFVIMAENGFL  894
ATNA$ARTSA  AMGIAGSDVSKQAADMILLDDNFASIVTGVEEGRLIFDNLKKSIAYTLTSNIPELSPFLMYILFDLPLAIGTVTILCIDLGTDMVPAISMAYEGPEADPRKPRDPVKEKLVNERLI-----SMAYGQIGVMQAFGGFFTYFVIMGENGFL  852

ATNA$TORCA  PIDLIGIREKWDELWTQDLEDSYGQQWTYEQRKIVEYTCHTAFFVSIVVVQWADLIICKTRRNSVFQQGMKNKILIFGLFEETALAAFLSYCPGTDIALRMYPLKPSWWFCAFPYSLIIFLYDEMRRFIIRRSPGGWVEQETYY  1022
ATN1$PIG    PIHLLGLRVDWDDRWINDVEDSYGQQWTYEQRKIVEFTCHTAFFVSIVVVQWADLVICKTRRNSVFQQGMKNKILIFGLFEETALAAFLSYCPGMGVALRMYPLKPTWWFCAFPYSLLIFVYDEVRKLIIRRRPGGWVEKETYY  1021
ATN1$SHEEP  PNHLLGIRVTWDDRWINDVEDSYGQQWTYEQRKIVEFTCHTAFFVSIVVVQWADLVICKTRRNSVFQQGMKNKILIFGLFEETALAAFLSYCPGMGVALRMYPLKPTWWFCAFPYSLLIFVYDEVRKLIIRRRPGGWVEKETYY  1021
ATN1$HUMAN  PIHLLGLRVDWDDRWINDVEDSYGQQWTYEQRKIVEFTCHTAFFVSIVVVQWADLVICKTRRNSVFQQGMKNKILIFGLFEETALAAFLSYCPGMGVALRMYPLKPTWWFCAFPYSLLIFVYDEVRKLIIRRRPGGWVEKETYY  1023
ATN1$RAT    PFHLLGIRETWDDRWINDVEDSYGQQWTYEQRKIVEFTCHTAFFVSIVVVQWADLVICKTRRNSVFQQGMKNKILIFGLFEETALAAFLSYCPGMGAALRMYPLKPTWWFCAFPYSLLIFVYDEVRKLIIRRRPGGWVEKETYY  1023
ATN1$CHICK  PSDLIGIRLQWDDRWINDVEDSYGQQWTFEQRKIVEFTCHTAFFVSIVVVQWADLIICKTRRNSVFQQGMKNKILIFGLFEETALAAFLSYCPGMDVALRMYPLKPSWWFCAFPYSFLIFVYDEIRKLIIRRRPGGWVEKETYY  1021
ATNA$DROME  PKHLIGIRKMWDSKAVNDLTDSYGQEWTYRDRKTLEYTCHTAFFISIVVVQWADLIICKTRRNSIFQQGMRNWALNFGLVFETVLAAFLSYCPGMEKGLRMYPLKLVWWFPAIPFALAIFIYDETRRFYLRRNPGGWLEQETYY  1038
ATNA$ARTSA  PNRLIGIRKAWESKAYNDLTDSYGQEWTWDARKQLEMTCHTAFFISIVVVQWADLIICKTRRLSIFQQGMRNGTLNFALVFETCVAAFLSYTPGMDKGLRMYPLKIWWWFPPMPFSLLILVYDECRKFIMRRNPGGFLERETYY  996
```

and Farley, 1979; Collins *et al.,* 1983) have provided some insight into the possible transmembrane orientation of the α-subunit. Under defined conditions, limited digestion of the native α-subunit by trypsin and chymotrypsin results in a specific digestion pattern that has been used to study the arrangement of the polypeptide in the membrane (Jorgensen, 1975, 1977; Castro and Farley, 1979). The analysis of digestion patterns has been greatly simplified because the native enzyme lacks extracellular trypsin or chymotrypsin sites (Giotta, 1975; Karlish and Pick, 1981). Therefore, all proteolytic digestion sites are necessarily at the cytoplasmic surface. Moreover, the susceptibility of the native protein to proteolytic digestion depends largely on the conformational state of the enzyme. While in the presence of Na and the E_1 conformation, trypsin digestion of the α-subunit releases into the cytoplasm 20 residues from the amino-terminal end (amino acids 1–20). Further digestion of the amino terminus causes cleavage between lysine 33 and glutamate 34 and results in an "invalid" Na,K-ATPase with several catalytic defects (Jorgensen, 1977; Castro and Farley, 1979; Jorgensen and Klodos, 1978; Jorgensen and Anner, 1979; Jorgensen and Collins, 1986). Another trypsin site, cleaved more slowly in the presence of Na, results in a fragment that remains associated with the membrane (Jorgensen, 1975). This fragment, which has an apparent molecular weight of 78,000 as determined by sodium dodecyl sulfate–polyacrylamide gel electrophoresis (SDS–PAGE), represents the carboxy-terminal region of the intact polypeptide (Castro and Farley, 1979). The amino-terminal residue of the 78-kDa fragment is isoleucine, placing this site between arginine 265 and isoleucine 266 and predicting a fragment with an actual molecular weight of 83,000. An adjacent chymotrypsin site, also accessible in the presence of Na, cleaves between leucine 266 and alanine 267. In the presence of K and in the E_2 conformation, a slightly different trypsin digestion pattern occurs. Trypsin initially cleaves the α-subunit at arginine 438 with subsequent cleavage at lysine 30. While in the E_2 conformation the α-subunit is not cleaved by chymotrypsin. The digestion studies of the native α-subunit have been useful in identifying cytoplasmic residues. To identify regions of the subunit that are extramembranous, the α-subunit has been exhaustively digested. Several soluble peptide fragments released after extensive proteolysis have been isolated and characterized (Ohta *et al.,* 1986a,b; Ovchinnikov *et al.,* 1986). These fragments should represent regions of the α-subunit not embedded or tightly associated with the membrane. The locations of the extramembra-

---

FIG. 1 Amino acid sequence of the Na,K-ATPase α-subunits. Alignment was performed using the GeneWorks multiple alignment software. Identical residues are boxed.

```
Na,K-ATPase  176  INAEEVVVGDLVEVKGGDRIPADLRIISAN**GCKVDNSSLTGESEPQTRS
H,K-ATPase   189  INADELVVGDLVEMKGGDRVPADIRILSAE**GCKVDNSSLTGESEPQTRS
Ca-ATPase    140  IKARDIVPGDIVEVAVGDKVPADIRILSIKSTTLRVDQSILTGESVSVIKH
H-ATPase     191  IEAPEVVPGDILQVEEGTIIPADGRIVTDD*AFLQVDQSALTGESLAVDKH
Cd-ATPase    229  IHVDDIAVGDIMIVKPGEKIAMDGIIVNGL***SAVNQAAITGESVPVSKA


                                                                     *
Na,K-ATPase  320  IGIIVANVPEGLLATVTVCLTLTAKRMARKNCLVKNLEAVETLGSTSTICSDKTGTLTQN
H,K-ATPase   333  MAIVVAYVPEGLLATVTVCLSLTAKRLASKNCVVKNLEAVETLGSTSVICSDKTGTLTQN
Ca-ATPase    300  VALAVAAIPEGLPAVITTCLALGTRRMAKKNAIVRSLPSVETLGCTSVICSDKTGTLTTN
H-ATPase     327  LAITIIGVPVGLPAVVTTTMAVGAAYLAKKKAIVQKLSAIESLAGVEILCSDKTGTLTKN
Cd-ATPase    364  LAVLVVGCPCALVISTPISIVSAIGNAAKKGVLVKGGVYLEKLGAIKTVAFDKTGTLTKG


                     ↓
Na,K-ATPase  500  LVMKGAPERILDRCSSILLHGKEQPLDEELKDAFQ
H,K-ATPase   514  LVMKGAPERVLERCSSILIKGQELPLDEQWREAFQ
Ca-ATPase    512  MFVKGAPEGVIDRCNYVRVGTTRVPMTGPVKEKIL
H-ATPase     471  TCVKGAPLFVLKTVEEDHPIPEEVDQAYKNKVAEF
Cd-ATPase    486  RGIKGIVNGTTYYIGSPKLFKELNVSDFSLGFENN


Na,K-ATPase  588  DPPRAAVPDAVGKCRSAGIKVIMVTGDHPITAK
H,K-ATPase   602  DPPRATVPDAVLKCRTAGIRVIMVTGDHPITAK
Ca-ATPase    601  DPRRKEVMGSIQLCRDAGIRVIMITGDNKGTAI
H-ATPase     534  DPPRHDTYKTVCEAKTLGLSIKMLTGDAVGIAR


                                       ___________________
Na,K-ATPase  682  EIVFARTSPQQKLIIVEGCQRQGAIVAVTGDGVNDSPALKKADIGVAMGIAGSDVSKQAAD
H,K-ATPase   696  EMVFARTSPQQKLVIVESCQRLGAIVAVTGDGVNDSPALKKADIGVAMGIAGSDAAKNAAD
Ca-ATPase    673  ACCFARVEPSHKSKIVEYLQSYDEITAMTGDGVNDAPALKKAEIGIAMGS*GTAVAKTASE
H-ATPase     604  ADGFAEVFPQHKYNVVEILQQRGYLVAMTGDGVNDAPSLKKADTGIAVEG*SSDAARSAAD
Cd-ATPase    590  SDIQSELMPQDKLDYIKKMQSEYDNVAMIGDGVNDAPALAASTVGIAMGGAGTDTAIETAD
```

FIG. 2 Regions of homology between the human α-subunit of the Na,K-ATPase and the rabbit Ca-ATPase, rat H,K-ATPase, *Neurospora* H-ATPase, and the *Staphylococcus* Cd-ATPase. An asterisk identifies the phosphorylated aspartic acid, an arrow the lysine labeled by FITC, and a bar the tryptic peptide labeled by FSBA. Numbering for the α-subunit is based on Figure 1.

nous peptides of the α-subunit are shown in a linear map in Fig. 3A (open bars).

Using reactive reagents that partition in the bilayer, several portions of the α-subunit presumably within or associated with the membrane have been labeled (Farley *et al.*, 1980; Nicholas, 1984). Upon exposure to light the lipophilic carbene precursor 1-tritiospiro [adamantane-4,3′-diazirine] labels several fragments of the Na,K-ATPase α-subunit (Nicholas, 1984). From apparent length, amino-terminal sequence, and amino acid composition, five labeled peptides could be assigned to residues, 263–342, 545–589, 842–880, 946–972, and 973–998 (Nicholas, 1984; Kyte *et al.*, 1987). The location of these peptides in the primary amino acid sequence is shown in Fig. 3A (hatched bars).

Specific labeling with ATP analogs, identification of the catalytic phosphorylation site, and comparison with other nucleotide binding proteins have given an indication of the cytoplasmic domains and residues that contribute to the ATP binding site. For example, the ATP analog 5′-(*p*-fluorsulfonyl)benzoyladenosine (FSBA) both labels and inhibits the Na,K-ATPase (Cooper and Winter, 1980; Cooper *et al.*, 1983; Ohta *et al.*, 1986a). The inactivation and labeling of the Na,K-ATPase by FSBA are specifically prevented by ATP (Ohta *et al.*, 1986a). Studies using radioactive FSBA have shown that FSBA is incorporated into two tryptic peptides (Ohta *et al.*, 1986a) (amino acids 658–667 and 704–722 in Fig. 1). The most likely residues labeled by FSBA are cysteine 656 and lysine 719. Another ATP analog, γ-[4-*N*-2-chloroethyl-*N*-methylamino)]benzolamide-ATP (Cl-ATP), covalently labels Asp 710, a residue near the lysine labeled by FSBA (Ovchinnikov *et al.*, 1987b). Another probe of the ATP binding site is fluorescein isothiocyanate (FITC). Fluorescein isothiocyanate reacts covalently to inhibit the Na,K-ATPase and this inhibition can be prevented by ATP (Karlish, 1979, 1980); FITC labels a lysine residue (504) in the tryptic peptide corresponding to residues 499–509 (Farley *et al.*, 1984; Kirley *et al.*, 1984).

As mentioned earlier, during the catalytic cycle an aspartate residue of the α-subunit is phosphorylated by ATP (Albers *et al.*, 1963; Post *et al.*, 1965). Determination of the amino acid sequence surrounding this site shows that aspartate 372 is phosphorylated (Bastide *et al.*, 1973). This location is consistent with the phosphorylation site of the Ca-ATPase (Allen and Green, 1976) and with the analysis of α-subunit peptide fragments (Castro and Farley, 1979). A comparison of the primary sequence of the α-subunit with other adenine nucleotide binding proteins has identified another region that may be involved in ATP binding. Table I shows a homologous region from several divergent proteins, all of which are involved in the binding of ATP. Notably, these proteins have a conserved sequence of R or K, X–X or X–X–X, G–X–X–X–L, followed by two or three hydrophobic residues (Walker *et al.*, 1982). As shown in Table I, the amino acid sequence of the rat α-subunit corresponding to residues 546 through 555 is homologous to this conserved region of the adenine nucleotide binding proteins. This region may exist as a hydrophobic β-sheet that binds to the nucleotide. Also, studies with phosphofructokinase have suggested that the conserved aspartic acid residue (boxed in Table I) is important for the binding of magnesium in the magnesium–ATP substrate complex (Walker *et al.*, 1982). Thus, there appear to be several regions of the Na,K-ATPase that interact with ATP. Not surprisingly, most of these regions are conserved in the other ion-transporting ATPases. As shown in Fig. 2, the phosphorylation site, the FITC binding region, and the tryptic peptide labeled with FSBA and Cl-ATP are in regions

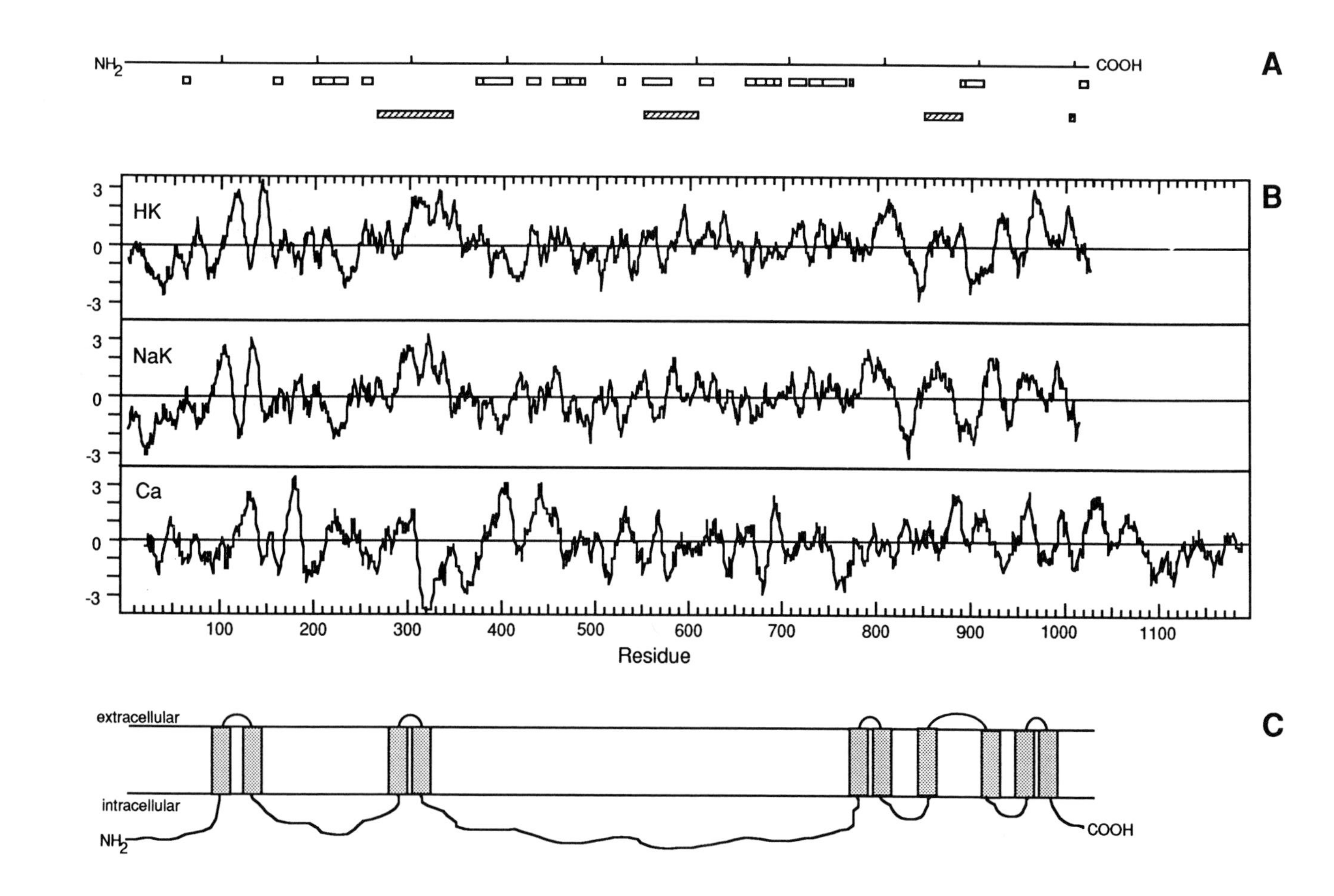
A
NH2
COOH
B
HK
NaK
Ca
3
0
-3
100
200
300
400
500
600
700
800
900
1000
1100
Residue
C
extracellular
intracellular
NH2
COOH

conserved in the P-type ATPases. These regions, though far apart in the primary sequence, may provide clues about the arrangement of the polypeptide chain around the ATP binding site. For example, the phosphorylated aspartate residue and the amino acids that are labeled by FSBA and Cl-ATP are presumed to be close to the terminal phosphate of ATP (Colman, 1983; Ovchinnikov *et al.,* 1987a). On the other hand, FITC is thought to bind to the portion of the Na,K-ATPase that is normally occupied by the adenine ring (Cantley *et al.,* 1982). Presumably the folding of the polypeptide is such that these regions must be in close proximity to form the tertiary structure necessary for the binding of ATP.

Figure 3B shows the hydropathy profiles of the α-subunits from the Na,K-ATPase, H,K-ATPase, and plasma membrane Ca-ATPase. The α-subunit from the sarcoplasmic reticulum Ca-ATPase, and the yeast and *Neurospora* H-ATPases have similar profiles. Hydrophobic regions are plotted above the x-axis and hydrophilic regions are below it. The hydropathy analysis, when combined with the other results, can be used to construct a model for the path of the Na,K-ATPase α-subunit in the membrane. As shown in Fig. 3C, these studies predict a cytoplasmic amino terminus followed by four transmembrane-spanning regions, which are followed by a large cytoplasmic domain consisting of roughly one-third of the polypeptide. The four transmembrane domains are predicted by hydropathy and secondary structure analysis assuming an α-helical stretch of at least 20 hydrophobic residues. As mentioned above, the large cytosolic domain contains the FITC and FSBA binding regions and the aspartate residue that is phosphorylated. Interestingly, one peptide (545–589) labeled with the lipophilic reagent 1-tritiospiro[adamantane-4,3′-diazirine] lies within the cytoplasmic domain. The isolation of soluble peptide fragments containing this region makes its involvement with the membrane unlikely. However, it has been suggested that the corresponding region of the Ca-ATPase has alternating β-strands and α-helicies that fold to give parallel β-sheets that are hydrophobic on both faces (MacLennan *et al.,* 1985). Furthermore, this region exhibits strong hydrophobic characteristics in high-pressure liquid chromatography (HPLC) (Pedemonte and Kaplan, 1990). Therefore, although this region is extramembranous, it may have some hydrophobic characteristics. The localization of several

---

FIG. 3 The Na,K-ATPase α-subunit. (A) The location of extramembranous peptides of the α-subunit subunit are identified as open bars and putative membrane domains as hatched bars. Peptide fragments determined from references 27, 37, 50 and 52. (B) Hydropathy plot of the α-subunits from the Na,K-, H,K- and Ca-ATPases generated by the method of Kyte and Doolittle. Hydrophobic regions are above the x-axis and hydrophilic regions below. (C) Model of the transmembrane orientation of the human Na,K-ATPase α-subunit.

TABLE I

Alignment of Homologous Sequences in Adenine Nucleotide Binding Proteins[a]

Protein	Residues	Sequences																	
Rat Na,K-ATPase	554–561	G	E	R	V	—	L	G	F	C	H	L	L	L	P	D	E	Q	F
Rabbit Ca-ATPase	612–628	L	C	R	Q	—	A	G	I	R	V	I	M	I	T	G	D	N	K
Rat H,K-ATPase	614–640	K	C	R	T	—	A	G	I	R	V	I	M	V	T	G	D	H	P
Bovine ATPase β	244–261	Y	F	R	D	Q	E	G	Q	D	V	L	L	F	I	D	N	I	F
E. coli ATPase β	230–246	K	F	R	D	—	E	G	R	D	V	L	L	F	V	D	N	I	Y
E. coli ATPase α	268–284	Y	F	R	D	—	R	G	E	D	A	L	I	I	Y	D	D	L	S
ATP/ADP translocase	277–294	V	L	R	G	M	G	G	A	F	V	L	V	L	Y	D	E	I	K
Adenylate kinase	105–121	F	E	R	K	—	I	G	Q	P	T	L	L	L	Y	V	D	A	G
Phosphofructokinase	88–104	Q	L	K	K	—	H	G	I	Q	G	L	V	V	I	G	G	D	G

[a] Residues enclosed in boxes are identical or conserved substitutions.

cytoplasmic antibody binding sites to tryptic and chymotryptic fragments of the α-subunit is consistent with the structure shown (Farley *et al.*, 1986).

Analysis of the carboxy-terminal portion of the enzyme suggests the presence of from two to six membrane-spanning regions (Shull *et al.*, 1985; Kawakami *et al.*, 1985, 1986b; Ovchinnikov *et al.*, 1986). Unfortunately little information from site-specific labeling or proteolytic digestion exists for this portion of the protein. However, three regions have been labeled with the lipophilic probe 1-tritiospiro[adamantane-4,3′-diazirine], and two soluble peptide fragments (Fig. 3A) have been isolated. These results are consistent with the presence of six transmembrane segments and is similar to the model suggested for the Ca and H,K-ATPases. The presence of six membrane-spanning domains would place the carboxy terminus and a potential phosphorylation site (R–R–N–S–V, 936–940 in Fig. 1) for a cAMP-dependent protein kinase at the cytoplasmic surface (Krebs and Beavo, 1979). It has been suggested that phosphorylation of the α-subunit at this site by either cAMP-dependent protein kinase or protein kinase C inhibits enzymatic activity by approximately 45% (Bertorello *et al.*, 1991). However, the specific kinase phosphorylation site and the exact membrane topology of this region of the Na,K-ATPase will have to be verified experimentally.

Comparison of the amino acid sequences from the various α-subunits and the other transport ATPases has provided other insights into Na,K-ATPase structure. As previously mentioned, when the enzyme is in the E_1 form, the amino terminus of the α-subunit is rapidly cleaved with trypsin. Cleavage of the α-subunit at this site appears to alter the equilibrium between the E_1 and the E_2 conformations of the enzyme. This tryptic site (K, 20) is in a conserved lysine rich region, which, surprisingly, is also found in the H,K-ATPase (Shull and Lingrel, 1986). It has been proposed that this region may be involved in the conformation shift that occurs during cation occlusion and may function as a movable ion-selective gate that controls the passage of ions to and from their binding sites (Shull *et al.*, 1986a).

The determination of the amino acid sequence of the α-subunit has made it possible to identify amino acids important in the binding of the cardiac glycosides. These drugs, which include digitalis and ouabain, bind to the extracellular surface of the α-subunit, resulting in the inhibition of enzymatic activity. The Na,K-ATPase is the sole receptor for the cardiac glycosides. Although the Na,K-ATPase from most species is sensitive to cardiac glycosides, there is one striking exception. The Na,K-ATPase from rodents is approximately 1000 times less sensitive to ouabain than the enzyme from humans, pigs, and sheep (Wallick *et al.*, 1980; Ahmed *et al.*, 1983). Also, the rat $\alpha 2$ and $\alpha 3$ isoforms of the Na,K-ATPase (discussed below) are much more sensitive to ouabain than the α1 form. Transfection of mouse or rat α-subunits into ouabain-sensitive cells confers ouabain resistance, demonstrating that the determinatants responsible for resistance reside solely on the α-subunit (Kent *et al.*, 1987). A chimeric α-subunit consisting of the amino-terminal half of the rat and the carboxy-terminal half of either the *Torpedo* or the sheep subunit is also resistant to inhibition by ouabain (Noguchi *et al.*, 1988; Price and Lingrel, 1988). The corresponding chimera containing the sheep N-terminus and the rodent C-terminus is not resistant, indicating that the determinants for resistance reside in the N-terminal portion of the subunit (Price and Lingrel, 1988). As shown in Table II there are only six amino acid differences in which the rat ouabain-insensitive α-subunit differs from the ouabain sensitive subunits. Of the six differences, only three are in the N-terminal half of the subunit. These residues are predicted to be in the extracellular region between the first and the second membrane-spanning domains. Conversion of the sheep residues to the corresponding charged residues of the rat (Arg 114 and Asp 125) causes the subunit to become ouabain insensitive. This mutated sheep α-subunit is identical with the rat α-subunit concerning Rb uptake and Na,K-ATPase activity. Changing only one sheep residue to the charged rat residue results in a subunit with intermediate sensitivity to ouabain (Price *et al.*, 1990). If the differences in ouabain sensitivity are caused by amino acid changes at the ouabain binding site, and not by differences at some location distant from ouabain binding, then presumably these amino acids should be involved in ouabain binding. Thus it appears that changes in single amino acids can significantly affect the sensitivity of the enzyme to cardiotonic steroids.

TABLE II

Amino Acid Differences between the Rat α1 (Ouabain-insensitive) and Rat α2, α3, *Artemia, Drosophila, Torpedo,* Chicken, Sheep, Pig, and Human (Ouabain-sensitive) Na,K-ATPase α-Subunits.

Residue	114	122	125	469	555	877
Rat α1	R	P	D	K	L	F
Rat α2	L	S	N	S	N	S
Rat α3	Q	S	N	R	Y	G
Artemia	M	L	N	—	E	N
Drosophila	Q	A	N	R	M	K
Torpedo	Q	A	N	R	K	I
Chicken	T	N	N	R	A	S
Sheep	Q	Q	N	R	M	N
Pig	Q	Q	N	R	F	I
Human	Q	Q	N	Q	F	I

III. Na,K-ATPase α-Subunit Isoforms

As first shown by Sweadner, two molecular forms of the α-subunit exist in mammals (Sweadner, 1979). Besides the renal form, now designated α1, another Na,K-ATPase α-subunit termed α(+) has been identified. The α(+) isoform (so named because of its larger molecular weight by SDS–gel electrophoresis) can be distinguished from the renal form by its higher affinity for ouabain (Sweadner, 1979; Mark and Seeds, 1978), an increased sensitivity to *N*-ethylmaleimide and pyrithiamin, and its resistance to digestion by trypsin (Sweadner, 1979; Matsuda *et al.*, 1984). Both isoforms are present in the brain, but it appears that the renal form is expressed in the glia cells, whereas both forms are expressed in the neurons (Sweadner, 1979; Specht and Sweadner, 1984; Atterwill *et al.*, 1984). In adipocytes and skeletal muscle both isoforms are present, but it appears that the α(+) form of the enzyme is selectively sensitive to stimulation by insulin (Lytton *et al.*, 1985). Determination of the amino-terminal sequences of the two isoforms has shown that they have distinct, although homologous, amino-terminal ends (Lytton, 1985). Recently, cDNAs coding for the α(+) isoform (termed α2) and a previously unidentified α-subunit, designated α3, have been isolated and sequenced (Shull *et al.*, 1986a; Takeyasu *et al.*, 1990). Figure 4 shows the complete amino acid sequences for the rat α1, α2, and α3, α-subunit isoforms. The degrees of sequence similarity among the rat and chicken isoforms are shown in Table III. Each rat or chicken α-subunit isoform differs from the other two isoforms by approximately 14%. However, the degree of similarity between identical isoforms across species is over 92%. The α3 isoforms exhibit the highest degree of similarity, with over 96% identity. Interestingly, the amount of amino acid sequence homology between the different α-subunit isoforms is approximately the same as that between the α1 subunits from the distantly related rat and *Torpedo*. If the degree of homology can be considered to be related to the time of divergence, then the divergence in the genes coding for the α-subunit isoforms must have been a distant evolutionary event.

As can be seen from Fig. 4, most of the amino acid divergence among the isoforms occurs in the amino terminus, the loop between the first and second membrane domains, and in the cytoplasmic region from amino acids 403 to 503. The greatest similarities occur around the phosphorylation site, the major hydrophobic regions, and the cytoplasmic region that includes the FSBA-reactive peptides (amino acids 589–785).

The determination of the nucleotide sequence and hybridization analysis of rat genomic DNA shows that each α-subunit isoform is the product of a different gene. Chromosomal mapping localizes the gene coding for the α1 isoform to chromosome 1p in the human and chromosome 3 in the

```
α1   MGKGV -1
α2   MGRGA
α3   *****

α1   GRDKYEPAAVSEHGD**KKS KKAK*KERDMDELKKEVSMD DHKLSLDELHRKYGTDLSRG LTPARPAEILARDGPNALTP PPTTPEWVKFCRQLFGGFSM
α2   GRE*YSPAATTAENGGGKK* *KQK**EKELDELKKEVAMD DHKLSLDELGRKYNVDLSKG LTNQRAQDILARDGPNALTP PPTTPEWVKFCRQLFGGFSI
α3   ****MGDKKDDKSSP**KKS *KAK*ERRDLDDLKKEVAMT EHKMSVEEVCRKYNTDCVQG LTHSKAQEILARDGPNALTP PPTTPEWVKFCRQLFGGFSI

α1   LLWIGAILCFLAYGIRSATE EEPPNDDLYLGVVLSAVVII TGCFSYYQEAKSSKIMESFK NMVPQQALVIRNGEKMSINA EDVVVGDLVEVKGGDRIPAD
α2   LLWIGALLCFLAYGILAAME DEPSNDNLYLGIVLAAVVIV TGCFSYYQEAKSSKIMDSFK NMVPQQALVIREGEKMQINA EEVVVGDLVEVKGGDRVPAD
α3   LLWIGAILCFLAYGIQAGTE DDPSGDNLYLGIVLAAVVII TGCFSYYQEAKSSKIMESFK NMVPQQALVIREGEKMQVNA EEVVVGDLVEIKGGDRVPAD

α1   LRIISANGCKVDNSSLTGESEP QTRSPDFTNENPLETRNIAF FSTNCVEGTARGIVVYTGDR TVMGRIATLASGLEGGQTPI AEEIEHFIHLITGVAVFL
α2   LRIISSHGCKVDNSSLTGESEP QTRSPEFTHENPLETRNICF FSTNCVEGTARGIVIATGDR TVMGRIATLASGLEVGQTPI AMEIEHFIQLITGVAVFL
α3   LRIISAHGCKVDNSSLTGESEP QTRSPDCTHDNPLETRNITF FSTNCVEGTARGVVVATGDR TVMGRIATLASGLEVGKTPI AIEIEHFIQLITGVAVFL

α1   GVSFFILSLILEYTWLEAVI FLIGIIVANVPEGLLATVTV CLTLTAKRMARKNCLVKNLE AVETLGSTSTICSDKTGTLT QNRMTVAHMWFDNQIHEADT
α2   GVSFFVLSLILGYSWLEAVI FLIGIIVANVPEGLLATVTV CLTLTAKRMARKNCLVKNLE AVETLGSTSTICSDKTGTLT QNRMTVAHMWFDNQIHEADT
α3   GVSFFILSLILGYTWLEAVI FLIGIIVANVPEGLLATVTV CLTLTAKRMARKNCLVKNLE AVETLGSTSTICSDKTGTLT QNRMTVAHMWFDNQIHEADT

α1   TENQSGVSFDKTSATWF*AL SRIAGLCNRAVFQANQENLP ILKRAVAGDASESALLKCIE VCCGSVMEMREKYTKIVEIP FNSTNKYQLSIHKNPNASEP
α2   TEDQSGATFDKRSPTWT*AL SRIAGLCNRAVFKAGQENIS VSKRDTAGDASESALLKCIE LSCGSVRKMRDRNPKVAEIP FNSTNKYQLSIHEREDSPQS
α3   TEDQSGTSFDK*SHTWVSAL SHIAGLCNRAVFKGGQDNIP VLKRDVAGDASESALLKCIE LSSGSVKLMRERNKKVAEIP FNSTNKYQLSIHETEDPNDN
```

```
α1   KHLLVMKGAPERILDRCSSI LLHGKEQPLDEELKDAFQNA YLELGGLGERVLGFCHLLLP DEQFPEGFQFDTDEVNFPVD NLCFVGLISMIDPPRAAVPD
α2   *HVLVMKGAPERILDRCSTI LVQGKEIPLDKEMQDAFQNA YMELGGLGERVLGFCQLNLP SGKFPRGFKFDTDELNFPTE KLCFVGLMSMIDPPRAAVPD
α3   RYLLVMKGAPERILDRCATI LLQGKEQPLDEEMKEAFQNA YLELGGLGERVLGFCHYYLP EEQFPKGFAFDCDDVNFTTD NLCFVGLMSMIDPPRAAVPD

α1   AVGKCRSAGIKVIMVTGDHP ITAKAIAKGVGIISEGNETV EDIAARLNIPVNQVNPRDAK ACVVHGSDLKDMTSEELDDI LRYHTEIVFARTSPQQKLII
α2   AVGKCRSAGIKVIMVTGDHP ITAKAIAKGVGIISEGNETV EDIAARLNIPVSQVNPREAK ACVVHGSDLKDMTSEQLDEI LRDHTEIVFARTSPQQKLII
α3   AVGKCRSAGIKVIMVTGDHP ITAKAIAKGVGIISEGNETV EDIAARLNIPVSQVNPRDAK ACVIHGTDLKDFTSQQIDEI LQNHTEIVFARTSPQQKLII

α1   VEGCQRQGAIVAVTGDGVND SPALKKADIGVAMGIVGSDV SKQAADMILLDDNFASIVTG VEEGRLIFDNLKKSIAYTLT SNIPEITPFLIFIIANIPLP
α2   VEGCQRQGAIVAVTGDGVND SPALKKADIGIAMGISGSDV SKQAADMILLDDNFASIVTG VEEGRLIFDNLKKSIAYTLT SNIPEITPFLLFIIANIPLP
α3   VEGCQRQGAIVAVTGDGVND SPALKKADIGVAMGIVGSDV SKQAADMILLDDNFASIVTG VEEGRLIFDNLKKSIAYTLT SNIPEITPFLLFIMANIPLP

α1   LGTVTILCIDLGTDMVPAIS LAYEQAESDIMKRQPRNPKT DKLVNERLISMAYGQIGMIQ ALGGFFTYFVILAENGFLPF HLLGIRETWDDRWINDVEDS
α2   LGTVTILCIDLGTDMVPAIS LAYEAAESDIMKRQPRNSQT DKLVNERLISMAYGQIGMIQ ALGGFFSYFVILAENGFLPS RLLGIRLDWDDRTTNDLEDS
α3   LGTITILCIDLGTDMVPAIS LAYEAAESDIMKRQPRNPRT DKLVNERLISMAYGQIGMIQ ALGGFFSYFVILAENGFLPG NLLGIRLNWDDRTVNDLEDS

α1   YGQQWTYEQRKIVEFTCHTA FFVSIVVVQWADLVICKTRR NSVFQQGMKNKILIFGLFEE TALAAFLSYCPGMGAALRMY PLKPTWWFCAFPYSLLIFVY
α2   YGQEWTYEQRKVVEFTCHTA FFASIVVVQWADLIICKTRR NSVFQQGMKNKILIFGLLEE TALAAFLSYCPGMGVALRMY PLKVTWWFCAFPYSLLIFIY
α3   YGQQWTYEQRKVVEFTEHTA FFVSIVVVQWADLIICKTRR NSVFQQGMKNKILIFGLFEE TALAAFLSYCPGMDVALRMY PLKPSWWFCAFPYSLLIFVY

α1   DEVRKLIIRRRPGGWVEKETYY
α2   DEVRKLILRRYPGGWVEKETYY
α3   DEVRKLILRRNPGGWVEKETYY
```

FIG. 4 Amino acid sequence of the rat Na,K-ATPase α-subunit isoforms. Differences from the $\alpha 1$ sequence are underlined.

mouse, whereas the genes coding for human $\alpha 2$ and $\alpha 3$ are on chromosomes 1q and 19, respectively (Kawakami *et al.*, 1986a; Yang-Fang *et al.*, 1988).

Analysis of mRNA of fetal and adult tissues has revealed several interesting and surprising results concerning the tissue-specific and developmental expression of the α-subunit isoforms. Ribonucleic acid hybridization analysis of α-subunit isoform expression in adult rat tissues identifies $\alpha 1$ mRNA in all tissues examined with the highest levels of $\alpha 1$ transcription in the kidney. Characterization of $\alpha 2$ transcription reveals $\alpha 2$ mRNA in the brain, skeletal muscle, diaphragm, heart, and spinal cord. Also, in these tissues two distinct mRNAs are present, one mRNA with a molecular weight corresponding to the $\alpha 1$ isoform and the other with a slightly larger molecular weight. The abundance of the two mRNA species is tissue-specific with the larger transcript predominating in the neural tissues and the smaller in the muscle. Whereas $\alpha 1$ and $\alpha 2$ isoforms exhibit a diverse expression pattern, $\alpha 3$ is found almost entirely in adult neural tissue. Cellular localization of $\alpha 3$ mRNA by *in situ* hybridization suggests that this isoform is predominately expressed in neurons, with high levels of expression in the large neurons of the cerebellum, cortex, and hippocampus; the Purkinje cells of the cerebellum; and the pyramidal cells of the cortex and hippocampus (Schneider *et al.*, 1988). The relative amount of Na,K-ATPase α isoform and β-subunit mRNA expression in various rat tissues is summarized in Table IV.

In addition to the tissue-specific regulation, the Na,K-ATPase α isoforms are regulated developmentally. For example, in the rat the multiple α isoform and β-subunit mRNAs appear to be coordinately regulated with maximum expression occurring between 15 and 25 days of age (Orlowski and Lingrel, 1988). As mentioned before, in the adult rat brain all three α isoforms are present in roughly equivalent amounts; however, in fetal brain $\alpha 3$ is the most abundant isoform. The levels of $\alpha 3$ mRNA increase over 10-fold during the first 7 days of development. In contrast, $\alpha 1$ and

TABLE III

Amino Acid Sequence Identity between Rat and Chicken Na,K-ATPase α-Subunit Isoforms.

	rat $\alpha 1$	rat $\alpha 2$	rat $\alpha 3$	chicken $\alpha 1$	chicken $\alpha 2$	chicken $\alpha 3$
rat $\alpha 1$	100	86	85	93	73	64
rat $\alpha 2$		100	86	73	93	87
rat $\alpha 3$			100	66	77	96
chicken $\alpha 1$				100	86	86
chicken $\alpha 2$					100	86
chicken $\alpha 3$						100

TABLE IV

Relative Tissue Distributions of Na,K-ATPase Isoform mRNAs in Adult Rat Tissues

Tissue	$\alpha 1$	$\alpha 2$	$\alpha 3$	$\beta 1$	$\beta 2$
Kidney	+++	−	−	+++	−
Heart	++	+	−	++	−
Brain	++	++	+++	++	+
Liver	+	−	−	+/−	−
Intestine	++	−	−	++	?
Skeletal muscle	+	+++	−	++	?
Smooth muscle	+	++	−	++	?

$\alpha 2$ mRNA increase more gradually, reaching levels similar to $\alpha 3$ at approximately 25 days of age (Orlowski and Lingrel, 1988). In skeletal muscle $\alpha 1$ is the major mRNA isoform in the fetus, remaining constant throughout development. The $\alpha 2$ mRNA levels are relatively low in the early stages of development; however, following birth, $\alpha 2$ levels rapidly rise over 10-fold to become the predominant isoform (Orlowski and Lingrel, 1988). In the heart the $\alpha 1$ mRNA represents the major α-subunit transcript; the other isoforms are expressed to a much lesser extent. In the fetal and neonatal rat there is a developmental switch between the $\alpha 2$ and the $\alpha 3$ isoforms. The $\alpha 3$ mRNA is expressed predominantly in the fetal and neonatal heart, decreasing to negligible levels in the adult. In contrast, $\alpha 2$ mRNA is low in the fetal and neonatal heart and increases in the juvenile and adult tissue (Orlowski and Lingrel, 1988).

Clearly Na,K-ATPase α-subunit mRNA expression is regulated in a tissue-specific fashion. The $\alpha 1$ isoform is found mainly in the kidney and transport epithelia, $\alpha 2$ is expressed largely in the heart and skeletal muscle, and $\alpha 3$ is found mostly in neuronal tissue. Also, each tissue exhibits a distinct pattern of Na,K-ATPase subunit expression during ontogenesis. It is easy to speculate that the α-subunit isoforms may have functional properties that are important in the regulation of Na,K-ATPase activity in the different tissues. However, the exact physiological functions of these different α-subunit isoforms have not been determined. Undoubtedly, an understanding of the significance of the α-subunit isoforms will provide insights to the function of the Na,K-ATPase.

IV. Structure of the β-Subunit

The complete amino acid sequences for the human (Kawakami *et al.*, 1986a), pig (Ovchinnikov *et al.*, 1986), sheep (Shull *et al.*, 1986b), rat

(Mercer *et al.*, 1986), chicken (Takeyasu *et al.*, 1987), dog (Brown *et al.*, 1987), *Xenopus* (Verrey *et al.*, 1989), and *Torpedo* (Noguchi *et al.*, 1986) β-subunits are shown in Fig. 5. The subunits consist of 302 to 305 amino acids and have an approximate molecular weight of 35,000. The mammalian β-subunits exhibit a high degree of amino acid homology ($\approx$95%) and are approximately 60% homologous with the nonmammalian protein. When comparing the mammalian β-subunits with the β-subunit from the *Torpedo,* two regions are highly conserved, a hydrophobic region consisting of 28 amino acids (35–63, 89% homology) and an uncharged region of 19 residues (233–250, $\approx$89% homology). As indicated by arrows in Fig. 5, the *Torpedo* β-subunit has four potential sites for N-linked glycosylation (Asn–X–Ser/Thr), whereas the mammalian subunits have three. A substitution in the mammalian sequences at amino acid 114 converts an asparagine residue to lysine, eliminating the most amino-terminal glycosylation site. Isolation and characterization of glycosylated peptide fragments (Ovchinnikov *et al.*, 1986) and *in vitro* expression of β-subunit cDNA (Gilmore-Hebert *et al.*, 1988) have clearly demonstrated that all gycosylation sites are used in the mammalian β-subunits.

It also appears that the β-subunit has several disulfide bonds that may be important for activity of the enzyme (Esmann, 1982; Kawamura and Nagano, 1984). The locations of all seven cysteine residues are conserved within all the β-subunits. All cysteines, except the cysteine predicted to be in the membrane-spanning domain of the polypeptide (Cys 45), are involved in disulfide bond formation (Miller and Farley, 1990). Isolation and characterization of β-subunit tryptic fragments have identified the disulfide bonds as between Cys 125–Cys 148, Cys 158–Cys 174, and Cys 212–Cys 275 (Ohta *et al.*, 1986b; Miller and Farley, 1990). Reduction of the β-subunit with 2-mercaptoethanol completely inhibits ATPase activity, suggesting that a change in disulfide bond formation can inactivate the enzyme (Kawamura and Nagano, 1984).

The β-subunit has at least one transmembrane segment since it possesses antigenic determinants at the cytoplasmic surface (Girardet *et al.*, 1981) and can be covalently labeled from the extracellular surface (Hall and Ruoho, 1980). Isolation of several soluble tryptic fragments corresponding to most of the exposed polypeptide demonstrates that most of the β-subunit is extramembranous (Ohta *et al.*, 1986b; Ovchinnikov *et al.*, 1986), and labeling studies using Bolton–Hunter reagent indicate that most of it is extracellular (Dzhandzhugazyan and Jorgensen, 1985). The hydropathy profiles and the predicted secondary structures for all the β-subunits are nearly identical, implying that the amino acid substitutions have not changed the general structural organization of the protein. The hydropathy profile suggests the presence of a single transmembrane domain corresponding to the highly conserved region consisting of residues

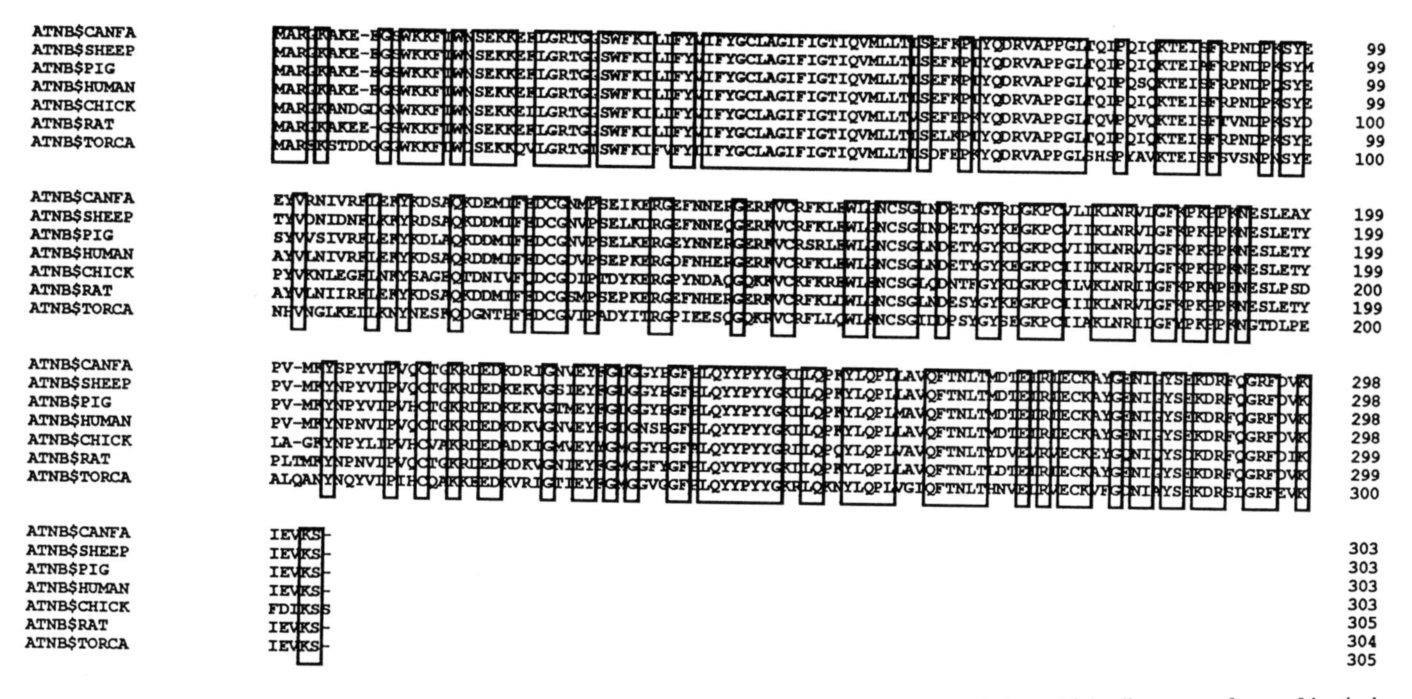

FIG. 5 Amino acid sequence of the Na,K-ATPase β-subunits. Alignment was performed using the GeneWorks multiple alignment software. Identical residues are boxed.

35 through 62. The presence of a single transmembrane domain would predict a β-subunit with a highly charged cytoplasmic amino terminus, followed by a single transmembrane region and a large extracellular carboxy-terminal domain. A model of the β-subunit showing the transmembrane orientation, location of the glycosylation sites, and position of the disulfide bonds is shown in Fig. 6. The conserved region consisting of residues 233 through 250 and comprising only uncharged amino acids may be associated with the membrane. Although this region probably does not span the membrane, it may be in contact with or embedded in the membrane or involved in hydrophobic interactions with the α-subunit.

In contrast to the α-subunit, the rat and human β-subunit gene encodes four distinct mRNA species that are expressed in a tissue-specific fashion. The β-subunit mRNA levels also vary considerably from tissue to tissue, suggesting that transcriptional control mechanisms may in part account for the differences in Na,K-ATPase activity. The highest levels of β mRNA are found in the kidney and brain, with much lower levels in the heart, muscle, lung, and liver (Mercer *et al.*, 1986; Young and Lingrel, 1987; Orlowski and Lingrel, 1988).

Also, different tissues exhibit distinctive preferences for individual transcripts. For example, rat kidney accumulates the largest β-subunit mRNA

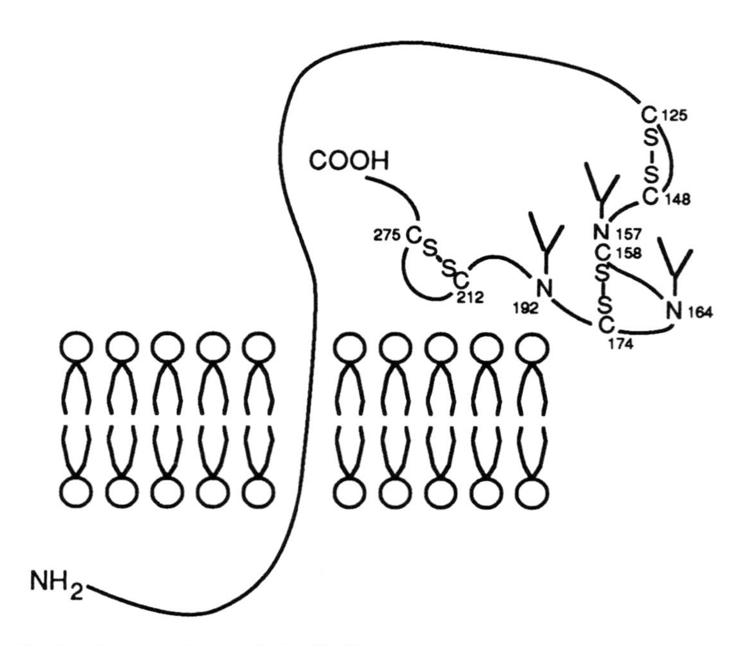

FIG. 6 Model of the β-subunit of the Na,K-ATPase. The three glycosylation sites and the position of the disulfide bridge are shown. Modified from Ohta, *et al.*, 1986b.

in preference to the other three, whereas different mRNAs are preferred in the bladder and brain. The different mRNAs, which produce a single protein product, are probably derived from the same gene through alternate splicing in the noncoding region (Young and Lingrel, 1987). In this regard it is interesting that the 3′-noncoding regions of the β-subunit mRNAs exhibit a remarkable degree of homology. For example, a region of over 140 nucleotides in the 3′-untranslated region of the rat mRNA differs by only one nucleotide when compared to the human sequence. This high degree of nucleotide sequence conservation suggests that the 3′-untranslated region plays an important role in β-subunit biogenesis.

As with the α-subunit, the β-subunit mRNA is coordinately regulated throughout development. However, although it seems clear that the α and β polypeptides are present in the Na,K-ATPase in equal amounts (see below), the α to β mRNA ratio varies greatly during development and in the different tissues. This implies that there is a translational or post-translational regulation of the subunits to maintain the stoichiometric amount of the subunits. Alternatively there may be other unidentified Na,K-ATPase isoforms that may contribute to the apparent differences.

V. Na,K-ATPase β-Subunit Isoforms

Recently a cDNA of a putative isoform of the β-subunit, termed $\beta2$, has been identified in rat brain and human liver (Martin-Vasallo *et al.*, 1989). As shown in Fig. 7 the rat $\beta2$ isoform exhibits 58% amino acid sequence similarity (34% identity, 24% favored substitutions) with the rat β-subunit (now termed $\beta1$). The putative transmembrane segment is more highly conserved; of the 28 residues, 16 are identical and 5 are conservative substitutions. There are seven potential N-linked glycosylation sites in the rat $\beta2$ subunit and eight in the human polypeptide. Three of these sites are conserved relative to the mammalian $\beta1$ subunit. It is not clear whether the other potential glycosylation sites are used. The human and rat $\beta2$ subunits are 98% identical. Analysis of rat tissue mRNA indicates that the $\beta2$ gene encodes a single 3.4-kilobase mRNA that is expressed in a tissue-specific fashion that is distinct from the $\beta1$ expression pattern. Whereas the highest levels of $\beta1$ mRNA are in the kidney, $\beta2$ is predominately expressed in the brain. High levels of $\beta2$ mRNA, relative to $\beta1$, are also found in the eye and spleen (Martin-Vasallo *et al.*, 1989). The high level of $\beta2$ expression in the brain is consistent with the finding that the mouse $\beta2$ polypeptide has been previously described as AMOG (adhesion molecule on glia), a protein that is involved in neuron–astrocyte adhesion (Gloor *et al.*, 1990).

```
rat ß1        MA-----RGKAKEEGSWKKFIWNSEKKEFLGRTGGSWFKILLFYVIFYGCLAGIFIGTIQVMLLTISELKPTYQD   70
rat ß2        MVIQKEKKSCGQVVEEWKEFVWNPRTHQFMGRTGTSWAFILLFYLVFYGFLTAMFTLTMWVMLQTVSDHTPKYQD   75
Xenopus ß3    MA-KEENKGSEQSGSDWKQFIYNPQKGEFMGRTASSWALILLFYLVFYGFLAGLFTLTMWVMLQTLDDSVPKYRD   74

rat ß1        RVAPPGLTQIPQIQKTEISFRPNDPKSYEAYVLNIIRFLEKYKDSAQ--KDDMIFED-CGSMPSE-----PKERG  137
rat ß2        RLATPGLMIRPKTENLDVIVNISDTESWDQHVQKLNKFLEPYNDSIQAQKNDVCRPGRYYEQPDNGVLNYPKRAC  150
Xenopus ß3    RVSSPGLMISPKSAGLEIKFSRSKTQSYMEYVQTLNTFLAPYNDSIQA-KNEFCPPGLYFDQDEE-V---EKKTC  144

rat ß1        EFNHERGERKVCRFKLDWLGNCSGLNDES-YGYKEGKPCIIIKLNRVLGFKPKPPKNESLETYPLTMKYNPNVLP  211
rat ß2        QFNRTQ------------LGNCSGIGDPTHYGYSTGQPCVFIKMNRVINFYAGANQSMNVTCVGKKDEDAENLGH  213
Xenopus ß3    QFNRTS------------LGICSGIEDPM-FGYGEGKPCVIVKINRIIGLKPEGNP--KINCTSKTED--VNL-Q  201

rat ß1        ---VQCTGKRD----EDKDKVGNIEYFGMGGFYGFPLQYYPYYGKLLQPKYLQP-LLAVQFTNLTLDTEIRIECK  278
rat ß2        FIMFPANGNIDLMYFPYYGKKFHVNYTQPLVAVKFLNVT-PNVEVNVECRINAA-NIATDDERDKFAARVAFKLR  286
Xenopus ß3    Y--FPDNGKIDLMYFPYYGKKTHVNYVQPVVAVKISPSNFTSEEIAVECKIHGSRNLKNEDERDKFLGRVTFKVK  274

rat ß1        AYGEN----------------------------IGYSEKDRFQGRFD-VKIEVKS                      304
rat ß2        I-NKA--------------------------------------------------                      290
Xenopus ß3    ITEKETQDLWPVVSLLLLLLFFFFFLIIFFYRPDLLFSDLEPSRRMIDWHMIQGST                     330
```

FIG. 7 Amino acid sequence of the Na,K-ATPase β-subunit isoforms. Alignment was performed using the GeneWorks multiple alignment software. Identical residues are boxed.

Another putative isoform of the β-subunit (β3) has been identified in *Xenopus* (Good et al., 1990). The deduced amino acid sequence for this β-subunit (Fig. 7) is distinct from the β1-subunit of *Xenopus;* moreover, comparison with the mammalian β2 suggests that it is not the frog homolog of the β2 subunit. For example, although the rat and *Xenopus* β1-subunits exhibit 67% sequence identity and differ in length by only one amino acid, the rat β2 and *Xenopus* β3 differ in size by 13 residues and are only 56% identical. In addition, the 3'-noncoding region of the β3 subunit does not share any sequence similarity with any other β-subunit 3'noncoding region. Analysis of mRNA by blot and *in situ* hybridization shows that the β3 transcript is a nervous-system-specific isoform that is regulated during early nervous system development (Good *et al.,* 1990).

Hydropathy analysis and the location of extracellular cysteine residues and potential N-linked glycosylation sites are similar between the β-subunits, suggesting that the tertiary structures of these polypeptides are conserved. It appears that the β2-subunit may preferentially associate with either the α2 or the α3 isoforms (Gloor *et al.,* 1990; Shyjan *et al.,* 1990; Schneider and Kraig, 1990); still, the possibility has not been excluded that it represents a β-like subunit from another P-type ion pump. The exact relationship of the β isoforms to the expression, regulation, and activity of the Na,K-ATPase has not yet been established.

VI. Structure of the γ-subunit

The γ-subunit is a hydrophobic peptide with a molecular weight of approximately 10,000. As mentioned before, the strongest evidence that the γ-subunit is associated with the Na,K-ATPase is that it can be photoaffinity-labeled with ouabain derivatives. These studies have shown that the binding site consists of not only the α- and β-subunits, but the γ-subunit as well (Forbush *et al.,* 1978; Rogers and Lazdunski, 1979b; Lowndes *et al.,* 1984). For example, the α- and γ-subunits are labeled by photoaffinity compounds with approximately equal probability when the reactive groups are 23–24 Å from the lactone ring of the cardiotonic steroid (Forbush, 1983). Quantitation of γ-subunit from lamb and shark Na,K-ATPase indicates about 1 γ-subunit per α-subunit, suggesting that the γ-subunit is present in the enzyme in stoichiometric amounts (Reeves *et al.,* 1980). When the lamb kidney γ-peptides are chromatographed in nonpolar solvents, two distinct proteolipid fractions, designated γ1 and γ2, can be identified. Determination of the amino acid compositions of the γ-peptides demonstrates that γ1 and γ2 have similar amino acid compositions, although γ1 has more hydrophobic and fewer charged residues (Reeves *et al.*, 1980). Although little is known about the role of the γ-subunit, the γ-subunits from lamb kidney and shark rectal gland have similar amino acid compositions, implying that the amino acid sequence is conserved and therefore important for Na,K-ATPase structure or function (Hardwicke and Freytag, 1981). Although small hydrophobic proteins are associated with the H-ATPase of mitochondria (Blondin, 1979), bacteria (Sierra and Tzagoloff, 1973), and the Ca-ATPase of sarcoplasmic reticulum (MacLennan *et al.*, 1972), the significance of these proteins has remained unclear. The exact role of the γ-subunit in Na,K-ATPase function is unknown, although its necessity for ATPase activity has been experimentally questioned (Hardwick and Freytag, 1981). For several reasons it has been suggested that the γ-polypeptide is a breakdown product of the α- and β-subunits; the amino acid composition of the γ-subunit and α-subunits is similar (Collins *et al.*, 1982), antibodies raised to γ cross-react with both the α- and β-subunits (Ball *et al.*, 1983), and a peptide derived from the γ-polypeptide fraction contains the N-terminus of the β-subunit (Harris and Stahl, 1988). Consequently, the labeling of the γ-subunit by ouabain derivatives could be explained if it was a portion of the γ- or β-subunits. However, partial peptide sequencing of HPLC-purified γ-subunit has clearly established that the polypeptide is different from the α- and β-subunits, and therefore cannot be a breakdown product of the larger subunits (Collins and Lesznk, 1987).

Recently, the cDNA coding for the γ-subunit of the NaK-ATPase has been isolated (Mercer *et al.*, 1991). The deduced amino acid sequences of the rat and sheep γ-subunits are shown in Fig. 8. The full-length rat kidney γ-subunit consists of 58 amino acids with a molecular weight of 6497. There are no sites for N-linked glycosylation. The amino acid sequences of the sheep and rat γ-subunits exhibit 91% sequence similarity. Hydropathy analysis of the γ-subunit suggests the presence of a cytoplasmic N-terminus, followed by a single membrane-spanning region and a highly charged extracellular domain. Immunological evidence suggests that the N-terminus of γ-subunit is cytoplasmic, although this has yet to be verified.

Ribonucleic acid blot hybridization indicates that the γ-subunit gene encodes two RNA species of approximately 1.5 and 0.8 kilobases in size. The smaller transcript is more abundant in the kidney, with low levels in spleen, lung, and heart. In contrast, the larger transcript is more abundant in the spleen and lung. Interestingly, in contrast to the α mRNA, there is

```
SHEEP        ENEDPFYYDYETVRNGGLIPAALAFIVGLMIILSKRFRCGAKKKHRQIPQDGL
RAT     MVAVQGTENPFEYDYETVRKGGLIPAGLAFVVGLLIILSKRFRCGGSKKHRQVNQDEL
```

FIG. 8 Amino acid sequence of the sheep and rat Na,K-ATPase γ-subunits. Identical residues are boxed.

little expression of γ-subunit mRNA in the brain. Thus, like the other Na,K-ATPase subunits, the γ-subunit is expressed in a tissue-specific fashion. In addition, the low levels of γ-subunit mRNA in the brain suggest that there may be other isoforms for the subunit.

Immunocytochemical studies (Minuth *et al.*, 1987) have localized the α-subunit of the Na,K-ATPase to the basolateral membrane of renal tubule epithelial cells. An antiserum specific to the γ-polypeptide also labels the basolateral membrane of the tubular epithelium (Mercer *et al.*, 1991). Moreover, double immunofluorescence using the γ-antiserum and an antibody to the α-subunit demonstrates that the γ-subunit codistributes with the α-subunit. These results are consistent with the view that the γ-subunit is specifically associated with and may be an important component of the Na,K-ATPase. However, speculation concerning the exact nature and function of the γ-peptide is best constrained until there is a better understanding of the subunit.

VII. Structure of the Na,K-ATPase

Although much information is known about the structure of the α- and β-subunits, relatively little is known concerning the subunit interactions of the Na,K-ATPase. The exact subunit structure of the enzyme and the minimum functional unit for the active transport of cations have not yet been determined. Uncertainty in the molecular weights of the α- and β-subunits has caused some ambiguity in the determination of the molar α/β ratio. Assuming molecular weights for the α- and β-subunits of 112,000 and 45,000, respectively, the quantization of the α/β mass ratio by several different methods generates an α/β ratio of approximately one (Jorgensen, 1974; Peterson and Hokin, 1980; Craig and Kyte, 1980; Peters *et al.*, 1981; Bonting *et al.*, 1983). Crosslinking studies (Craig and Kyte, 1980), high-pressure gel chromatography (Hayashi *et al.*, 1983), and measurements of the ratio of the amino-terminal residues for the subunits (Cantley, 1981) also support a 1:1 ratio. Thus, several lines of evidence seem to support a model for the Na,K-ATPase in which the number of α- and α-subunits is equal.

Although there is general agreement for a 1:1 α/β ratio, there is much less agreement on whether the αβ unit represents the active form of the enzyme (reviewed in Glynn, 1985; Jorgensen, 1982). However, there is some evidence that suggests that the αβ unit has enzymatic activity. In 1981, Brotherus *et al.* demonstrated that solubilization of pig kidney Na,K-ATPase in the *n*-dodecyl octaethylene glycol monoether ($C_{12}E_8$) results in predominantly αβ units that retain ATPase activity (Brotherus *et al.*,

1981). In addition it has been shown that each soluble $\alpha\beta$ unit can bind one molecule of ATP (Jensen and Ottolenghi, 1983), undergo transitions between the E_1 and the E_2 conformation of the α-subunit (Jorgensen and Andersen, 1986), and when reconstituted into phospholipid vesicles can transport Na and K (Brotherus *et al.*, 1983). However, it is not clear whether the $\alpha\beta$ protomer by itself can form the pathway for the transport of cations across the membrane or whether active transport requires an association betwen $\alpha\beta$ units in the membrane. Indeed there are several lines of evidence which suggest that the active unit of the Na,K-ATPase involves an aggregation of the $\alpha\beta$ protomers. Molecular weight determinations using analytical ultracentrifugation (Esmann *et al.*, 1979, 1980), low-angle laser light scattering (Ellory *et al.*, 1979), radiation inactivation (Ellory *et al.*, 1979; Ottolenghi *et al.*, 1983; Nørby and Jensen, 1989), and evidence from crosslinking (Askari and Huang, 1980; Askari *et al.*, 1980; Huang and Askari, 1981) and electron microscopy (Skriver *et al.*, 1981; Hebert *et al.*, 1982) support oligomeric models for Na,K-ATPase-mediated transport. Thus, although it is possible that the $\alpha\beta$ protomer has independent catalytic activity, it cannot be excluded that the transport of Na and K across the cellular membrane requires the formation of oligomeric complexes.

VIII. Conclusion

Tremendous strides have been made in the understanding of the structure of the Na,K-ATPase. The isolation of the DNA coding for the Na,K-ATPase subunits will undoubtedly provide opportunities for further understanding this important enzyme. In view of the recent advances it may be possible to correlate some of the functional studies of the Na,K-ATPase with the structural aspects of the enzyme to obtain some fundamental observations on the molecular mechanisms of transport.

References

Ahmed, K., Rohrer, D. C., Fullerton, D. S., Detto, T., Kitatsuji, E., and From, A. H. L. (1983). *J. Biol. Chem.*, **258**, 8092–8097.
Albers, R. W., Fahn, S., and Koval, G. J. (1963). *Proc. Natl. Acad. Sci. U.S.A.* **50**, 474–481.
Allen, G., and Green, N. M. (1976). *FEBS Lett.* **63**, 188–192.
Askari, A., and Huang, W.-H. (1980). *Biochem. Biophys. Res. Commun.* **93**, 448–453.
Askari, A., Huang, W.-H., and Antieau, J. M. (1980). *Biochemistry* **19**, 1132–1140.
Atterwill, C. K., Cunningham, V. J., and Balazs, R. (1984). *J. Neurochem.* **43**, 8–18.

Ball, W., Jr., Collins, J. H., Land, L., and Schwartz, A. (1983). *Curr. Top. Membr. Transp.* **19,** 781–785.

Bastide, F., Meissner, G., Fleischer, S., and Post, R. L. (1973). *J. Biol. Chem.* **248,** 8385–8391.

Baxter-Lowe, L. A., Guo, J. Z., Bergstrom, E. E., and Hokin, L. E. (1989). *FEBS Lett.* **257,** 181–187.

Bertorello, A. M., Aperia, A., Walaas, S. I., Nairn, A. C., Greengard, P. (1991). *Proc. Natl. Acad. Sci. U.S.A.* **88,** 11359–11362.

Blondin, G. A. (1979). *Biochem. Biophys. Res. Commun.* **87,** 1087–1094.

Bonting, S. L., Swarts, H. G. P., Peters, W. H. M., Schuurmans Stekhoven, F. M. A. H., and De Pont, J. J. H. H. M. (1983). *Curr. Top. Membr. Transp.* **19,** 403–424.

Brotherus, J. R., Moller, J. V., and Jorgensen, P. L. (1981). *Biochem. Biophys. Res. Commun.* **100,** 146–154.

Brotherus, J. R., Jacobsen, L., and Jorgensen, P. L. (1983). *Biochim. Biophys. Acta* **731,** 290–303.

Brown, T. A., Horowitz, B., Miller, R. P., McDonough, A. A., and Farley, R. A. (1987). *Biochim. Biophys. Acta* **912,** 244–253.

Cantley, L. C. (1981). *Curr. Top. Bioenerg.* **11,** 201–237.

Cantley, L. C., Carilli, C. T., Farley, R. A., and Perlman, D. M. (1982). *Ann. N. Y. Acad. Sci.* **402,** 289–291.

Castro, J., and Farley, R. A. (1979). *J. Biol. Chem.* **254,** 2221–2228.

Collins, J. H., and Lesznk, J. (1987). *Biochemistry* **26,** 8665–8668.

Collins, J. H., Forbush, B., Lane, L. K., Ling, E., Schwartz, A., and Zot, A. (1982). *Biochim. Biophys. Acta* **686,** 7–12.

Collins, J. H., Zot, A. S. Ball, W. J., Jr., Lane, L. K., and Schwartz, A. (1983). *Biochim. Biophys. Acta* **742,** 358–365.Colman, R. F. (1983). *Annu. Rev. Biochem.* **52,** 67–91.

Cooper, J. B., and Winter, C. G. (1980). *J. Supramol. Struct.* **13,** 165–174.

Cooper, J. B., Johnson, C., and Winter, C. G. (1983). *Curr. Top. Membr. Transp.* **19,** 367–370.

Craig, W. S., and Kyte, J. (1980). *J. Biol. Chem.* **255,** 6262–6269.

De Tomaso, A. W., Zdankiewicz, P., and Mercer, R. W. (1991). *In* "The Sodium Pump" (P. De Weer and J. H. Kaplan, eds.), pp. 69–73. Rockefeller Univ. Press, New York.

Dzhandzhugazyan, K. N., and Jorgensen, P. L. (1985). *Biochim. Biophys. Acta* **817,** 165–173.

Ellory, J. C., Green, J. R., Jarvis, S. M., and Young, J. D. (1979). *J. Physiol. (London)* **295,** 10P-11P.

Esmann, M. (1982) *Biochim. Biophys. Acta* **668,** 251–259.

Esmann, M., Skou, J. C., and Christiansen, C. (1979). *Biochim. Biophys. Acta* **567,** 410–420.

Esmann, M., Christiansen, C., Karlsson, K. A., Hansson, G. C., and Skou, J. C. (1980). *Biochim. Biophys. Acta* **603,** 1–12.

Farley, R. A., Goldman, D. W., and Bayley, H. (1980). *J. Biol. Chem.* **255,** 860–864.

Farley, R. A., Tran, C. H., Carilli, C. T., Hawke, D., and Shrively, J. E. (1984). *J. Biol. Chem.* **259,** 9532–9535.

Farley, R. A., Ochoa, G. T., and Kudrow, Y. (1986). *Am. J. Physiol.* **250,** C896–C906.

Forbush, B., III (1983). *Curr. Top. Membr. Transp.* **19,** 167–201.

Forbush, B., Kaplan, J. H., and Hoffman, J. F. (1978). *Biochemistry* **17,** 3667–3676.

Gilmore-Hebert, M., Mercer, R. W., Schneider, J. W., and Benz, E. J., Jr. (1988). *In* "Na,K-ATPase" (J. C. Skou and J. G. Nørby, eds.), pp. 71–76. Alan R. Liss, New York.

Giotta, G. J. (1975). *J. Biol. Chem.* **250,** 5159–5164.

Girardet, M., Geering, K., Frantes, J. M., Geser, D., Rossier, B. C., Kraehenbuhl, J.-P., and Bron, C. (1981). *Biochemistry* **20,** 6684–6691.

Gloor, S., Antonicek, H., Sweadner, K. J., Pagliusi, SD., Frank, R., Moos, M., and Schacher, M. (1990). *J. Cell Biol.* **110,** 165–174.

Glynn, I. M. (1985). *In* "The Enzymes of Biological Membrane" (A. Martonosi, ed.), 2nd Ed., Vol. 3, pp. 35–114. Plenum, New York.
Good, P. J., Richter, K., and Dawid, I. G. (1990). *Proc. Natl. Acad. Sci. U.S.A.* **87,** 9088–9092.
Hager, K. M., Mandala, S. M., Davenport, J. W., Speicher, D. W., Benz, E. J., Jr., and Stayman, C. W. (1986). *Proc. Natl. Acad. Sci. U.S.A.* **83,** 7693–7697.
Hall, C., and Ruoho, A. (1980). *Proc. Natl. Acad. Sci. U.S.A.* **77,** 4529–4533.
Hardwicke, P. M. D., and Freytag, J. W. (1981). *Biochem. Biophys. Res. Commun.* **102,** 250–257.
Harris, W. E., and Stahl, W. L. (1988). *Biochim. Biophys. Acta* **942,** 236–244.
Hayashi, Y., Takagi, T., Maezawa, S., and Matsui, H. (1983). *Biochim. Biophys. Acta* **748,** 153–167.
Hebert, H., Jorgensen, P., Skriver, E., and Maunsbach, A. B. (1982). *Biochim. Biophys. Acta* **689,** 571–574.
Hesse, J. S., Wieczorek, L., Altendorf, K., Reicin, A. S., Dorus, E., and Epstein, W. (1984). *Proc. Natl. Acad. Sci. U.S.A.* **81,** 4746–4750.
Hokin, L. E., Dahl, J. L., Duupree, J. D., Dixon, J. F., Hackney, J. F., and Perdue, J. F. (1973). *J. Biol. Chem.* **248,** 2593–2605.
Horowitz, B., Eakle, K. A., Scheiner-Bobis, G., Randolph, G. R., Chen, C. Y., Hitzeman, R. A., and Farley, R. A. (1991). *J. Biol. Chem.* **265,** 4189–4192.
Huang, W.-H., and Askari, A. (1981). *Biochim. Biophys. Acta* **645,** 54–58.
Jensen, J., and Ottolenghi, P. (1983). *Biochim. Biophys. Acta* **731,** 282–289.
Jorgensen, P. L. (1974). *Biochim. Biophys. Acta* **356,** 36–52.
Jorgensen, P. L. (1975). *Biochim. Biophys. Acta* **401,** 399–415.
Jorgensen, P. L. (1977). *Biochim. Biophys. Acta* **466,** 97–108.
Jorgensen, P. L. (1982). *Biochim. Biophys. Acta* **694,** 27–68.
Jorgensen, P. L., and Andersen, J. P. (1986). *Biochemistry* **25,** 2889–2897.
Jorgensen, P. L., and Anner, B. M. (1979). *Biochim. Biophys. Acta* **555,** 485–492.
Jorgensen, P. L., and Collins, J. H. (1986). *Biochim. Biophys. Acta* **860,** 570–576.
Jorgensen, P. L., and Klodos, I. (1978). *Biochim. Biophys. Acta* **507,** 8–16.
Karlish, S. J. D. (1979). *In* "Na,K-ATPase Structure and Kinetics" (J. C. Skou and J. G. Nørby, eds.) pp. 115–128. Academic Press, New York.
Karlish, S. J. D. (1980). *J. Bioenerg. Biomembr.* **12,** 111–136.
Karlish, S. J. D., and Pick, U. (1981). *J. Physiol. (London)* **312,** 505–529.
Kawakami, K., Noguchi, S., Noda, M., Takahashi, H., Ohta, T., Kawamura, M., Nojima, H., Nagano, K., Hirose, T., Inayama, S., Hayashida, H., Miyata, T., and Numa, S. (1985). *Nature (London)* **316,** 733–736.
Kawakami, K., Nojima, H., Ohta, T., and Nagano, K. (1986a). *Nucleic Acids Res.* **14,** 2833–2844.
Kawakami, K., Ohta, T., Nojima, H., and Nagano, K. (1986b). *J. Biochem. (Tokyo)* **100,** 389–397.
Kawamura, M., and Nagano, K. (1984). *Biochim. Biophys. Acta* **774,** 188–192.
Kent, R. B., Emanuel, J. R., Neriah, Y. B., Levenson, R., and Housman, D. E. (1987). *Science* **237,** 901–903.
Kirley, T. L., Wallick, E. T., and Lane, L. K. (1984). *Biochem. Biophys. Res. Commun.* **125,** 767–773.
Krebs, E. G., and Beavo, J. A. (1979). *Annu. Rev. Biochem.* **48,** 923–959.
Kyte, J., Xu, K., and Bayer, R. (1987). *Biochemistry* **26,** 8350–8360.
Le, D. T. (1986). *Biochemistry* **25,** 2379–2386.
Lebovitz, R. M., Takeyasu, K., and Fambrough, D. M. (1989). *EMBO J.* **8,** 193–202.
Lowndes, J. M., Hokin-Neaverson, M., and Ruoho, A. E. (1984). *J. Biol. Chem.* **259,** 10533–10538.

Lytton, J. (1985). *Biochem. Biophys. Res. Commun.* **132,** 764–769.
Lytton, J., Lin, J. C., and Guidotti, G. (1985). *J. Biol. Chem.* **260,** 1177–1184.
MacLennan, D. H., Yip, C. C., Iles, G. H., and Seeman, P. (1972). *Cold Spring Harbor Symp. Quant. Biol.* **37,** 469-477.
MacLennan, D. H ., Brandl, C. J., Korczak, B., and Green, N. M. (1985). *Nature (London)* **316,** 696–700.
Marks, M. J., and Seeds, N. W. (1978). *Life Sci.* **23,** 2735–2744.
Martin-Vasallo, P., Dackowski, W., Emanuel, J. R., and Levenson, R. (1989). *J. Biol. Chem.* **264,** 4613–4618.
Matsuda, T., Iwata, H.,and Cooper, J. R. (1984). *J. Biol. Chem.* **259,** 35858–3863.
McDonough, A. A., Hiatt, A., and Edelman, I. S. (1982). *J. Membr. Biol.* **69,** 13–22.
Mercer, R. W., Schneider, J. W., Savitz, A., Emanuel, J., Benz, E. J., Jr., and Levenson, R. (1986). *Mol. Cell. Biol.* **6,** 3884–3890.
Mercer, R. W., Biemesderfer, D., Bliss, D. P., Jr., Collins, J. H., and Forbush, B., III (1991). *In* "The Sodium Pump" (P. De Weer and J. H. Kaplan, eds.), pp. 37–41. Rockefeller Univ. Press, New York.
Miller, R. P., and Farley, R. A. (1990). *Biochemistry* **29,** 1524–1532.
Minuth, W. W., Gross, P., Gilbert, P., and Kashgarian, M. (1987). *Kidney Int.* **31,** 1104–1112.
Nicholas, R. A. (1984). *Biochemistry* **23,** 888–898.
Nørby, J. G., and Jensen, J. (1989). *J. Biol. Chem.* **264,** 19548–19558.
Noguchi, S., Noda, M., Takahashi, H., Kawakami, K., Ohta, T., Nagano, K., Hirose, T., Inayama, S., Kawamura, M., and Numa, S. (1986). *FEBS Lett.* **196,** 315–319.
Noguchi, S., Mishina, M., Kawamura, M., and Numa, S. (1987). *FEBS Lett.* **225,** 27–32.
Noguchi, S., Ohta, T., Takeda, K., Ohtsubo, M., and Kawamura, M. (1988). *Biochem. Biophys. Res. Commun.* **155,** 1237–1243.
Ohta, T., Nagano, K., and Yoshida, M. (1986a). *Proc. Natl. Acad. Sci. U.S.A.* **83,** 2071–2075.
Ohta, T., Yoshida, M., Hirano, K., and Kawamura, M. (1986b). *FEBS Lett.* **204,** 297–301.
Orlowski, J., and Lingrel., J. B. (1988). *J. Biol. Chem.* **263,** 10436–10442.
Ottolenghi, P., Ellory, J. C., and Klein, R. (1983). *Curr. Top. Membr. Transp.* **19,** 139–143.
Ovchinnikov, Y. A., Modyanov, N. N., Broude, N. E., Petrukhin, K. E., Grishin, A. V., Arzamazova, N. M., Aldanova, N. A., Monastyrskaya, G. S., and Sverdlov, E. D. (1986). *FEBS Lett.* **201,** 237–245.
Ovchinnikov, Y. A., Dzhandzugazyan, K. N., Lutsenko, S. V., Mustayev, A. A., and Modyanov, N. N. (1987a). *FEBS Lett.* **217,** 111–116.
Ovchinnikov, Y. A., Monastyrskaya, G. S., Broude, N. E., Modyanov, N. N., Allikmets, R. L., Ushkaryov, Y. A., Melkov, A. M., Smirnov, Y. V., Malyshev, I. V., Dulubova, I. E., Petrukhin, K. E., Gryshin, A. V., Sverdlov, V. E., Kiyatkin, N. I., Kostina, M. B., Modyanov, N. N., and Sverdlov, E. D. (1987b). *FEBS Lett.* **213,** 73–80.
Pedemonte, C. H., and Kaplan, J. H. (1990). *Am. J. Physiol.* **258,** C1–C23.
Pedersen, P. L., and Carafoli, E. (1987). *Trends Biochem. Sci.* **12,** 186–189.
Peters, W. H. M., De Pont, J. J. H. H. M., Koppers, A., and Bonding, S. L. (1981). *Biochim. Biophys. Acta* **641,** 55–70.
Peterson, G. L., and Hokin, L. E. (1980). *Biochem. J.* **192,** 107–118.
Post, R. L., Sen, A. K., and Rosenthal, A. S. (1965). *J. Biol. Chem.* **240,** 1437–1445.
Price, E. M., and Lingrel, J. B. (1988). *Biochemistry* **27,** 8400–8408.
Price, E. M., Rice, D. A., and Lingrel, J. B. (1990). *J. Biol. Chem.* **264,** 6638–6641.
Reeves, A. S., Collins, J. H., and Schwartz, A. (1980). *Biochem. Biophys. Res. Commun.* **95,** 1591–1598.
Rogers, T. B., and Lazdunski, M. (1979a). *Biochemistry,* **18,** 135–140.
Rogers, T. B., and Lazdunski, M. (1979b). *FEBS Lett.* **98,** 373–376.
Schneider, B. G., and Kraig, E. (1990). *Exp. Eye Res.* **51,** 553–564.

Schneider, J. W., Mercer, R. W., Caplan, M., Emanuel, J. R., Sweadner, K. J., Benz, E. J., Jr., and Levenson, R. (1985). *Proc. Natl. Acad. Sci. U.S.A.* **82,** 6357–6361.

Schneider, J. W., Mercer, R. W., Gilmore-Hebert, M., Utset, M. F., Lai, C., Green, A., and Benz, E. J., Jr. (1988). *Proc. Natl. Acad. Sci. U.S.A.* **85,** 284–288.

Serrano, R., Kielland-Brandt, M. C., and Fink, G. R. (1986). *Nature (London)* **319,** 689–693.

Shull, G. E., and Lingrel, J. B. (1986). *J. Biol. Chem.* **261,** 16788–16791.

Shull, G. E., Schwartz, A., and Lingrel, J. B. (1985). *Nature (London)* **316,** 691–695.

Shull, G. E., Greeb, J., and Lingrel, J. B. (1986a). *Biochemistry* **25,** 8125–8132.

Shull, G. E., Lane, L. K., and Lingrel, J. B. (1986b). *Nature (London)* **321,** 429–431.

Shyjan, A. W., Cena, V., Klein, D. C., and Levenson, R. (1990). *Proc. Natl. Acad. Sci. U.S.A.* **87,** 1178–1182.

Sierra, M. F., and Tzagoloff, A. (1973). *Proc. Natl. Acad. Sci. U.S.A.* **70,** 3155–3159.

Silver, S., Nucifora, G., Chu, L., and Misra, T. K. (1989). *Trends Biochem. Sci.* **14,** 76–80.

Skou, J. C. (1957). *Biochim. Biophys. Acta* **23,** 394–401.

Skriver, E., Maunsbach, A. B., and Jorgensen, P. L. (1981). *FEBS Lett.* **131,** 219–222.

Specht, S. C., and Sweadner, K. J. (1984). *Proc. Natl. Acad. Sci. U.S.A.* **81,** 1234–1238.

Sweadner, K. J. (1979). *J. Biol. Chem.* **254,** 6060–6067.

Takeyasu, K., Tamkun, M. M., Siegel, N. R., and Fambrough, D. M. (1987). *J. Biol. Chem.* **262,** 10733–10740.

Takeyasu, K., Lemas, V., and Fambrough, D. M. (1990). *Am. J. Physiol.* **259,** C619–C630.

Verrey, F., Kairouz, P., Schaerer, E., Fuentes, P., Geering, K., Rossier, B. C., and Kraehenbuhl, J. P. (1989). *Am. J. Physiol.* **256,** PF1034–PF1043.

Walker, J. E., Saraste, M., Runswick, M. J., and Gay, N. J. (1982). *EMBO J.* **1,** 945–951.

Wallick, E. T., Pitts, B. J. R., Lane, L. K., and Schwartz, A. (1980). *Arch. Biochem. Biophys.* **202,** 442–449.

Yang-Fang, T. L., Schneider, J. W., Lindgren, V., Shull, M. M., Benz, E. J., Jr., Lingrel, J. B., and Francke, U. (1988). *Genomics* **2,** 128–138.

Young, R. M., and Lingrel, J. B. (1987). *Biochem. Biophys. Res. Commun.* **145,** 52–58.

The Mouse Multidrug Resistance Gene Family: Structural and Functional Analysis

Philippe Gros and Ellen Buschman
Department of Biochemistry, McGill University, Montreal, Quebec, Canada H3G 1Y6

I. Introduction

One major limitation to the successful chemotherapeutic treatment of many types of human tumors is the emergence and outgrowth of subpopulations of drug-resistant cells. These populations of tumor cells often display cross-resistance to cytotoxic compounds that share little or no structural and functional similarities, and to which they have not been previously exposed (Moscow and Cowan, 1988). Drugs that form the multidrug resistance (MDR) spectrum include anthracyclines (such as adriamycin), *Vinca* alkaloids (vincristine, vinblastine), colchicine, actinomycin D, etoposides (VP-16, VM-21), topoisomerase inhibitors (amsacrine), and many others (see, e.g., Gupta *et al.*, 1988). There are only few structural and functional similarities between these compounds: they are small, biplanar, hydrophobic molecules that have a basic nitrogen atom (hydrophobic cations) and that enter the cell by passive diffusion across the membrane lipid bilayer (Gupta *et al.*, 1988; Zamora *et al.*, 1988; Pearce *et al.*, 1989; Beck, 1990).

Since clinical tumor specimens are often heterogeneous and composed of normal, drug-resistant and drug-sensitive neoplastic cells present in unknown proportions, and since it is virtually impossible to obtain serial specimens of the same tumor at different times after initiation of treatment, the MDR phenotype has been inherently difficult to study *in vivo*. Most of our understanding of this phenomenon has been derived over the years from the *in vitro* study of multidrug-resistant cell lines obtained by continuous exposure to increasing concentrations of a given cytotoxic MDR drug (stepwise selection)(Biedler and Riehm, 1970; Ling and Thompson, 1974; Fojo *et al.*, 1985; Lemontt *et al.*, 1988). The phenotypic characteristics of independently derived MDR cell lines are remarkably similar despite the diversity of their tissue origin and *in vitro* selection protocols used in their production (for comprehensive reviews, see Bradley *et al.*, 1988; Endicott and Ling, 1989; Roninson, 1991). First, MDR cell lines display a strikingly similar pattern of cross-resistance to the group of natural products that form the MDR spectrum, although they tend to express higher levels of resistance to the compound used in the initial drug selection (Tsuruo *et al.*, 1983; Conter and Beck, 1984; Scotto *et al.*, 1986). Second, the emergence of multidrug resistance is linked to a decreased degree of intracellular drug accumulation and to a concomitant increase in cellular drug efflux, both of which are strictly dependent on the presence of intact intracellular nucleotide triphosphate pools (e.g., ATP)(Dano, 1973; Skovsgaard, 1978). Third, the very high levels of resistance detected in certain MDR cell lines are unstable, are lost upon prolonged culture of the cells in drug-free medium, and are often associated with unstable genetic elements such as double minute chromosomes and homogeneously staining chromosomal regions (hallmarks of gene amplification), suggesting that overexpression of a specific polypeptide(s) from amplified gene copies may be responsible for the MDR phenotype (Roninson *et al.*, 1984, 1986). Fourth, although the levels of several polypeptides such as cytoplasmic calcium binding proteins (Koch *et al.*, 1986), enzymes of the glycosylation pathway (Peterson *et al.*, 1979), and membrane glycoproteins (Peterson *et al.*, 1983; Richert *et al.*, 1985) have been found to be modulated in MDR cell lines, the most ubiquitous phenotypic marker of MDR is the overexpression of a membrane protein called P-glycoprotein (P-gp). P-glycoproteins are a heterogeneous group of membrane phosphoglycoproteins of apparent molecular weight of 170,000 to 200,000 first described by Juliano and Ling (1976) in a Chinese hamster cell line (C5) highly resistant to colchicine. P-glycoprotein has since been shown to bind photoactivatable analogs of ATP (Cornwell *et al.*, 1987a; Schurr *et al.*, 1989) and cytotoxic drugs (Cornwell *et al.*, 1986; Safa *et al.*, 1986, 1989) and has been shown to possess ATPase activity (Hamada *et al.*, 1988), leading to the proposal that P-gp functions as an ATP-dependent drug efflux pump to reduce the intracellular accumulation of cytotoxic compounds of the MDR spectrum. Finally, the MDR phenotype can be reversed in drug-resistant cells overexpressing P-gp by a heterogeneous group of compounds that include verapamil (channel blocker: Tsuruo *et al.*, 1981), trifluoperazin (calmodulin inhibitor: Tsuruo *et al.*, 1982), cyclosporin A (immunosuppressor: Slater *et al.*, 1986), quinidine (Tsuruo *et al.*, 1984), reserpine (Pearce *et al.*, 1989), and several others (reviewed in Beck, 1990). Although the mechanism of MDR reversal by these compounds is not yet fully understood, it appears that some of them behave as competitive inhibitors for drug binding sites on P-gp (Cornwell *et al.*, 1987b; Safa *et al.*, 1987; Akiyama *et al.*, 1988; Safa, 1988; Kamiwatari *et al.*, 1989; Pearce *et al.*,

1989; Tamai and Safa, 1990). Some of these compounds have shown promise, *in vivo,* for the treatment of drug-resistant tumors (Dalton *et al.*, 1989) and several (verapamil, cyclosporin) are in clinical trials.

The mechanism by which P-gp can apparently recognize and prevent the intracellular accumulation of a large group of structurally and functionally unrelated compounds remains mysterious. In particular, P-gp domains and discrete amino acid residues implicated in substrate recognition and transport have yet to be identified. In addition, the mechanism(s) by which so-called "reversal" agents can apparently inhibit P-gp function or bypass its action is only partly elucidated. These questions are only but a few that need to be answered in order to clarify the functional role of P-gp in drug-resistant tumor cells, and to design experimental protocols and compounds susceptible to bypass its action. Finally, the parallel study of the physiological role and substrates of P-gp in normal differentiated cell types should also prove critical in achieving these goals. Over the past 5 years, the successful isolation of molecular clones encoding several P-gp isoforms has shed considerable light on the proposed structure of this group of proteins and has provided new analytical tools for the study of its function. In addition, molecular cloning experiments have shown that P-gps are part of a large group of structurally and functionally related membrane proteins that have been highly conserved during evolution, from bacterial periplasmic transport proteins to the transporter of mating pheromone in yeast cells, the chloroquine efflux pump of the malarial parasite, peptide pumps in lymphocytes, and chloride channels in airway epithelial cells. This review will be limited to a summary of current work from our own laboratory and that of others addressing the structure/function analysis of P-gp, using a molecular genetic approach. The reader is referred to excellent reviews and books for more complete discussions of other related aspects of MDR or P-gp, including the isolation and biochemical characterization of MDR cell lines, the pharmacology of multidrug resistance and drug transport, and the clinical implications of P-gp expression in human tumors (Bradley *et al.*, 1988; Endicott and Ling, 1989; Roninson, 1991; Nooter and Herweijer, 1991).

II. Cloned *mdr* Genes and Predicted P-gp Polypeptides

Molecular clones corresponding to the human, mouse, and hamster P-gps have been independently obtained by different experimental approaches, including cloning of genomic DNA commonly amplified in independently derived MDR cells (Roninson *et al.*, 1984; Gros *et al.*, 1986a), screening of cDNA expression libraries with specific anti-Pgp antibodies (Kartner *et al.*, 1985), and the isolation of cDNA clones corresponding to genes overexpressed in MDR cells (Van der Bliek *et al.*, 1986). These experiments have shown that Pgps are encoded by a small group of closely related genes, termed *mdr*,and full-length cDNA clones corresponding to the cellular transcripts of two human *MDR* genes *MDR1*, *MDR2* (C. J. Chen *et al.*, 1986; Van der Bliek *et al.*, 1987) and three rodent *mdr* genes (*mdr1*, *mdr2*, *mdr3*) have been isolated and sequenced from mouse (Gros *et al.*, 1986b, 1988; Devault and Gros, 1990; Hsu *et al.*, 1990) and hamster (Ng *et al.*, 1989). Nucleotide and predicted amino acid sequence analyses indicate that these genes encode highly homologous P-gps. The prototype P-gp is composed of 1276 amino acids (mouse) and can be subdivided into two halves, each half encoding a large hydrophilic domain and a large hydrophobic domain. Each hydrophobic region contains six highly hydrophobic segments characteristic of membrane-associated domains, which can be arranged into three transmembrane (TM) loops (Fig. 1). Each hydrophilic segment contains a classical consensus sequence for nucleotide binding (NB site) that has been described in a variety of ATP binding proteins and kinases (Higgins *et al.*, 1986). These sequence analyses are in agreement with previous biochemical characterizations of P-gp and suggest that P-gp is a membrane glycoprotein capable of binding ATP. Considering that a cluster of predicted N-linked glycosylation sites would be located between the first and the second TM domain (likely to be extracellular), and since the predicted NB sites are likely to be located on the intracellular side of the membrane, a model for the membrane arrangement of P-gp has been proposed (Fig. 1) (Gros *et al.*, 1986b; C. J. Chen *et al.*, 1986; Gottesman and Pastan, 1988). This proposed membrane organization of P-gp is in agreement with the results of epitope mapping using specific antibodies directed against synthetic P-gp peptides of known sequence, or against fusion proteins containing P-gp subfragments (Bruggemann *et al.*, 1989; Yoshimura *et al.*, 1989). The two halves of P-gp share up to 38% identical residues (62% overall homology), and it has been proposed that at least part of the protein evolved from the duplication of an ancestral unit. The conservation of the position of specific introns within each of the two predicted NB sites (Raymond and Gros, 1989; Chen *et al.*, 1990) sustains this contention. Sequence homology between the two halves of individual P-gps is highest within the predicted NB domains, decreases within the predicted TM domains, and disappears in the extreme 5′ end of each half, which corresponds to the amino terminus of the protein and the highly charged so-called "linker" domain which links the two halves together. Surprisingly, P-gp segments overlapping the NB sites were found to be homologous (30%) to a group of bacterial periplasmic transport proteins such as *HisP* (Higgins *et al.*, 1982), *MalK* (Gilson *et al.*, 1982), *OppD* (Ames, 1986), and *PstB* (Surin *et al.*, 1985) implicated in the

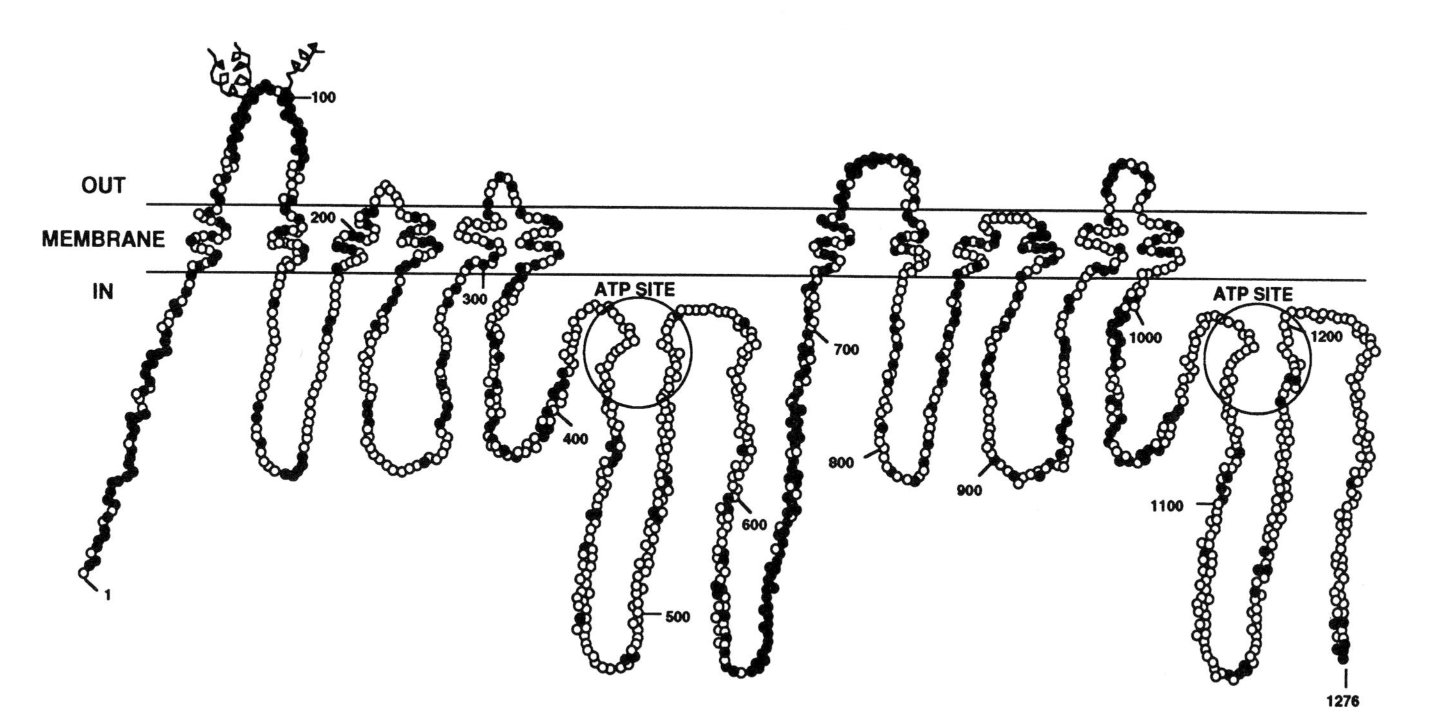

FIG. 1 Schematic representation of P-glycoproteins encoded by the mouse *mdr* gene family. The proposed membrane topology of P-glycoprotein is shown. The position of the predicted nucleotide binding folds (ATP site) is indicated. Open circles indicate either identical residues or conserved substitutions among *mdr1*, *mdr2*, and *mdr3*. Filled circles represent nonconservative substitutions among any of the three sequences. This drawing is a modification of that presented by Gottesman and Pastan (1988).

high-affinity transmembrane transport of histidine, maltose, oligopeptides, and phosphates into *Escherichia coli*. An even stronger homology (50%) is detected with a second group of bacterial proteins such as *HlyB* (Felmlee *et al.*, 1985), *ChvA* (Cangelosi *et al.*, 1989), *CyaB* (Glaser *et al.*, 1988), *ArsA* (C. M. Chen *et al.*, 1986), and *LktB* (Stanfield *et al.*, 1988), implicated in the ATP-dependent export of hemolysin, toxins, and arsenate in Gram-negative bacteria. In the case of *HlyB*, the homology to P-gp overlaps not only the NB sites but also involves the TM domains, and both proteins share almost superimposable hydropathy profiles (Gros *et al.*, 1988). A partial listing of bacterial proteins sharing sequence homology and predicted secondary structure with P-gp is presented in Table I. The predicted structural features of P-gp as well as its remarkable sequence homology to prokaryotic transport proteins implicated in extracellular export of specific substrates are also in agreement with the proposition that P-gp functions as an ATP-dependent efflux pump.

In humans and rodents, the gradient of sequence conservation between individual members of the *mdr* gene family is similar to that detected between the two homologous halves of each P-gp; e.g., it is highest within the NB sites (over 95% homology), decreases in the TM regions (70% homology), and is lowest in the amino terminus and linker domains of the proteins (20% homology) (Fig. 1). In rodents, sequence analyses indicate that the three genes arose from a common ancestor by two successive gene duplication events, the most recent one producing *mdr1* and *mdr3*, which are more homologous to each other than they are to *mdr2* (Devault and Gros, 1990). In humans, a single gene duplication event would have generated the two genes, *MDR1* and *MDR2* (Chin *et al.*, 1989; Ng *et al.*, 1989).

III. The *mdr* Super Gene Family

The *mdr* gene family is itself part of a larger supergene family of sequence-related members whose structural and functional aspects have been preserved throughout evolution. Over the past 2 years, *mdr* homologs and *mdr*-like genes from this family have been identified in a number of distantly related organisms and found associated with some remarkable physiological events. The amino acid sequence homology between *mdr* genes and these *mdr*-like homologs is modest (20 to 40%) and is clustered mostly around the predicted NB sites, and at a lower level within the membrane-associated domains of the encoded proteins. However, the predicted polypeptides have very similar length and identical hydropathy profiles and, consequently, share highly similar predicted secondary structures. In gen-

TABLE I

Proteins That Share Sequence Homology with P-gp (*mdr* and *mdr*- Homologs)

Protein	Organism	Substrate	Reference
Import (prokaryotes)			
AraG	*E. coli*	Arabinose	Scripture *et al.* (1987)
BtuD	*E. coli*	Vitamin B12	Friedrich *et al.* (1986)
ChlD	*E. coli*	Molybdate	Johann and Hinton (1987)
FecE	*E. coli*	Fe hydroxamate	Staudenmaier *et al.* (1989)
FhuC	*E. coli*	Fe citrate	Coulton *et al.* (1987)
GlnQ	*E. coli*	Glutamine	Nohno *et al.* (1986)
HisP	*S. typhimurium*	Histidine	Higgins *et al.* (1982)
MalK	*E. coli*	Maltose	Gilson *et al.* (1982)
OppD	*S. typhimurium*	Peptides	Ames (1986)
OppF	*S. typhimurium*	Peptides	Ames (1986)
PstB	*E. coli*	Phosphate	Surin *et al.* (1985)
RbsA	*E. coli*	Ribose	Ames (1986)
Export (prokaryotes)			
ArsA	*E. coli*	Arsenate	C. M. Chen *et al.* (1986)
HlyB	*E. coli*	Hemolysin	Felmlee *et al.* (1985)
CyaB	*B. pertussis*	Hemolysin	Glaser *et al.* (1988)
LktB	*R. meliloti*	Cyclic sugars	Stanfield *et al.* (1988)
ChvA	*A. tumefaciens*	Cyclic sugars	Cangelosi *et al.* (1989)
DrrA	*S. peucetius*	Anthracyclines	Guifoile and Hutchison (1991)
Others (prokaryotes)			
FtsE	*E. coli*	Cell division	Higgens *et al.* (1986)
NodI	*R. leguminosarum*	Nodulation	Higgins *et al.* (1986)
UvrA	*E. coli*	DNA repair	Doolittle *et al.* (1986)
Eukaryotes			
White	*D. melanogaster*	Pteridine	O'Hare *et al.* (1984)
Brown	*D. melanogaster*	Pteridine	Dreesen *et al.* (1988)
Mdr (49, 65)	*D. melanogster*	?	Wu *et al.* (1991)
Cepgp (1,2,3)	*C. elegans*	?	P. Borst and C. Lincke, personal communication
Ltpgp (A,B,C,D,E)	*L. tarentolae*	Multiple drugs	Ouellette *et al.* (1990)
Pfmdr (1,2)	*P. falciparum*	Chloroquine	Foote *et al.* (1989)
Ehpgp (1,2,3)	*E. histolytica*	Emetine	Samuelson *et al.* 1990
STE-6	*S. cerevisiae*	Pheromone	McGrath *et al.* (1989)
MpbX	*M. polymorpha*	?	Ohyama *et al.* (1986)
Mammals			
MDR (1,2)	Human	Multiple drugs	C. J. Chen *et al.* (1986)
mdr (1,2,3)	Mouse	Multiple drugs	Gros *et al.* (1986c)

continues

TABLE I *Continued*

Protein	Organism	Substrate	Reference
CFTR	Human	Halides	Riordan *et al.* (1989)
PMP70	Human	?	Kamijo *et al.* (1990)
Ham (1,2)	Mouse	Peptides	Monaco *et al.* (1990)
RING (4,11)[a]	Human	Peptides	Trousdale *et al.*, 1990; Spies *et al.*, 1990; Glynne *et al.*, 1991
mtp (1,2)	Rat	Peptides	Deverson *et al.*, 1990; Powis *et al.*, 1991

[a] RING genes have recently been renamed TAP (1,2) (Bodmer *et al.*, 1992).

eral, these genes have been isolated by the polymerase chain reaction technique, using synthetic oligonucleotides corresponding the highly conserved NB sites or using specific antibodies directed against these or other P-gp sites. In the yeast *Saccharomyces cerevisiae*, an *mdr* homolog called *STE-6* has been discovered (McGrath and Varshavsky, 1989) and found responsible for the transport of the mating pheromone *a* factor across the membrane of *a* cells (Kuchler *et al.*, 1989). In the malarial parasite *Plasmodium falciparum*, the widespread emergence of resistance to chloroquine (CLQ) is a major health threat to human populations of the third world. Choroquine resistance in *P. falciparum* is caused by an increase in drug efflux from the parasite that can be competed by CLQ analogs, but surprisingly, also by vinblastine and verapamil, two inhibitors of mammalian P-gp (Martin *et al.*, 1987; Krogstad *et al.*, 1987; Bitoni *et al.*, 1988). An *mdr* homolog called *pfmdr 1* has been cloned from *P. falciparum* and has been found amplified and/or overexpressed and/or mutated in CLQ-resistant isolates of the parasite (Wilson *et al.*, 1989; Foote *et al.*, 1989). Likewise, in the kinetoplastidae *Leishmania tarentolae*, a group of five *mdr* homologs designated *ltpgp* has been isolated, with some members of this family found amplified and overexpressed in arsenate and methotrexate-resistant isolates of this parasite (Elenberger and Beverley, 1989; Ouellette *et al.*, 1990). Also, the emergence of emetine resistance in the enteric protozoan *Entamoeba histolytica* has been found associated with the overexpression of at least three *mdr*-related mRNA species (Samuelson *et al.*, 1990). Finally, three *mdr* homologs have been cloned and sequenced from the nematode *Caenorhabditis elegans* (P. Borst and C. Lincke, personal communication), and two *mdr* homologs have been cloned form the fly *Drosophila melanogaster* (Wu *et al.*, 1991). The function of these *mdr* homologs has not been elucidated in these organisms, although their overexpression appears to confer resistance to emetine and colchicine, respectively. Since the creation of null alleles by homologous recombination is relatively easy to achieve in these organisms, these experimental systems may prove useful to elucidate the function of these *mdr* homologs in the normal physiology of these lower eukaryotes. Finally, a chloroplast gene of yet unknown function but encoding a protein sharing sequence homology with the bacterial MalK and HisP proteins has been identified in *Marchantia polymorpha* (Ohyama *et al.*, 1986).

In humans, at least three *mdr*-like genes have been recently cloned and shown to participate in key physiological events. Cystic fibrosis (CF) is an inherited disorder characterized by a defect in chloride transport, and for which a candidate gene designated CFTR (for cystic fibrosis transmembrane conductance regulator) has been isolated by reverse genetics (Riordan *et al.*, 1989). The CFTR gene is an *mdr* homolog that is mutated in CF patients, with the vast majority of mutations found within the predicted NB sites of CFTR (Kerem *et al.*, 1989; Cutting *et al.*, 1990). Direct transfection experiments of wild-type or mutant CFTR cDNAs in cells otherwise not expressing this gene very strongly suggest that CFTR is indeed the membrane-bound chloride channel that is defective in CF patients (Rich *et al.*, 1990; Drumm *et al.*, 1990; Anderson *et al.*, 1991). The identification in CF patients of several inactivating, naturally occuring mutations in the CFTR gene has pointed out key functional residues in the protein and facilitated its structure–function analysis in *in vitro* transfection systems (Rich *et al.*, 1991; Tabcharani *et al.*, 1991; Cheng *et al.*, 1991). A pair of closely related *mdr* homologs have been discovered within the major histocompatibility complex (MHC) of both humans and rodents. The chromosomal region carrying these two genes is either delated or rearranged in T-cell lines that fail to properly present antigens to their surface in association with the class I MHC antigens. The two genes, designated *HAM* (Monaco *et al.*, 1990), *mtp* (Deverson *et al.*, 1990), *RING* (Trowsdale *et al.*, 1990), and *PSF* (Spies *et al.*, 1990), encode for proteins exactly half the size of P-gp and containing only six TM domains and one NB sites. It has been proposed that these two proteins work in concert to transport short soluble peptides from proteolyzed antigens across the reticulum endoplasmic membrane, and eventually to the cell surface in association with class I MHC antigens for presentation to other cells of the immune system (Spies and De Mars, 1991). Finally a protein sharing sequence homology with P-gp has been identified in the membrane of peroxysomes (Kamijo *et al.*, 1990). This protein, termed PMP70, has an apparent molecular mass of 70,000 and shows a structure similar to that of the *HAM* gene products and half the size of P-gp. The function of this protein is still uncertain, but it has been proposed to be part of the import machinery of peroxysomes in general, and possibly an acyl CoA carrier, in particular (Kamijo *et al.*, 1990).

Taken together, these observations indicate that the *mdr* super gene family is composed of an increasing number of genes encoding proteins in which the basic structural unit is conserved (at least at the predicted secondary structure level), and that carry out membrane-associated transport of small lipophilic compounds (chemotherapeutic drugs), peptides (mating pheromone, antigenic peptides), and even ions (chloride channel). Identifying the protein domains implicated in (*a*) the common mechanism of transport and (*b*) the recognition of specific substrates by these transporters are important research goals in the study of this gene family.

IV. *mdr* Gene Expression in Multidrug-Resistant Cells

To determine the specific contribution of individual members of the *mdr* gene family to the emergence of MDR, the level of expression of the three mouse and two human genes has been measured at the RNA and protein levels in MDR cell lines of either species. We have used gene-specific hybridization probes to monitor the degree of amplification and the levels of mRNA expression of the three mouse *mdr* genes in 12 independently derived multidrug-resistant cell lines (Raymond *et al.*, 1990). These gene-specific probes were derived from the short linker domain, which shows minimum sequence homology between the three mouse P-gps. Ribonucleic acid blotting experiments show that the *mdr1* gene encodes a unique mRNA of 5 kb in size, whereas the *mdr2* gene produces a unique 4.5-kb transcript and the *mdr3* gene encodes two mRNA species of 5 and 6 kb in size, which differ in their 3′ untranslated region through the alternate use of distinct polyadenylation signals (Croop *et al.*, 1989; Hsu *et al.*, 1990). Long-range physical mapping by pulse field gel electrophoresis indicate that the three genes are contiguous on a genomic DNA fragment of approximately 625 kb in size (Raymond *et al.*, 1990) located on mouse chromosome 5 (Martinsson and Levan, 1987). The gene order and direction of transcription of the three genes have been tentatively assigned to be (5′) *mdr3* (3′)–(5′) *mdr1* (3′)–(3′) *mdr2* (5′). By comparison, the two human *MDR* genes, *MDR1* and *MDR2*, are contiguous on a 230-kb fragment of human chromosome 7 (Chin *et al.*, 1989), with both genes in the same transcriptional orientation (Lincke *et al.*, 1991). Southern blotting analyses of genomic DNAs from a panel of independently derived multidrug-resistant mouse cell lines identified *mdr* gene amplification in 10 of the 12 lines tested. In individual MDR cell lines showing gene amplification, the copy number of each *mdr* gene was found to be identical, suggesting that the three genes are amplified together as part of a single amplicon. Multidrug resistance in these cells was linked to the independent overexpression of either *mdr1* or *mdr3*, but not *mdr2*, suggesting that *mdr1* and *mdr3* but not *mdr2* can independently confer multidrug resistance to cultured cells (Raymond, *et al.*, 1990). The independent analysis of series of multidrug-resistant derivatives of the mouse macrophage line J774 produced identical results (Hsu *et al.*, 1989). Interestingly, increased *mdr1* expression was always linked to gene amplification, whereas *mdr3* overexpression was detected in the absence of *mdr3* gene amplification. In addition, amplification of a particular *mdr* gene did not necessarily correlate with its level of expression. This discrepancy was linked neither to the degree of drug resistance nor the drug used for *in vitro* selection, but was linked to the specific tissue type of the cell. These observations suggest that *mdr1* and *mdr3* gene expression is controlled by tissue-specific factors that may show nonoverlapping distributions in the cell lines tested. Indeed, recent studies of the promotor region of the *mdr1* gene reveals that its expression is strictly controlled in normal tissues by the activity of cell-specific cellular factors (Raymond and Gros, 1990). In humans the situation is very similar to that observed in the mouse model: both *MDR1* and *MDR2* are found amplified in MDR cells, whereas only *MDR1* is overexpressed at detectable levels (Chin *et al.*, 1989), suggesting that *MDR1* but not *MDR2* can convey multidrug resistance.

V. *mdr* Gene Expression in Normal Tissues

Although the precise functional role of *mdr* genes in normal physiological events remains unknown, the organ and cellular distribution of *mdr* mRNA transcripts and polypeptides appears to be tightly regulated in a tissue- and cell-specific fashi n. In addition, subcellular distribution of the discrete P-gp isoforms also appears to be narrowly restricted to the plasma membrane, in particular in polarized epithelia. These have been measured at the RNA level by Northern or slot blotting (Fojo *et al.*, 1987; Croop *et al.*, 1989) or by *in situ* hybridization using RNA probes (Arceci *et al.*, 1988), or by the polymerase chain reaction (Chin *et al.*, 1989). Specific polypeptides encoded by individual members of the *mdr* gene family have been detected by Western blotting or by immunohistochemistry using generic or isoform-specific anti-Pgp antibodies (Georges *et al.*, 1990; Buschman *et al.*, 1992). This type of analysis has been carried out for both the human and the rodent genes (hamster, mouse) and has shown that the tissue-specific distribution of human *MDR1* overlaps that of mouse *mdr1* and *mdr3*, whereas human *MDR2* and mouse *mdr2* share similar tissue distribution. The subcellular distribution of discrete P-gp isoforms in each respective tissues of both species is very similar. In the case of human tissues, RNA analysis indicates that *MDR1* is expressed at highest levels in the adrenal glands, kidneys, jejunum, and colon (Fojo *et al.*, 1987), whereas *MDR2* is expressed mostly in the liver (Chin *et al.*, 1989). In normal mouse tissues, *mdr1* is expressed at high levels in the pregnant uterus and adrenals, whereas lower levels are found in the kidneys and placenta (Croop *et al.*, 1989; Arceci *et al.*, 1988). *mdr2* is expressed almost exclusively in the liver, and *mdr3* is expressed in the intestine and detected at lower levels in the heart and brain (Croop *et al.*, 1989). Immunohistochemical localization of P-gp in epithelial cells with a polarized secretory function has shown that it is expressed on the apical surface of cells opposed to a lumen: it is found on the canalicular face of hepatocytes; the apical surface of biliary ductules; and the apical surface of columnar epithelial cells of the intestine, brush border of renal proximal tubules, and pancreatic ductule cells (Thiebaut *et al.*, 1987; Bradley *et al.*, 1990; Georges *et al.*, 1990). P-glycoprotein is also expressed on endothelial cells of the blood–brain barrier (Cordon-Cardo *et al.*, 1989; Thiebaut *et al.*, 1989). In the adrenal glands, P-gp is diffusely distributed on the cell surface in the zona glomerulosa and fasciculata of the cortex, with lower expression on the cell surface of all medullary cells (Sugarawa *et al.*, 1988). Finally, P-gp (*mdr1*) is expressed at very high levels on the luminal surface of glandular epithelial cells of the endometrium, late during pregnancy (Arceci *et al.*, 1988). The role of P-gp in this tissue is unknown but its expression is tightly regulated by estrogen and progesterone (Arceci *et al.*, 1990). Finally, a recent study has detected high levels of P-gp (*MDR1*) expression in the hematopoietic system, in particular in a population of proposed pluripotent stem cells of the bone marrow (Chaudhary and Roninson, 1991). Overall, the precise substrates of P-gp in normal tissues remain to be identified: It has been proposed that it could act as a normal detoxifying mechanism that would protect the normal environment against toxic insults by xenobiotics and that would simply become overexpressed in drug-resistant tumor cells. The high levels of expression and specific subcellular localization of P-gp in the intestine, kidney, endothelial cells of the blood–brain barrier, and, in particular, pluripotent precursor stem cells of the bone marrow would certainly argue in favor of this proposal. Others have proposed that chemotherapeutic drugs really represent fraudulent substrates for P-gp that would under normal physiological conditions transport separate classes of yet to be identified substrates. P-glycoprotein expression detected in adrenals and pregnant uterus together with the observation that P-gp can directly bind (Qian and Beck, 1990) and interact with (Arceci *et al.*, 1990; Yang *et al.*, 1990) progesterone would favor the later proposal.

VI. Functional Analysis of *mdr* Genes in Transfected Cells

The functional role of *mdr* genes in the establishment of multidrug resistance has been addressed in cell clones transfected and stably expressing individual members of the mouse (Gros *et al.*, 1986c) and human (Ueda *et al.*, 1987) *mdr* family. These studies have also provided a starting point and general model system to carry out functional analysis of wild type, chimeric, and mutant P-gp. The mammalian expression vector we have used in our laboratory to overexpress individual *mdr* genes (p91023b, pMT2) utilizes strong viral promotor/enhancer sequences located in the early regions of SV40 and adenovirus, to direct high levels of expression of cloned cDNAs (Gros *et al.*, 1986c; Devault and Gros, 1990). Retroviral-based constructs have also been used by us and others to express the mouse and human *MDR* genes, respectively (Ueda *et al.*, 1987; Guild *et al.*, 1988). The drug-sensitive cells we have used as recipients in our transfection assays were Chinese hamster LR73 ovary cells, which show undetectable levels of expression of endogenous P-gp. Full-length cDNAs for individual *mdr* genes were introduced directly in these cells or after cotransfection with a dominant selectable marker such as the Tn5 (*neo*) gene (Southern and Berg, 1982), followed by selection in genetycin. In this assay, the mouse *mdr1* cDNA can directly confer multidrug resistance to LR73 cells, as drug-resistant colonies can be directly selected from transfected cells in culture medium containing drugs of the MDR spectrum (Gros *et al.*, 1986c). Transfection and overexpression of *mdr1* in these cells confer readily detectable levels of resistance to adriamycin, colchicine, vinblastine, actinomycin D, and gramicidin D, but not gramicidin S or bleomycin, two non-MDR drugs (Fig. 2) (Devault and Gros, 1990). Drug-resistant *mdr1* transfectants express a 180- to 200-kDa membrane glycoprotein that is highly phosphorylated and capable of combining photoactivatable analogs of ATP and known P-gp substrates (Schurr *et al.*, 1989). The binding of ATP is specific and restricted to two discrete tryptic peptides detected by two-dimensional eletrophoresis and chromatographic analysis (Schurr *et al.*, 1989). The emergence of multidrug resistance in these *mdr1* transfectants is linked to a decreased intracellular drug accumulation and an increased drug efflux, both of which are dependent on the presence of intact intracellular nucleotide triphosphate pools, and can be inhibited by combined treatment with deoxyglucose and rotenone (Hammond *et al.*, 1989). By contrast, transfection and overexpression of the mouse *mdr2* cDNA fail to confer an MDR phenotype (Gros *et al.*, 1988). Clones stably transfected with *mdr2* overexpress a 160 kDa glycoprotein that is enriched in the membrane fraction (E. Buschman and P. Gros, unpublished observations). However, *mdr2* overexpression fails to confer any detectable increase in resistance of LR73 cells to classical MDR drugs such as those listed in Fig. 2, but also to non-MDR drugs such as 5-fluorouracil, *cis*-platinum, cyclophosphamide, or chlorambucil (P. Gros, unpublished observations). Finally, a full-length cDNA for the *mdr3* gene has also been shown to be capable of conferring MDR to LR73 cells. Interestingly, the apparent MDR phenotype conferred by *mdr3* seemed to

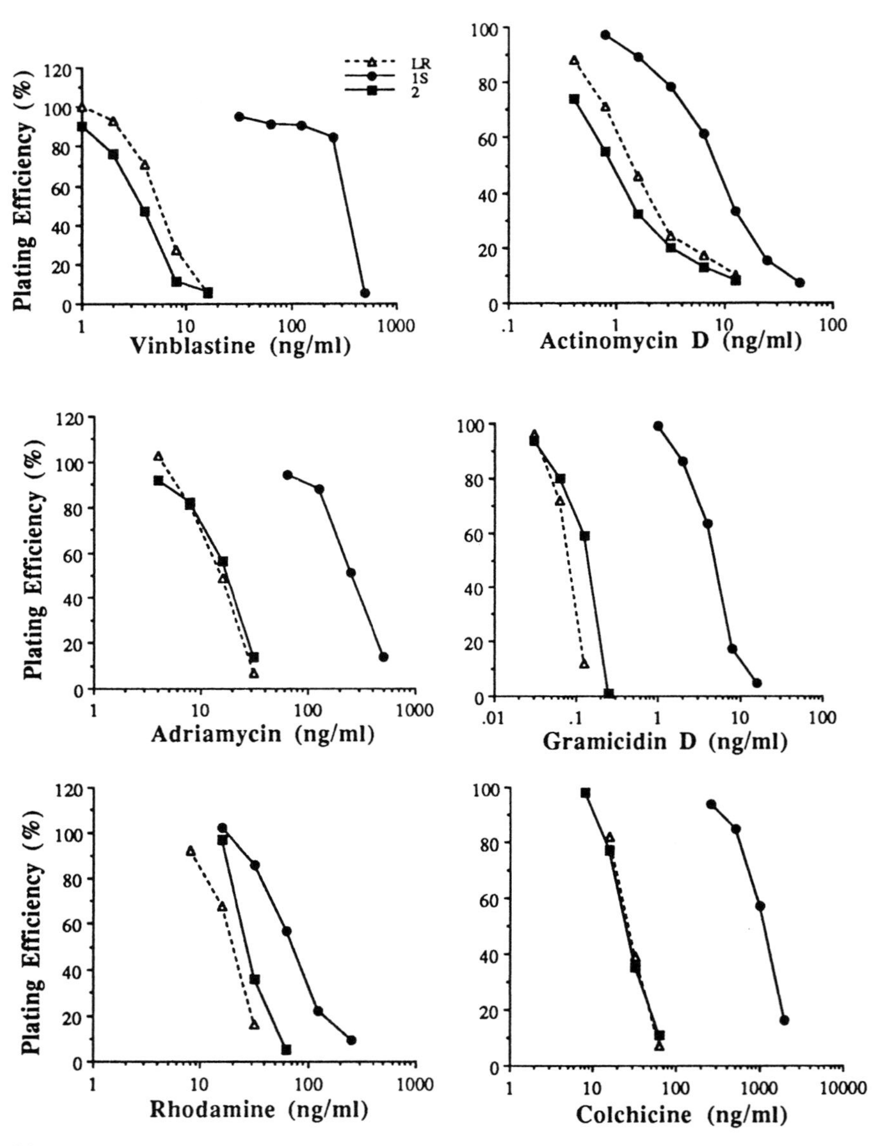

FIG. 2 Drug survival characteristics of Chinese hamster LR73 cells (LR) and cell clones stably transfected and overexpressing P-glycoproteins encoded by either mouse *mdr1* (1S) or *mdr2* (2) cells. Control and transfected cells (10^4 cells/well) were plated in increasing concentrations of cytotoxic drugs and incubated for 72 hr. Cells were fixed and stained with sulforhodamine B, and the staining was quantitated by spectrophotometry. The plating efficiency of individual clones was calculated by dividing the optical density value measured at a given drug concentration by that measured in control wells lacking drug, and is expressed as a percentage. Each point represents the average of dupiicate measurements.

be qualitatively distinct from that encoded by the other biologically active *mdr* gene, *mdr3:* In independent cell clones expressing similar amounts of both polypeptides, it appeared that although both cDNAs could confer similar degrees of resistance to vinblastine, *mdr1* conferred levels of colchicine and adriamycin resistance 10-fold greater than those conveyed by *mdr3*. In the case of actinomycin D the situation was reversed, as *mdr3* conveyed higher levels of resistance to this drug that did *mdr1* (Devault and Gros, 1990). Taken together, these results clearly indicate strong functional differences (*mdr1/3* vs *mdr2*) between individual members of the mouse *mdr* gene family, despite considerable sequence homologies. In addition, our results indicate that although *mdr1* and *mdr3* are clearly capable of conferring MDR, sequence divergence has produced two drug efflux pumps showing overlapping but distinct substrate specificities.

Similar findings have been obtained from the independent analysis of the human *MDR* gene family. Although the *MDR1* gene can confer multidrug resistance (Ueda *et al.*, 1987), transfection and overexpression of the protein encoded by the *MDR2* gene fail to alter the drug-sensitive phenotype of recipient human melanoma BRO cells (Schinkel *et al.*, 1991).

VII. Analysis of Chimeric Genes Obtained by Exchanging Homologous Domains of *mdr1* and *mdr2*

To initiate a structure/function analysis of the mouse *mdr* gene family, we have taken advantage of the strong sequence homology and striking functional differences detected in transfection experiments between *mdr1* and *mdr2* to identify the domains of *mdr1* that are essential for multidrug resistance and that may be functionally distinct in *mdr2*. We have constructed chimeric cDNA molecules in which discrete domains of *mdr2* have been introduced into the homologous region of *mdr1* and analyzed these chimeras for their capacity to transfer multidrug resistance (Fig. 3) (Buschman and Gros, 1991). For this, we used restriction enzyme sites that were conserved between the two cDNAs or we introduced new sites by site-directed mutagenesis at the homologous position in both clones. The two predicted NB sites of *mdr2* share almost complete sequence identity with those of *mdr1*. Both *mdr2* NB sites were found to be functional, since either could independently complement the biological activity of *mdr1*. Thus, it appears that the NB domains of *mdr2* are functional and it is likely that these sites are involved in the common mechanism of action of the two proteins, most likely the energy coupling component of the transport system. Likewise, a chimeric molecule in which the highly se-

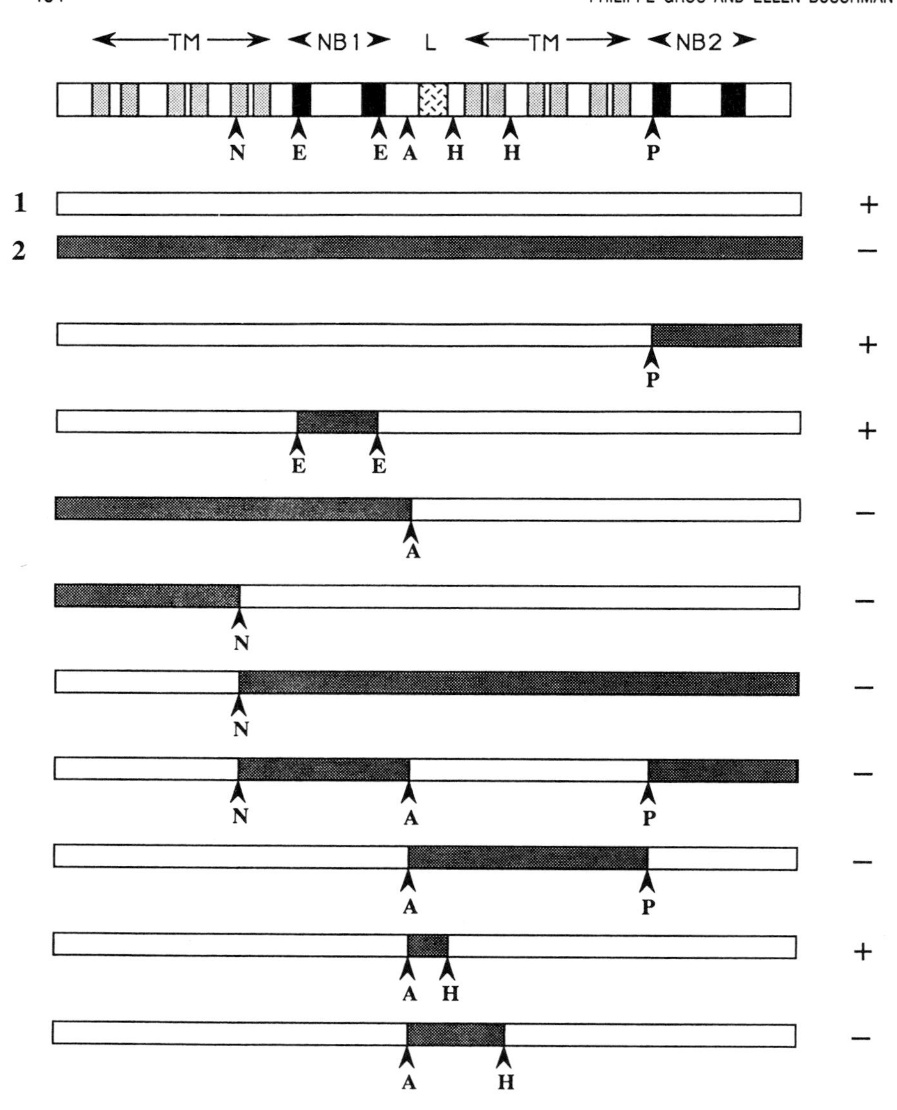

FIG. 3 Functional analysis of *mdr1/mdr2* chimeric cDNA clones. Schematic representation of wild-type *mdr* cDNA and some of the predicted structural features of the protein, including proposed transmembrane domains (TM), nucleotide binding folds 1 and 2 (NB1,NB2), and linker domain (L). Restriction enzyme sites (N,*Nsi*I;E,*Eco*0109;A,*Acc*I;H,*Hin*dIII;P,*Pst*I) used to construct the chimeras are identified by arrows below the cDNA. The wild-type *mdr1* (open box), *mdr2* (shaded box), and chimeric constructs are shown, along with their phenotypes in transfection experiments, either positive (+) or negative (-) for drug resistance.

quence divergent linker domain of *mdr2* had been introduced at the homologous position of *mdr1* could also confer resistance to adriamycin and colchicine. It is interesting to note that this linker domain contains in *mdr1* the only two consensus sites for cyclic AMP-dependent protein kinase A found in *mdr1*. Since these sites are not conserved in *mdr2*, it is unlikely that they play a key mechanistic or regulatory role in the drug efflux function of *mdr1*.However, the replacement of either the amino- or the carboxy-terminus membrane-associated domains (transmembrane domain, intra- and extracellular loops) of *mdr1* by the homologous segments of *mdr2* resulted in inactive chimeras. The replacement of as few as two TM domains from either the amino (TM5-6) or the carboxy (TM7-8) half of *mdr1* by the homologous *mdr2* region was sufficient to destroy the activity of *mdr1*. These results suggest that the TM regions of *mdr1* are essential for its capacity to confer multidrug resistance and are functionally distinct in *mdr2*. The data also suggest that the functional differences detected between the two proteins are not encoded by a single TM segment but rather involve protein domains present in both homologous halves of each protein. Finally, the observation that *mdr2* differs from *mdr1* and *mdr3* only be a few nonconserved residues within the TM5-6 and TM7-8 intervals suggests that functional differences detected between *mdr1* and *mdr2* in these regions may be encoded by a very small number of amino acid residues. Taken together, these findings are compatible with the proposal that *mdr1* and *mdr2* encode for membrane associated transport proteins that act by the same mechanism on perhaps distinct sets of substrates. It is interesting to speculate that NB domains represent functional sites common to both transport systems and that TM domains may be involved in specific substrate recognition and transport. This hypothetical model has been tested in experiments described in the two following sections.

VIII. Mutational Analysis of the Predicted Nucleotide Binding Domains of *mdr1*

The capacity of *mdr1* and *mdr3* to prevent intracellular accumulation of cytotoxic drugs in stably transfected cells is strictly ATP dependent. Analysis of the predicted amino acid sequence of P-gps encoded by these two genes reveal the presence of two putative nucleotide binding folds, originally compiled by Walker *et al*. (1982) for a number of ATP binding proteins and ATPases, which are highly conserved in *mdr1* and *mdr3* but also in *mdr2* and in other members of the *mdr* super gene family such as *STE-6*, *pfmdr*, and *CFTR*. Each predicted NB site is formed by two consensus motifs, a G–(X)4–G–K–(T)–(X)6–I/N segment at positions 419 and 1061 (known as motif A), and a hydrophobic pocket (known as motif B) of sequence R/K–(X)3–G–(X)3–L–(hydrophobic)4–D, located 100 residues downstream of motif A at positions 542 and 1186. The A motif is believed to form a flexible loop between a β strand and an α helix. Through conformational changes induced by ATP binding, this loop would control access to the nucleotide binding domain (Fry *et al.*, 1986). The conserved lysine residue is thought to interact directly with one of the phosphate groups and would be essential for ATPase activity. On the other hand, the B motif is thought to be close to the glycine-rich flexible loop, and to form a homing pocket for the adenine moiety of the ATP molecule, possibly conferring the nucleotide specificity to the binding site. The aspartate residue is believed to interact directly with the magnesium atom of the nucleotide triphosphate complex. Theoretical three-dimensional models for this ATP binding cassette have been independently proposed (Hyde *et al.*, 1990; Mimura *et al.*, 1991). To assess the functional importance of these NB sites in the drug resistance phenotype encoded by *mdr* genes, we have introduced discrete amino acid sbustitutions within the core consensus sequence for nucleotide binding, GX<u>GK</u>ST (motif A). Mutants bearing the sequence GX<u>A</u>KST or GXG<u>R</u>ST at either of the two NB sites of *mdr1* and a double mutant harboring the sequence GXG<u>R</u>ST at both NB sites were generated (Azzaria *et al.*, 1989). The integrity of the two NB sites was essential for the biological activity of *mdr1,* since all five mutants were, unlike the wild-type *mdr1*, unable to confer drug resistance to drug-sensitive LR73 hamster cells in transfection experiments. Conversely, a lysine-to-arginine substitution outside the core consensus sequence (position 1100) had no effect on the activity of *mdr1*. The loss of activity detected in cell clones stably expressing individual *mdr1* mutants was paralleled by a failure to reduce the intracellular accumulation of radiolabeled vinblastine. However, the ability to combine the photoactivatable analog of ATP, 8-azido ATP, was retained in the five inactive *mdr1* mutants, suggesting that a step subsequent to ATP binding, possibly ATP hydrolysis, was impaired in the *mdr1* mutants (Azzaria *et al.*, 1989). We have observed that the mutant *mdr1* carrying a lysine-to-arginine substitution in both NB sites has indeed lost its ATPase activity (F. Sharom and P. Gros, unpublished observations). Taken together, these results indicate that both NB sites are essential for the overall expression of multidrug resistance by *mdr1*. Moreover, these two sites do not appear to function independently of each other since mutating either site completely abolishes the biological activity of *mdr1*. This suggests that the dual nature of the P-gp encoded by *mdr1* is pivotal for drug effux, not merely by duplicating the biological activity of a functional monomer but rather suggesting cooperative interactions between the two halves of the molecule. The nature of these cooperative interactions is unknown and can only be speculated upon: it could involve either concerted or sequential ATP binding and hydrolysis at both sites. Concerted hydrolysis of two ATP molecules may be required to produce a conformational change responsible for efflux; alternatively, sequential mechanisms could involve ATP hydrolysis at one site to effect efflux and at a second site to regain the original conformation.

IX. Analysis of *mdr1* and *mdr3* Mutants in Predicted Transmembrane Domain 11

The drug resistance profiles of transfected cells expressing a cloned cDNA for the *mdr3* gene isolated from a pre-B cell cDNA library (Devault and Gros, 1990) differ qualitatively and quantitatively from that of multidrug-resistant J7-V3-1 macrophages overexpressing the endogenous *mdr3* (designated *mdr1a* in that study) gene (Yang *et al.*, 1990). A comparison of the amino acid sequences predicted for *mdr3* (Devault and Gros, 1990) and *mdr1a* (Hsu *et al.*, 1990) clones identifies a single amino acid difference between the two lones, located within the predicted TM11 of the protein (a Ser-to-Phe substitution at position 939 of *mdr3*). The Ser residue at this position within TM11 is conserved in all human and rodent P-gps sequenced to date. Direct DNA sequencing from control RNA specimens from drug-sensitive and -resistant cells in that region of *mdr3* reveals that the Phe^{939} in our clone results from a polymerase error during cDNA construction (Gros *et al.*, 1991). Surprisingly, this mutation maps at the homologous position of TM11 of two nonconservative amino acid substitutions present in the mutant 7G8 allele of the *pfmdr1* gene detected in a large proportion of chloroquine-resistant isolates of *P. falciparum* (Foote *et al.*, 1990). To directly test the functional role of this Ser-to-Phe substitution on *mdr* function, site-directed mutagenesis was used to convert the Phe^{939} to Ser^{939} in *mdr3* and to replace the wild-type Ser^{941} by Phe^{941} at the homologous position of *mdr1* (Fig. 4). The biological activity of wild-type and mutant *mdr1* and *mdr3* clones was tested and compared in transfected cell clones expressing similar amounts of each protein. In *mdr3*, the Phe^{939} (*mdr3F*)-to-Ser^{939} (*mdr3S*) substitution caused a considerable and general increase in drug resistance levels conferred by this protein (Fig. 5). Interestingly, the increase in resistance was not identical for all drugs tested and was small for vinblastine (3-fold) and large for adriamycin and colchicine (15- and 30-fold, respectively). These results suggest that the mutation at position 939 alters the overall activity of *mdr3* and possibly, its substrate specificity. To te t this possibility further a Ser (*mdr1S*)-to-Phe (*mdr1F*) replacement was introduced at the homologous position (941) in TM11 of

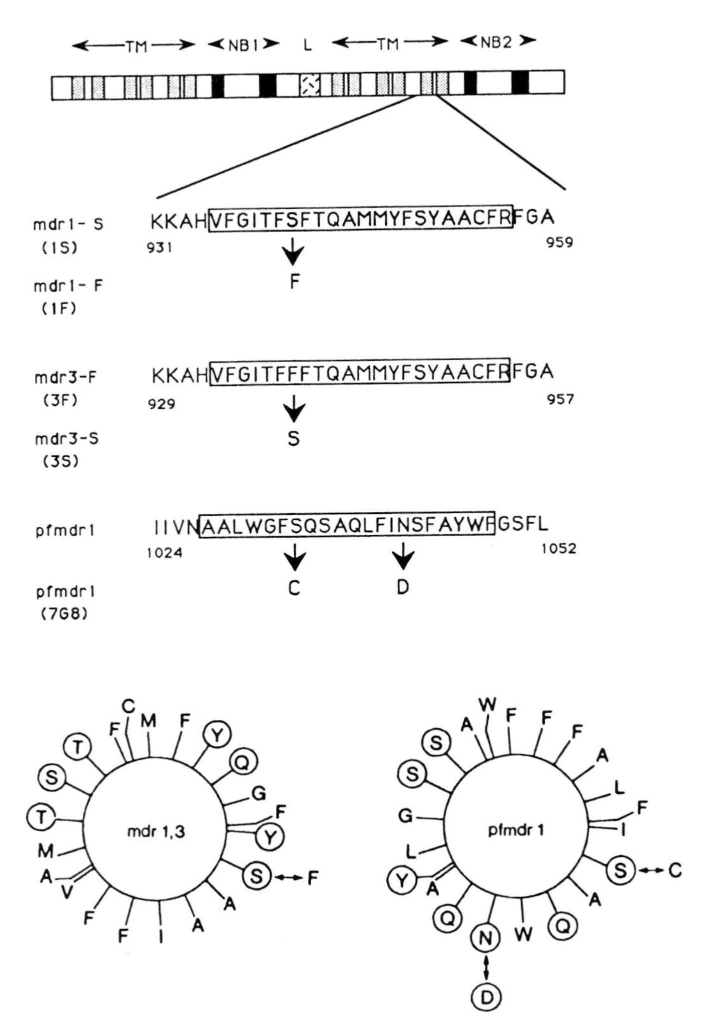

FIG. 4 Site-directed mutagenesis of *mdr* cDNA clones. A schematic representation of a wild-type *mdr* cDNA clone is shown (see legend to Fig. 3). The segment overlapping transmembrane domain 11 has been enlarged, and the amino acid sequences of wild-type and mutant *mdr* cDNA clones are shown. The amino acid sequence of the corresponding region of *pfmdr1* and its mutant 7G8 allele is also shown. At the bottom, helical wheel projections of the protein overlapping TM11 segments for *mdr1*/*mdr3* and *pfmdr1* are shown. Hydrophilic residues assigned by the algorithm of Eisenberg *et al.* (1984) are circled and the position of the mutated residues is indicated by arrows.

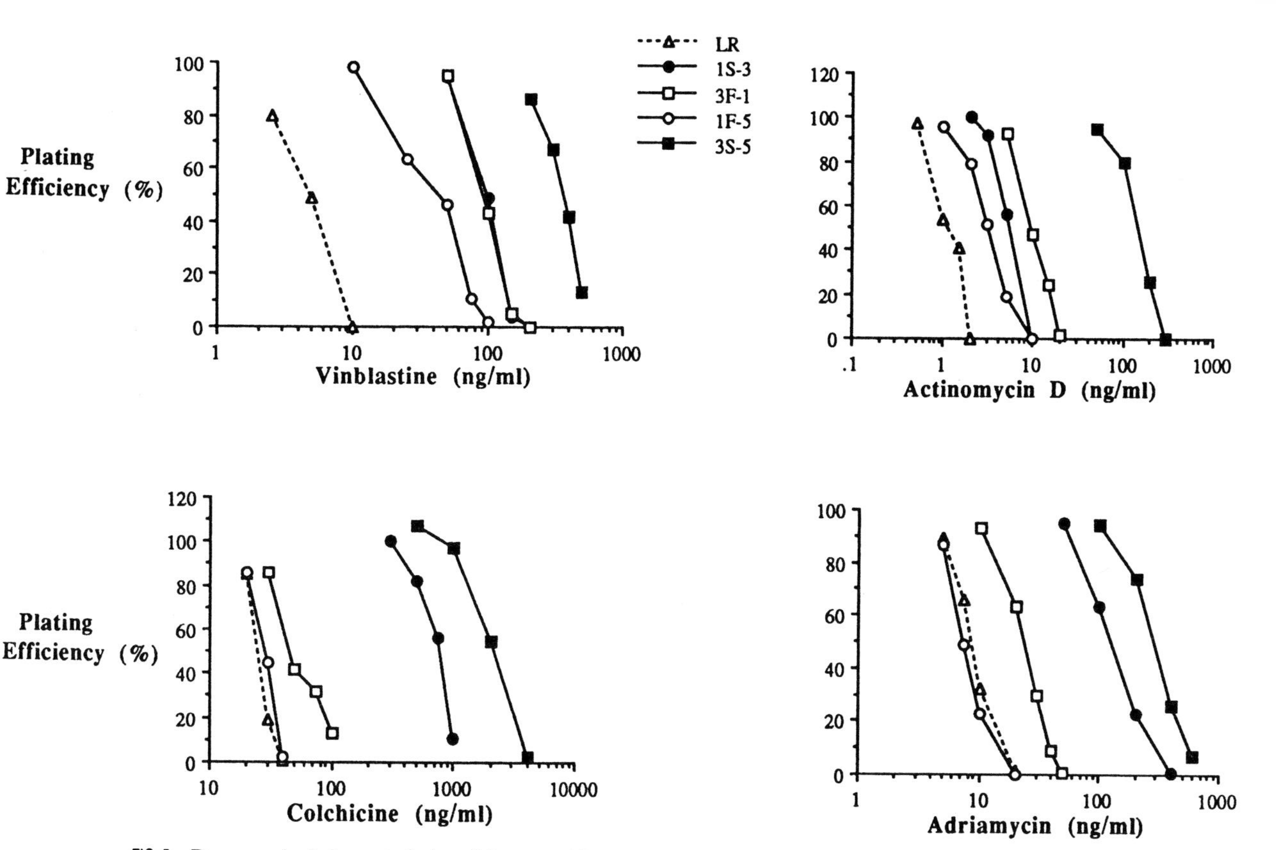

FIG. 5 Drug survival characteristics of drug sensitive Chinese hamster LR73 (LR) cells and cell clones stably transfected and overexpressing wild-type (1S,*mdr1S*;3S,*mdr3S*) or mutant (1F,*mdr1F*;1S,*mdr1S*) *mdr* cDNAs. Data are tabulated and presented as described in the legend to Fig. 3.

mdr1, and the activity of each protein tested in transfected clones. The Ser-to-Phe substitution at position 941 of *mdr1* resulted in a general decrease in the activity of the protein. As noted earlier for *mdr3*, the modulating effect of this mutation was very different for the four drugs tested: it was strongest for colchicine and adriamycin (33- and 15-fold decreases in levels of resistance) and was smallest for vinblastine (2-fold reduction). In fact, *mdr1F* encodes a unique P-gp that can clearly confer resistance to vinblastine but has lost the capacity to confer resistance to adriamycin and colchicine. The segregation of vinblastine and colchicine/adriamycin resistance in the same *mdr* mutant suggests the interesting possibility that these compounds have distinct binding sites on P-gp. The combined analysis of these *mdr1* and *mdr3* mutants identifies $Ser^{939/941}$ in TM11 as a critical residue for the overall activity and substrate specificity of the efflux pumps encoded by the two genes (Gros *et al.*, 1991). In addition, the finding of mutations at the homologous position of TM11 in proteins encoded by distantly related members of the *mdr* super gene family (in this case *pfmdr1*) suggests that this specific protein domain underlies a common structural and functional determinant of transport in the drug efflux by *mdr* proteins and that of chloroquine by *pfmdr1*. Besides their high degree of hydrophobicity, no sequence homology is detected between TM11 of *mdr* and *pfmdr1*. However, both segments are capable of forming amphiphilic helices and the mutations in both proteins map at the interface of the hydrophobic and hydrophilic segments of this domain (Fig. 4).

X. General Discussion and Conclusion

The precise mechanism of action of proteins encoded by members of the *mdr* and *mdr*-like family remains unknown. However, the combined biochemical and genetic analyses of these distantly related proteins have allowed the identification of common structural and functional aspects of these transport systems. These studies have prompted the formulation of hypotheses and working models for the mechanism of action of these proteins. Our simple working hypothesis has been that proteins encoded by the *mdr* and *mdr*-like gene family transport different types of substrates according to the same mechanism, and that evolutionary pressure has acted to diversify the substrate specificity of these systems while preserving their common mechanistic basis. We would like to speculate that the predicted NB sites are the protein domains responsible for the common mechanism of action, whereas the membrane-associated domains are implicated in substrate recognition and hence specificity of the individual transport systems.

A large body of experimental evidence supports the contention that NB sites represent the major determinants in the proposed mechanism of action common to these proteins. First, these segments are the most highly conserved at the primary amino acid sequence level among *mdr* and *mdr*-like genes (Riordan *et al.*, 1989; Foote *et al.*, 1989; Kuchler *et al.*, 1989; Hyde *et al.*, 1990). This high degree of evolutionary conservation undoubtedly indicates that these sites are functionally significant and that very few amino acid replacements can be tolerated in these regions. Second, our studies with *mdr1*/*mdr2* chimeras identify those domains as functionally interchangeable and capable of functioning in the context of different membrane-associated domains (Buschman and Gros, 1991). This suggests that the NB domains can carry out the same functional steps in perhaps distantly related proteins. Third, mutational analysis *in vitro* of the mouse *mdr1* gene clearly demonstrates that even highly conservative amino acid substitutions can ot be tolerated at either of these sites and completely abrogates the activity of this protein (Azzaria *et al.*, 1989). Finally, the vast majority of naturally occurring mutations in the CFTR gene are within the predicted NB domains of the proteins (Kerem *et al.*, 1989; Riordan *et al.*, 1989; Cutting *et al.*, 1990), again consistent with the notion that these sites are key functional determinants of this protein superfamily.

Biochemical and genetic evidence also suggests that the membrane-associated domains of these proteins (transmembrane segment, extra- and intracellular loops) participate in substrate interactions and may confer substrate specificity to the transport system. In the case of P-gps, the hydrophobic nature of the substrates, their ability to partition in the lipid phase of the membrane, and their mode of cell entry by passive diffusion intuitively suggest that P-gp segments embedded in the membrane are the primary sites of substrate interaction. This proposition is sustained by two sets of biochemical data: First, mapping of the binding site(s) of photoactivatable P-gp ligands by tryptic digestion of photolabeled P-gp followed by epitope mapping with specific antibodies identifies the membrane-associated segments as primary sites of photolabeling (Bruggemann *et al.*, 1989; Yoshimura *et al.*, 1989); Second, P-gp can be directly labeled by daunomycin by energy transfer from a photoactivatable probe, 5-[^{125}I]iodonaphthalene-1-azide (Raviv *et al.*, 1990). Indirect evidence from the yeast *mdr* homolog *STE-6* also suggests a role of TM domains in substrate recognition: *STE-6* shares with *STE-3*, the receptor for the *a* mating pheromone expressed on α cells, short segments of sequence homology near TM7 and TM12, possibly implicated in *a* factor binding (Kuchler *et al.*, 1989). A number of mutations have been found to have a modulating effect on the substrate specificity of proteins of the *mdr* or *mdr*-like genes. Invariably, these mutations have been found within the membrane-associated segments of the respective proteins. A single Val-to-Gly substitution near TM3 (pst 185) in the human *MDR1* gene strongly modulates the degree of colchicine resistance conferred by this gene in transfected cells (Choi *et al.*, 1988). Chloroquine resistance in *P. falciparum* is associated with two alleles at the *pfmdr1* gene, the K1 allele is a single Asn-to-Tyr substitution near TM1 (pst 86), and the 7G8 allele is Ser^{1034}/Asn^{1042} to Cys^{1034}/Asp^{1042} within TM11 (Foote *et al.*, 1990). A single Ser-to-Phe substitution in TM11 of *mdr1* and *mdr3* modulates the overall activity but also substrate specificity of the two mouse P-gps, and in the case of *mdr1* uncouples colchicine and anthracycline resistance from vinblastine re istance (Gros *et al.*, 1991). Finally, mutating charged residues within TM1, TM6, and TM10 of the CFTR protein has been found to modulate the halide specificity of the channel (Anderson *et al.*, 1991).

The parallel study of these various P-gp and P-gp-like proteins in their respective assay systems has allowed us to postulate some common mechanistic aspects of transport. It appears likely that the different substrates are recognized in association with the membrane lipid bilayer or directly interact with the membrane-associated segments of these proteins. It also appears that ATP binding and hydrolysis occur at both predicted NB sites, in either a concerted or a sequential fashion. This ATP hydrolysis could then transduce a signal(s) to membrane-associated domains to mediate transport of the various substrates. High-resolution crystal structures have been obtained for NB domains containing Walker's A and B motifs in adenylate kinase (ADK) and other soluble ATP binding proteins and other ATPases. The most important difference between the NB site of ADK and *mdr*-like proteins is the presence of an extra α-helical loop that has been proposed (Hyde *et al.*, 1990) to be responsible for signal transduction to TM domains. It is interesting to note that the major CF mutation, del F508, maps within this loop (Riordan *et al.*, 1989). Finally, it has been recently observed that deletion of the R domain in CFTR suppresses the inactivating effect of a mutation in NB2, but not NB1, providing direct evidence that the R domain interacts specifically with NB2 (Rich *et al.*, 1991). The intramolecular interactions between NB sites and other segments of these proteins, and the details of signal transduction from these sites, will be most easily d ciphered probably in lower eukaryotes such as the yeast (*STE-6* gene), where these types of suppressor mutations are most easily generated and analyzed.

References

Akiyama, S. I., Cornwell, M. M., Kuwano, M., Pastan, I., and Gottesman, M. M. (1988). *Mol. Pharmacol.* **33,** 144–147.

Ames, G. F. L. (1986). *Annu. Rev. Biochem.* **55,** 397–425.

Anderson, M. P., Gregory, R. J., Thompson, S., Souza, D. W., Sucharita, P., Mulligan, R. C., Smith, A. E., and Welsh, M. J. (1991). *Science* **253,** 202–205.
Arceci, R. J., Croop, J. M., Horwitz, S. B., and Housman, D. E. (1988). *Proc. Natl. Acad. Sci. U.S.A.* **85,** 4350–4354.
Arceci, R. J., Baas, F., Raponi, R., Horwitz, S. B., Housman, D. E., and Croop, J. (1990). *Mol. Reprod. Dev.* **25,** 101–109.
Azzaria, M., Schurr, E., and Gros, P. (1989). *Mol. Cell. Biol.* **9,** 5289–5297.
Beck, W. T. (1990). *Eur. J. Cancer.* **26,** 513–515.
Biedler, J. L., and Riehm, H. (1970). *Cancer. Res.* **30,** 1174–1184.
Bitoni, A. J., Sjoerdsma, A., McCann, P. P., Kyle, D. E., Oduola, A. M. J., Rossan, A. N., Milhous, W. K., and Davidson, D. E. (1988). *Science.* **242,** 1301–1303.
Bodmer, J. G., Marsh, S. G. E., Albert, E. D., Bodmer, W. F., Dupont, B., Erlich, H. A., Mach, B., Mayr, W. R., Parham, P., Sasazuki, T., Schreuder, G. M. T., Strominger, J. L., Svejgaard, A., and Terasaki, P. I. (1992). *Tissue Antigens* **39,** 161–173.
Bradley, G., Juranka P. F., and Ling, V. (1988). *Biochim. Biophys. Acta* **948,** 87–128.
Bradley, G., Georges, E., and Ling, V. (1990). *J. Cell. Physiol.* **145,** 398–408.
Bruggemann, E. P., German, U. A., Gottesman, M. M., and Pastan, I. (1989). *J. Biol. Chem.* **264,** 15483–15488.
Buschman, E., and Gros, P. (1991). *Mol. Cell. Biol.* **11,** 595–603.
Buschman, E., Arceci, R. J., Croop, J. M., Che, M., Arias, I. M., Housman, D. E., and Gros, P. (1992). *J. Biol. Chem.* **267,** in press.
Cangelosi, G. A., Martinetti, G., Leigh, J. A., Lee, C. C, Theines, C., and Nester, E. W. (1989). *J. Bacteriol.* **171,** 1609–1615.
Chaudhary, P. M., and Roninson, I. B. (1991). *Cell* **66,** 85–94.
Chen, C. J., Chin, J. E., Ueda, K., Clark, D. P., Pastan, I., Gottesman, M. M., and Roninson, I. B. (1986). *Cell* **47,** 381–389.
Chen, C. J., Clark, D., Ueda, K., Pastan, I., Gottesman, M. M., and Ronsinson, I. B. (1990). *J. Biol. Chem.* **265,** 506–514.
Chen, C. M., Misra T. K., Silver, S., and Rosen, B. P. (1986). *J. Biol. Chem.* **261,** 15030–15038.
Cheng, S. H., Rich, D. P., Marshall, J., Gregory, R. J., Welsh, M. J., and Smith, A. E. (1991). *Cell* **66,** 1027–1036.
Chin, J. E., Soffir, R. Noonan, K. E., Choi, K., and Roninson, I. B. (1989). *Mol. Cell. Biol.* **9,** 3808–3820.
Choi, K., Chen, C. J , Kriegler, M., and Roninson, I. B. (1988). *Cell* **53,** 519–529.
Conter, V., and Beck, W. T. (1984). *Cancer Treat Rep.* **68,** 831–839.
Cordon-Cardo, C., O'Brien, J. P., Casals, D., Rittman-Grauer, L., Biedler, J. L., Melamed, M. R., and Bertino, J. R. (1989). *Proc. Natl. Acad. Sci. U.S.A.* **86,** 695–698.
Cornwell, M. M., Safa, A. R., Felsted, R. L., Gottesman, M. M., and Pastan, I. (1986). *Proc. Natl. Acad. ci. U.S.A.* **83,** 3847–3850.
Cornwell, M. M., Tsuruo, T., Gottesman, M. M., and Pastan, I. (1987a). *FASEB J.* **1,** 51–54.
Cornwell, M. M., Pastan, I., and Gottesman, M. M. (1987b). *J. Biol. Chem.* **262,** 2166–2170.
Coulton, J. W., Mason, P., and Allatt, D. D. (1987). *J. Bacteriol.* **169,** 3844–3849.
Croop, J. M., Raymond, M., Haber, D., Devault, A., Arceci, R., Gros, P., and Housman, D. E. (1989). *Mol. Cell. Biol.* **9,** 1346–1350.
Cutting, G. R., Kash, L. M., Rosenstein, B. J., Zielenski, J., Tsui, L. C., Antonarakis, S. E., and Kazazian, H. H. (1990). *Nature (London)* **346,** 366–369.
Dalton, W. S., Grogan, T. M., Meltzer, P. S., Scheper, P. S., Durie, B. G. M., Taylor, C. W., Miller, T., and Salmon, S. E. (1989). *J. Clin. Oncol.* **7,** 415–424.
Dano, K. (1973). *Biochim. Biophys. Acta* **323,** 446–483.
Devault, A., and Gros, P. (1990). *Mol. Cell. Biol.* **10,** 1652–1663.

Deverson, E. V., Gow, I. R., Coadwell, J., Monaco, J. J., Butcher, G. W., and Howard, J. C. (1990). *Nature (London)* **348,** 738–741.

Doolittle, R. F., Johnson, M. S., Husain, I., Van Houten, B., Thomas, D. C., and Sancar, A. (1986). *Nature (London)* **323,** 451–453.

Dreesen, T. D., Johnson, D. H., and Henikoff, S. (1988). *Mol. Cell. Biol.* **8,** 5206–5215.

Drumm, M. L., Pope, H. A., Cliff, W. H., Rommens, J. M., Marvin, S. A., Tsui, L. C., Collins, F. S., Frizzell, R. A., and Wilson, J. M. (1990). *Cell* **62,** 1227–1233.

Eisenberg, D., Schwartz, E., Komaromy, M., and Wall, R. (1984). *J. Mol. Biol.* **179,** 125–142.

Ellenberger, T. E., and Beverley, S. M. (1989). *J. Biol. Chem.* **264,** 15094–15103.

Endicott, J. A., and Ling, V. (1989). *Annu. Rev. Biochem.* **58,** 137–171.

Felmlee, T., Pellett, S., and Welch, R. A. (1985). *J. Bacteriol.* **163,** 94–105.

Fojo, A., Akiyama, S., Gottesman, M. M., and Pastan, I. (1985). *Cancer Res.* **45,** 3002–3007.

Fojo, A., Ueda, K., Slamon, D. J., Poplack, D. G., Gottesman, M. M., and Pastan, I. (1987). *Proc. Natl. Acad. Sci. U.S.A.* **84,** 265–269.

Foote, S. J., Thompson, J. K., Cowman, A. F., and Kemp, D. J. (1989). *Cell* **57,** 921–930.

Foote, S. J., Kyle, D. E., Martin, R. K., Oduola, A. M. J., Forsyth, K., Kemp, D. J., and Cowman, A. F. (1990). *Nature (London)* **345,** 255–258.

Friedrich, M. J., Deveaux, L. C., and Kadner, R. J. (1986). *J. Bacteriol.* **167,** 928–934.

Fry, D. C., Kub , S. A., and Mildvan, A. S. (1986). *Proc. Natl. Acad. Sci. U.S.A.* **83,** 907–911.

Georges, E., Bradley, G., Gariepy, J., and Ling, V. (1990). *Proc. Natl. Acad. Sci. U.S.A.* **87,** 152–156.

Gilson, E., Higgins, C. F., Hofnung, M., Ames, G. F. L., and Nikaido, H. (1982). *J. Biol. Chem.* **257,** 9915–9918.

Glaser, P., Sakamoto, H., Bellalou, J., Ullmann, A., and Danchin, A. (1988). *EMBO. J.* **7,** 3997–4004.

Glynne, R., Povis, S. H., Beck, S., Kelly, A., Kerr, L.-A., and Trowsdale, J. (1991). *Nature,* **353,** 351–360.

Gottesman, M. M., and Pastan, I. (1988). *J. Biol. Chem.* **263,** 12163–12166.

Gros, P., Croop, J. C., Roninson, I. B., Varshavsky, A., and Housman, D. E. (1986a). *Proc. Natl. Acad. Sci. U.S.A.* **83,** 337–341.

Gros, P., Croop, J., and Housman, D. E. (1986b). *Cell* **47,** 371–380.

Gros, P., Ben Neriah, Y., Croop, J. C., and Housman, D. E. (1986c). *Nature (London)* **323,** 728–731.

Gros, P., Raymond, M., Bell, J., and Housman, D. E. (1988). *Mol. Cell. Biol.* **8,** 2770–2778.

Gros, P., Dhir, R., Croop, J. M., and Talbot, F. (1991). *Proc. Natl. Acad. Sci. U.S.A.* **88,** 7289–7293.

Guifoile, P. G., and Hutchinson, C. R. (1991). *Proc. Natl. Acad. Sci. U.S.A.* **88,** 8553–8557.

Guild, B. C., Mulligan, R. C., Gros, P., and Housman, D. E. (1988). *Proc. Natl. Acad. Sci. U.S.A.* **85,** 1595–1599.

Gupta, R. S., Murray, W., and Gupta, R. (1988). *B. J. Cancer* **58,** 441–447.

Hamada, H., and Tsuruo, T. (1988). *J. Biol. Chem.* **263,** 1454–1458.

Hammond, J., Johnstone, R. M. J., and Gros, P. (1989). *Cancer Res.* **49,** 3867–3871.

Higgins, C. F., Haag, P. D., Nikaido, K., Ardeshir, F., Garcia, G., and Ames, G. F. L. (1982). *Nature (London)* **298,** 723–727.

Higgins, C. F., Hiles, I. D., Salmond, G. P. C., Gill, D. R., Downie, J. A., Evans, I. J., Holland, I. B. Gray, L., Buckel, S. D., Bell, A. W., and Hermodson, M. A. (1986). *Nature (London)* **323,** 448–450.

Hsu, S. I. H., Lothstein, L., and Horwitz, S. B. (1989). *J. Biol. Chem.* **264,** 12053–12062.

Hsu, S. I. H., Cohen, D., Kirschner, L. S., Lothstein, L., Hartstein, M., and Horwitz, S. B. (1990). *Mol. Cell. Biol.* **10,** 3596–3606.

Hyde, S. C., Emsley, P., Hartshorn, M. J., Mimmack, M. M., Gileadi, U., Pearce, S. R., Gallagher, M. P., Gill, D. R., Hubbard, R. E., and Higgins, C. F. (1990). *Nature (London)* **346,** 362–366.

Johann, S., and Hinton, S. M. (1987). *J. Bacteriol* **169,** 1911–1916.
Juliano, R. L., and Ling, V. (1976). *Biochim. Biophys. Acta* **455,** 152–162.
Kamijo, K., Taketani, S., Yokota, S., Osumi, T., and Hasimoto, T. (1990). *J. Biol. Chem.* **265,** 4534–4540.
Kamiwatari, M., Nagata, Y., Kikuchi, H., Yoshimura, A., Sumizawa, T., Shudo, N., Sakoda, R., Seto, K., and Akiyama, S. I. (1989). *Cancer Res.* **49,** 3190–3195.
Kartner, N., Evernden-Porelle, D., Bradley, G., and Ling, V. (1985). *Nature (London)* **316,** 820–823.
Kerem, B. S., Romens, J. M., Buchanan, J. A., Markiewicz, D., Cox, T. K., Chakravarti, A., Buchwald, M., and Tsui, L. C. (1989). *Science* **245,** 1073–1080.
Koch, G., Smith, M., Twentyman, P., and Wright, K. (1986). *FEBS Lett.* **195,** 275–279.
Krogstad, D. J., Gluzman, I. Y., Kyle, D. E., Oduola, A. M. J., Martin, S. K., Milhous, W. K., and Schlesinger, P. H. (1987). *Science* **238,** 1283–1285.
Kuchler, K., Sterne, R. E., and Thorner, J. (1989). *EMBO J.* **8,** 3973–3984.
Lemontt, J., Azzaria M., and Gros, P. (1988). *Cancer Res.* **48,** 6348–6353.
Lincke, C. R., Smit, J. J. M., Van der Velde-Koerts, T., and Borst, P. (1991). *J. Biol. Chem.* **266,** 5303–5310.
Ling, V., and Thompson, L. H. (1974). *J. Cell. Physiol.* **83,** 103–111.
Martin, S. K., Oduola, A. M. J., and Milhous, W. K. (1987). *Science* **235,** 899–901.
Martinsson, T., and Levan, G. (1987). *Cytogenet. Cell. Genet.* **45,** 99–101.
McGrath, J. P., and Varshavsky, A. (1989). *Nature (London)* **340,** 400–404.
Mimura, C. S., Holbrook, S. R., and Ames, G. F. L. (1991). *Proc. Natl. Acad. Sci. U.S.A.* **88,** 84–88.
Monaco, J. J., Cho, S., and Attaya, M. (1990). *Science* **250,** 1723–1726.
Moscow, J. A., and Cowan, K. H. (1988). *J. Natl. Cancer Inst.* **80,** 14–16.
Ng, W. F., Sarangi, F., Zastawny, R. L., Veinot-Drebot, L., and Ling, V. (1989). *Mol. Cell. Biol.* **9,** 1224–1232.
Nohno, T., Saito, T., and Hong, J. S. (1986). *Mol. Gen. Genet.* **205,** 260–269.
Nooter, K., and Herweijer, H. (1991). *Br. J. Cancer* **63,** 663–669.
O'Hare, K., Murphy C., Levis, R., and Rubin, G. M. (1984). *J. Mol. Biol.* **180,** 437–455.
Ohyama, K., Fukuzawa, H., Kohchi, T., Shirai, H., Sano, T., Sano, S., Umesono, K., Shiki, Y., Takeuchi, M., Chang, Z., Aoto, S. I., Inokuchi, H., and Ozeki, H. (1986). *Nature (London)* **322,** 572–574.
Ouellette, M., Fase-Fowler, F., and Borst, P. (1990). *EMBO J.* **9,** 1027–1033.
Pearce, H. L., Safa, A. R., Bach, N. J., Winter, M. A., Cirtain, M. C., and Beck, W. T. (1989). *Proc. Natl. Acad. Sci. U.S.A.* **86,** 5128–5132.
Peterson, R. H., Beutler, W. J., and Biedler, J. L. (1979). *Biochem. Pharmacol.* **28,** 579–582.
Peterson, R. H., Meyers, M. B., Spengler, B. A., and Biedler, J. L. (1983). *Cancer Res.* **43,** 222–228.
Powis, S. J., Townsend, A. R. M., Deverson, E. V., Bastin, J., Butcher, G. W., and Howard, J. C. (1991). *Nature* **354,** 528–531.
Qian, X. D., and Beck, W. T. (1990). *J. Biol. Chem.* **265,** 18753–18756.
Raviv, Y., Pollard, H. B., Bruggemann, E. P., Pastan, I., and Gottesman, M. M. (1990). *J. Biol. Chem.* **265,** 3975–3980.
Raymond, M., and Gros, P. (1989). *Proc. Natl. Acad. Sci. U.S.A.* **86,** 6488–6492.
Raymond, M., and Gros, P. (1990). *Mol. Cell. Biol.* **10,** 6036–6040.
Raymond, M., Rose, E., Housman, D. E., and Gros, P. (1990). *Mol. Cell. Biol.* **10,** 1642–1651.
Rich, D. P., Anderson, M. P., Gregory, R. J., Cheng, S. H., Paul, S., Jefferson, D. M., McCann, J. D., Klinger, K. W., Smith, A. E., and Welch, M. J. (1990). *Nature (London)* **347,** 358–363.
Rich, D. P., Gregory, R. J., Anderson, M. P., Manavalan, P., Smith, A. E., and Welsh, M. J. (1991). *Science* **253,** 205–207.

Richert, N., Akiyama, S., Shen, D., Gottesman, M. M., and Pastan, I. (1985). *Proc. Natl. Acad. Sci. U.S.A.* **82,** 2330–2333.

Riordan, J. R., Rommens, J. M., Kerem, B. S., Alon, N., Rozmahel, R., Grzelczack, Z., Zielenski, J., Lok, S., Plavsic, N., Chou, J. L., Drumm, M. L., Iannuzzi, M. C., Collins, F. S., and Tsui, L. C. (1989). *Science* **245** 1066–1073.

Roninson, I. B. (991). "Molecular and Cellular Biology of Multidrug Resistance in Tumor Cells." Plenum, New York.

Roninson, I. B., Abelson, H. T., Housman, D. E., Howell, N., and Varshavsky, A. (1984). *Nature (London)* **309,** 626–628.

Roninson, I. B., Chin, J. E., Choi, K., Gros, P., Housman, D. E., Shen, D. W., Gottesman, M. M., and Pastan, I. (1986). *Proc. Natl. Acad. Sci. U.S.A.* **83,** 4538–4542.

Safa, A. R. (1988). *Proc. Natl. Acad. Sci. U.S.A.* **85,** 7187–7191.

Safa, A. R., Glover, C. J., Meyers, M. B., Biedler, J. L., and Felsted, R. L. (1986). *J. Biol. Chem.* **261,** 6137–6140.

Safa, A. R., Glover, C. J., Sewell, J. L., Meyers, M. B., Biedler, J. L., and Felsted, R. L. (1987). *J. Biol. Chem.* **262,** 7884–7888.

Safa, A. R., Metha, N. D., and Agresti, M. (1989). *Biochem. Biophys. Res. Commun.* **162,** 1402–1408.

Samuelson, J., Ayala, P., Oroczo, E., and Wirth, D. (1990). *Mol. Biochem. Parasitol.* **38,** 281–290.

Schinkel, A. H., Roelofs, M. E. M., and Borst, P. (1991). *Cancer Res.* **51,** 2628–2635.

Schurr, E., Raymond, M., Bell, J. C., and Gros, P. (1989). *Cancer Res.* **49,** 2729–2734.

Scotto, K. W., Biedler, J. L., and Mellera, P. (1986). *Science* **232,** 751–755.

Scripture, J. B., Voelker, C., Miller, S., O'Donnell, R. T., Polgar, L., Rade, J., Hrazdovsky, B. F., and Hogg, R. W. (1987). *J. Mol. Biol.* **197,** 37–46.

Skovsgaard, T. (978). *Cancer Res.* **39,** 4722–4727.

Slater, L., Sweet, P., Stupecky, M., and Gupta, S. (1986). *J. Clin. Invest.* **77,** 1405–1408.

Southern, P. J., and Berg, P. (1982). *J. Mol. Appl. Genet.* **1,** 327–341.

Spies, T., and De Mars, R. (1991). *Nature (London)* **351,** 323–324.

Spies, T., Bresnahan, M., Bahram, S., Arnold, D., Blanck, G., Mellins, E., Pious, D., and De Mars, R. (1990). *Nature (London)* **348,** 744–747.

Stanfield, S. W., Ielpi, L., O'Brochta, D., Helinski, D. R., and Ditta, G. S. (1988). *J. Bacteriol.* **170,** 3523–3530.

Staudenmaier, H., Van Hove, B., Yaraghi, Z., and Braun, V. (1989). *J. Bacteriol.* **171,** 2626–2633.

Sugarawa, I., Nakahama, M., Hamada, H., Tsuruo, T., and Mori, S. (1988). *Cancer Res.* **48,** 4611–4614.

Surin, B. P., Rosenberg, H., and Cox, G. B. (1985). *J. Bacteriol.* **161,** 189–198.

Tabcharani, J. A., Chang, X. B., Riordan, J. R., and Hanrahan, J. W. (1991). *Nature (London)* **352,** 28–631.

Tamai, I., and Safa, A. R. (1990). *J. Biol. Chem.* **265,** 16509–16513.

Thiebaut, F., Tsuruo, T., Hamada, H., Gottesman, M. M., Pastan, I., and Willingham, M. C. (1987). *Proc. Natl. Acad. Sci. U.S.A.* **84,** 7735–7738.

Thiebaut, F., Tsuruo, T., Hamada, H., Gottesman, M. M., Pastan, I., and Willingham, M. C. (1989). *J. Histochem. Cytochem.* **37,** 159–164.

Trowsdale, J., Hanson, I., Mockridge, I., Beck, S., Townsend, A., and Kelly, A. (1990). *Nature (London)* **348,** 741–743.

Tsuruo, T., Iida, H., Tsukagoshi, S., and Sakurai, Y. (1981). *Cancer Res.* **41,** 1967–1972.

Tsuruo, T., Iida, H., Tsukagoshi, S., and Sakurai, Y. (1982). *Cancer Res.* **42,** 4730–4733.

Tsuruo, T., Iida, H., Ohkochi, E., Tsukagoshi, S., and Sakurai, Y. (1983). *Gann* **74,** 751–758.

Tsuruo, T., Iida, H., Kitatani, Y., Yokota, K., Tsukagoshi, S., and Sakurai, Y. (1984). *Cancer Res.* **44,** 4303–4307.

Ueda, K., Cardarelli, C., Gottesman, M. M., and Pastan, I. (1987). *Proc. Natl. Acad. Sci. U.S.A.* **84,** 3004–3008.
Van der Bliek, A. M. Van der Velde-Koerts, T., Ling, V., and Borst, P. (1986). *Mol. Cell. Biol.* **6,** 1671–1678.
Van der Bliek, A. M. Baas, F., Ten Houte de Lange, T., Kooiman, P. M., Van der Velde-Koerts, T., and Borst, P. (1987). *EMBO J.* **6,** 3325–3331.
Walker, J. E., Sraste, M., Runswick, M. J., and Gay, N. J. (1982). *EMBO J.* **1,** 945–951.
Wilson, C. M., Serrano, A. E., Wasley, A., Bogenshutz, M. P., Shankar, A. H., and Wirth, D. F. (1989). *Science* **244,** 1184–1186.
Wu, C. T., Budding, M., Griffin, M. S., and Croop, J. (1991). *Mol. Cell. Biol.* **11,** 3940–3948.
Yang, C. P. H., Cohen, D., Greenberger, L. M., Hsu, S. I. H., and Horwitz, S. B. (1990). *J. Biol. Chem.* **265,** 10282–10288.
Yoshimura, A., Kuwazuru, Y., Sumizawa, T., Ichikawa, M., Ikeda, S. I., Ueda, T., and Akiyama, S. I. (1989). *J. Biol. Chem.* **264,** 16282–16291.
Zamora, J. M., Pearce, H. L., and Beck, W. (1988). *Mol. Pharmacol.* **33,** 454–462.

Molecular and Kinetic Aspects of Sodium–Calcium Exchange

Kenneth D. Philipson and Debora A. Nicoll
Departments of Medicine and Physiology, and the Cardiovascular Research Laboratory, UCLA School of Medicine, Los Angeles, California, 90024

I. Introduction

Na^+–Ca^{2+} exchange was first described in squid axon (Baker *et al.*, 1969) and in guinea pig atria (Reuter and Seitz, 1968) almost 25 years ago. Due to the importance of Ca^{2+} movements in intracellular signaling mechanisms, Na^+–Ca^{2+} exchange has attracted intense interest. In recent years, two international symposia have been devoted to this topic (Allen *et al.*, 1989; Blaustein *et al.*, 1991). Na^+–Ca^{2+} exchange is present in membranes from a wide variety of tissues and many speculative roles for Na^+–Ca^{2+} exchange have been proposed. This diversity has attracted investigators using many approaches at different levels. In general, it has been difficult to unequivocally prove specific roles for Na^+–Ca^{2+} exchange. The major experimental problem has been the presence of multiple Ca^{2+} transport pathways in most cell types. The complexity of Ca^{2+} movements often precludes assignment of importance to a particular pathway. Nevertheless, in some cell types, the exchanger is able to rapidly transport large quantities of Ca^{2+} and functional roles are becoming more clear.

Cardiac myocytes and the outer segments of photoreceptor rod cells (ROS) have especially high Na^+–Ca^{2+} exchange activity and have been the starting materials for the most extensive molecular studies. Some background on the physiological importance of exchange in these tissues will be given before molecular aspects are reviewed.

II. Physiological Significance

A. Cardiac Muscle

Cardiac muscle contracts in response to a rise in intracellular Ca^{2+}. The source of this Ca^{2+} is release from the sarcoplasmic reticulum and influx across the sarcolemma (plasma membrane) through voltage-sensitive Ca^{2+} channels. To bring about muscle relaxation and to maintain Ca^{2+} homeostasis, Ca^{2+} efflux must equal Ca^{2+} influx. In the past, the relative roles of the sarcolemmal ATP-dependent Ca^{2+} pump and the Na^+–Ca^{2+} exchanger in cardiac Ca^{2+} efflux had been controversial. Recent progress has clearly demonstrated, however, that Na^+–Ca^{2+} exchange is by far the dominant Ca^{2+} efflux mechanism (Bers and Bridge, 1989; Bridge *et al.*, 1990). It is difficult to substantiate any role for the sarcolemmal Ca^{2+} pump in cardiac muscle. Recently, Na^+–Ca^{2+} exchange has been postulated to contribute to cardiac Ca^{2+} influx (Leblanc and Hume, 1990). Recognition that Na^+–Ca^{2+} exchange has a major regulatory influence on cardiac contractility has intensified interest in this area.

B. Rod Outer Segments

In the outer segment of the vertebrate rod photoreceptor cell, calcium regulates a number of the key enzymes involved in the responses of the cell to light (reviewed in Yau and Baylor, 1989). The pathways for Ca^{2+} movement across the outer segment plasma membrane are well defined. Influx occurs through the cGMP-gated cation channel and efflux occurs via the exchanger. The concerted actions of these two proteins are ultimately responsible for the cellular responses to light.

Perhaps because it is the sole mechanism for calcium extrusion, the ROS exchanger utilizes not only the Na^+ but also the K^+ gradient as an energy source. The ROS exchanger is more correctly referred to as a Na^+–(Ca^{2+}+K^+) exchanger since it operates with a stoichiometry of 4 Na^+:(1 Ca^{2+}+1 K^+) (see below).

C. Other Tissues

Smooth muscle and neural tissue have substantial Na^+–Ca^{2+} exchange activity. The importance of Na^+–Ca^{2+} exchange as a regulator of smooth muscle contractile state is highly controversial and no concensus opinion has arisen. A hypothesis links the Na^+–Ca^{2+} exchanger of vascular smooth muscle to essential hypertension (Blaustein *et al.*, 1986). According to this hypothesis, an endogenous Na^+ pump inhibitor can cause a rise in the internal Na^+ concentration of vascular smooth muscle cells. This rise in Na^+ could then increase internal Ca^{2+} via Na^+–Ca^{2+} exchange and lead to increased muscle tone. The validity of this hypothesis is under active investigation and an endogenous ouabain-like factor has recently been isolated (Hamlyn *et al.*, 1989).

The Na^+–Ca^{2+} exchange of neural tissue was initially described for the squid giant axon but has now been described in several preparations. Activity is moderately high, compared with cardiac tissue, and may account for the majority of Ca^{2+} efflux from nerve terminals following excitation (Sanchez-Armass and Blaustein, 1987). The importance of the neuronal Na^+–Ca^{2+} exchanger may have been underestimated in the past. It will be of interest to quantitiate the amount of Na^+–Ca^{2+} exchanger in different regions of the brain and during different stages of development.

Na^+–Ca^{2+} exchange activity has been detected in almost all other tissues though generally with low activity. It is enticing to assume an importance for Na^+–Ca^{2+} exchange in tissues such as kidney, intestine, and adrenal medulla but physiologic proof is lacking. The role of Na^+–Ca^{2+} exchange in these tissues will become better understood as studies of Ca^{2+} fluxes and Ca^{2+} signaling become more sophisticated.

III. Kinetic Characterization

A. General Considerations

Kinetic characterization of the Na^+–Ca^{2+} exchanger comes from a variety of preparations using a variety of techniques. The preparations range from intact tissues to the isolated, reconstituted exchanger protein. The majority of studies have employed radioisotope flux or electrophysiological measurements. Generally, these approaches have resulted in consensus opinion on the overall workings of the exchanger, but there is still disagreement on many of the details.

At a molecular level, there appear to be two distinct forms of Na^+–Ca^{2+} exchangers: the cardiac-type, defined by a 120-kDa protein, which is K^+-independent, and the ROS-type, which is defined by a 220-kDa protein that cotransports K^+. The cardiac sarcolemmal exchanger has been cloned and there is now a report that the ROS exchanger has also been cloned (Achilles *et al.*, 1992). There is surprisingly little sequence similarity between the two proteins (see below). Nevertheless, there are many functional similarities, and we will attempt to point out both similarities and differences in this chapter.

There is, of course, Na^+–Ca^{2+} exchange activity in many other tissues, but it appears that these exchangers are of the cardiac type. Due to low activity, the exchanger has generally not been as well characterized in these tissues. However, those functional properties of exchangers from other tissues that have been measured, appear to be similar to the cardiac exchanger. Also, with the recent development of molecular probes to the cardiac exchanger (antibodies and oligonucleotides), it appears that other tissues contain cardiac-type exchanger proteins. Undoubtedly, many isoforms remain to be discovered (e.g., alternatively spliced products, separate gene products, alternatively processed proteins). Since the most detailed kinetic studies have been done with cardiac tissue, we will use these studies to describe the prototypical Na^+–Ca^{2+} exchanger.

The Na^+–Ca^{2+} exchange activity of the ROS plasma membrane is probably higher than that of cardiac sarcolemma. Recognition of the importance and high activity of the exchanger in the ROS has been a relatively recent occurrence. The ROS exchanger has thus received less attention than the cardiac exchanger, but this situation is rapidly changing.

There is also a Na^+–Ca^{2+} exchanger in the mitochondrial inner membrane (Crompton, 1990). This exchanger is difficult to study and has generally resisted molecular analysis. It is clearly a different exchanger from that present in plasma membranes in that it is not electrogenic and Li^+ can substitute for Na^+. It is hoped that molecular comparisons of the mitochondrial and plasma membrane exchangers will be possible and instructive in the not too distant future. The mitochondrial exchanger will not be further addressed here.

B. Methodological Considerations

A large mass of data has accumulated from isotope flux studies using isolated plasma membrane vesicles. As first described by Reeves and Sutko (1979) for cardiac sarcolemmal vesicles, this approach allows Na^+-dependent Ca^{2+} fluxes to be easily and unequivocally identified. A variety of control experiments can demonstrate convincingly that fluxes are specifically due to a Na^+–Ca^{2+} exchange mechanism. Cardiac sarcolemmal membranes are especially amenable to this approach with initial rates as high as 25 nmol Ca^{2+}/mg protein/sec. Measurements with membranes from other tissues are sometimes not as straightforward because of low activity.

The use of isolated vesicles has some technical problems. First, some regulatory properties of the exchanger appear to be lost. In intact preparations, the cardiac Na^+–Ca^{2+} exchanger has a high-affinity Ca^{2+} regulatory site (Kimura *et al.*, 1986; Noda *et al.*, 1988; Hilgemann, 1990) and is stimulated by ATP (Hilgemann, 1990). Neither of these modulatory influences can be observed in sarcolemmal vesicles. The ATP effect on exchange may be indirectly due to an effect on lipid asymmetry (see below). How this effect should extrapolate to vesicles is not clear. However, the lack of a high-affinity Ca^{2+} regulatory site on the vesicular exchanger remains to be explained.

Second, the very small size of plasma membrane vesicles (about 0.1 μm) causes intravesicular ion levels to change drastically following initiation of ion transport reactions. This confounds analysis of kinetic mechanisms. One recent approach to circumvent this problem has been to study Ca^{2+} uptake (in the presence of EGTA) using proteoliposomes prepared from solubilized sarcolemmal vesicles (Khananshvili, 1990). The EGTA prevents a rise in intravesicular Ca^{2+} allowing true initial rates to be measured.

Third, sarcolemmal vesicle preparations contain a mixture of inside-out and right-side-out vesicles. Attempts to physically separate these two populations of vesicles have not been successful. Since the Na^+–Ca^{2+} exchanger has ion binding sites on both sides of surface membranes, this creates difficulties in the interpretation of some data. We have developed a technique for measuring exchange in only the inside-out vesicles in a mixed population (Philipson and Nishimoto, 1982a). Inside-out vesicles were preloaded with Na^+ via the ATP-dependent Na^+ pump. The uptake of Ca^{2+}, exchanging for the pumped Na^+, could then be quantitated. The data suggested that the affinities of the exchanger for Ca^{2+} at the two surfaces of the sarcolemmal membrane were identical and were in the low micromolar range.

In light of more recent data, however, we have reinterpreted these symmetry experiments. From inhibitor studies and from data on Ca^{2+} affinities measured electrophysiologically, it appears that right-side-out sarcolemmal vesicles do not normally participate in vesicular Na^+–Ca^{2+} exchange reactions. That is, even in a mixed population of vesicles, only the inside-out vesicles are responsible for the observed Na_i^+-dependent Ca^{2+} uptake. The reason for this is as follows: Vesicular exchange is usually measured isotopically using $^{45}Ca^{2+}$. To maintain high specific activities, low Ca^{2+} concentrations are preferred and inside-out vesicles, with a moderately high-affinity Ca^{2+} transport site facing outward, will take up Ca^{2+}. In contrast, right-side-out vesicles, with a low affinity Ca^{2+} transport site facing outward (see below), will not be activated. This argument is presented in greater detail in Li *et al.* (1991).

The sophistication of electrophysiological approaches for studying Na^+–Ca^{2+} exchange has increased substantially in the last few years. The combination of patch clamp techniques and the use of dissociated myocytes has eliminated some problems such as inadequate voltage control and poor diffusion in intercellular clefts. By blocking other currents and transport pathways, Na^+–Ca^{2+} exchange currents can be identified and analyzed. A useful approach has been the whole-cell voltage clamp in combination with intracellular perfusion (Kimura *et al.*, 1986). Simultaneous use of fura-2 to monitor internal Ca^{2+} (Barcenas-Ruiz *et al.*, 1987) and rapid solution changes (Bridge *et al.*, 1990) has further increased the power of the technique. Potential drawbacks of these electrophysiological approaches include the possibility that other ion pathways contribute to the measured currents, slow diffusion of pipet contents into the cell, and the inconvenience of cellular contraction during Ca^{2+} influx. Many of these drawbacks have been eliminated by the use of giant excised patches to study exchanger activity. This technique, which was developed by Hilgemann (1989), allows exchange currents to be easily quantitated. Regulatory (Hilgemann, 1990) and mechanistic (Hilgemann *et al.,* 1991) studies have taken advantage of the approach as discussed below.

The discussion of techniques above has focused on studies with cardiac sarcolemmal vesicles and on recent electrophysiology. This emphasis reflects the authors' bias; major contributions have also come from other systems and techniques and will be referred to at appropriate places below.

C. Stoichiometry

Early in Na^+–Ca^{2+} exchange research, it was thought that the stoichiometry of exchange was 2 Na^+ for 1 Ca^{2+} (Reuter and Seitz, 1968). There is now a concensus that, with the exception of the ROS exchanger (see below), the stoichiometry of exchange is 3 Na^+ for 1 Ca^{2+}. The stoichiometry of exchange has profound physiological importance. An electrogenic 3 Na^+/1 Ca^{2+} exchanger will be affected by membrane potential and will also generate its own current. These factors are important to excitable tissues such as heart. Also, the stoichiometry, through thermodynamic considerations, sets a lower limit on the intracellular free Ca^{2+} level attainable by Na^+–Ca^{2+} exchange alone. A stoichiometry of 3 to 1 appears to function admirable for cardiac muscle both theoretically and empirically.

Many studies have confirmed the 3 Na^+/1 Ca^{2+} stoichiometry. Using cardiac sarcolemmal vesicles, Reeves and Hale (1984) applied a thermodynamic approach by measuring the Na^+ gradient required to block net Ca^{2+} movements at different membrane potentials. A coupling ratio of 3 was calculated from the results. A more direct measurement was that of Rasgado-Flores and Blaustein (1987), who isotopically measured the countermovement of Na^+ and Ca^{2+} in barnacle muscle fibers and found a stoichiometry of 3 to 1. Kimura *et al.* (1987), Bridge *et al.* (1990), and Crespo *et al.* (1990) also determined a coupling ratio of 3 in recent studies using the patch clamp technique on isolated cardiac myocytes. In the absence of evidence to the contrary, a tight coupling ratio of 3 Na^+/1 Ca^{2+} has become accepted as the correct stoichiometry for the "cardiac type" Na^+–Ca^{2+} exchanger. Whether certain conditions permit other stoichiometries is not known.

It seemed reasonable to expect that the ROS exchanger would have the same stoichiometry as the cardiac exchanger. Nevertheless, Hodgkin *et al.* (1985), Schnetkamp *et al.* (1988), and Schwartz (1985) had suggested that the ROS exchanger operates with a different stoichiometry, involving K^+, to explain the low intracellular Ca^{2+} observed in ROS. Subsequently, two groups (Cervetto *et al.*, 1989; Schnetkamp *et al.*, 1989) have described the stoichiometry of the ROS exchanger to be 4 Na^+ ions in exchange for 1 Ca^{2+} and 1 K^+ ion. Cervetto *et al.* (1989) measured Na^+–Ca^{2+} exchanger currents and examined the relative changes of Ca^{2+}, Na^+, and K^+ required to maintain equilibrium. From thermodynamic arguments, they deduced a stoichiometry of 4 Na^+/1 Ca^{2+}+1 K^+. Schnetkamp *et al.* (1989) used more direct measurements of cation fluxes to reach a similar conclusion. This stoichiometry allows the exchanger, by using the energy of the outward K^+ gradient, to maintain free internal Ca^{2+} levels lower than that attainable by a 3 Na^+/1 Ca^{2+} exchange. Although a K^+ dependence of the ROS exchanger had been noted previously (Hodgkin *et al.*, 1985), these were the first studies to demonstrate that the ROS exchanger could transport K^+, in marked contrast to the "cardiac-type" exchanger.

An interesting characteristic of the ROS system is an apparent variable stoichiometry. In the absence of an electrical shunt to dissipate the charge buildup associated with exchanger activity, the exchanger switches to a nonelectrogenic mode with a stoichiometry of 3 Na^+/1 Ca^{2+}+1 K^+ (Schnetkamp *et al.*, 1989, 1991). Under physiological conditions, the ROS exchanger would apparently operate in the 4 Na^+/1 Ca^{2+}+1 K^+ mode, and other stoichiometries may have no functional importance (Schnetkamp *et al.*, 1991). Schnetkamp *et al.* (1992b) have also observed that the ROS exchanger can operate in a K^+ independent mode.

D. Ion Dependencies

1. Ca^{2+} Dependence

The cardiac-type Na^+–Ca^{2+} exchangers have a relatively high-affinity Ca^{2+} transport site at the intracellular membrane surface and a low-affinity Ca^{2+} transport site at the extracellular surface. The internal Ca^{2+} site has attracted the most interest because of the physiological importance of regulating intracellular Ca^{2+}. A wide variety of K_m values have been measured for the internal Ca^{2+} site, ranging from about 1 to 40 μM. The reasons for this large range are unclear but may reflect different regulatory states of the cardiac exchanger (see below). The easiest system with which to measure K_m values is the sarcolemmal vesicle preparation. Nevertheless, even with this preparation, a wide range of values has been obtained (reviewed in Reeves and Philipson, 1989). "Typical values," however, are about 20 μM Ca^{2+}; K_m (Ca^{2+}) values obtained with plasma membrane vesicles from other tissues are also generally in this range. As described above, the data are complicated by the presence of both inside-out and right-side-out vesicles. However, if only the inside-out vesicles participate in the usual Na^+–Ca^{2+} exchange assay (Li *et al.*, 1991), then the $K_m(Ca^{2+})$ values obtained should reflect interactions of Ca^{2+} with the exchanger at the intracellular surface only.

Other systems for which K_m values for the intracellular Ca^{2+} site are available include perfused squid axons and barnacle muscle fibers, whole-cell patch clamped cardiac cells, and giant excised patches from cardiac cells. In squid axons and barnacle muscle (Blaustein, 1984), the $K_m(Ca^{2+})$ is changed by ATP by an unresolved mechanism. In the absence of ATP, the $K_m(Ca^{2+})$ for the squid axon is about 12 μM but decreases to 1–3 μM in the presence of ATP (DiPolo and Beauge, 1988).

Electrophysiological determinations of the internal $K_m(Ca^{2+})$ values for the Na^+–Ca^{2+} exchanger of cardiac cells have yielded quite variable results. Miura and Kimura (1989) determined a K_m value of 0.6 μM Ca^{2+} for patch clamped myocytes. In contrast, Barcenas-Ruiz *et al.* (1987) and Berlin *et al.* (1988) found that exchanger currents were linear with internal Ca^{2+} up to at least 1 and 2.5 μM, respectively. The implication is that the $K_m(Ca^{2+})$ must be substantially above these values since the exchanger activity had not yet begun to saturate. Study of the effects of intracellular Ca^{2+} is more direct with the giant excised sarcolemmal patch than with the whole-cell patch procedure. Hilgemann *et al.* (1991) have found the $K_m(Ca^{2+})$ to be variable from about 1 to 10 μM depending on the extracellular Na^+ concentration (see below).

In ROS, the $K_m(Ca^{2+})$ at the internal site appears to be in the 1–2 μM range (Lagnado *et al.*, 1988; Schnetkamp, 1989). As pointed out by Schnetkamp (1989), the measurement is very dependent on the occupation of other exchanger binding sites by other ions and on the presence of competing cations.

A $K_m(Ca^{2+})$ of approximately 1 μM (or larger) for both the ROS and the cardiac-type exchangers raises an important physiological question. How does Na^+–Ca^{2+} exchange lower free internal Ca^{2+} to 0.1 μM if the K_m is substantially above this value? The answer seems to be that even at low Ca^{2+} concentrations, there is enough exchange activity to lower the free Ca^{2+} concentration to significantly below the K_m value. In cardiac muscle, the ATP-dependent Ca^{2+} pump may contribute to Ca^{2+} extrusion at low Ca^{2+}. In contrast, the exchanger is the only Ca^{2+} efflux mechanism of ROS; in this system, the exchanger alone is clearly capable of lowering Ca^{2+} to physiological levels.

There is a striking difference in the symmetry of Ca^{2+} affinities between the ROS exchanger and exchangers from other tissues. The cardiac and squid axon Na^+–Ca^{2+} exchangers are highly asymmetric, whereas the

ROS exchanger is symmetric with similar $K_m(Ca^{2+})$ values in the micromolar range at both surfaces (Schnetkamp, 1989). The squid giant axon has a K_m (Ca^{2+}) at the external surface of about 3.0 m*M*, whereas the external K_m for cardiac cells has been measured to be 0.15–0.4 m*M* Ca^{2+} (Wakabayashi and Goshima, 1981; Vemuri *et al.*, 1989; Miura and Kimura, 1989). These values are much greater than the $K_m(Ca^{2+})$ values at the intracellular surfaces.

2. Na^+ Dependence

The Na^+ dependence of Na^+–Ca^{2+} exchange has been measured in a variety of systems. Generally, cardiac K_m (Na^+) values are found to be 20–30 m*M*, with Hill coefficients of 2–3 (reviewed in Reeves and Philipson, 1989; Philipson, 1990a; Reeves, 1990). In contrast to the K_m (Ca^{2+}), there is little apparent asymmetry in K_m (Na^+) values for the intra- and extracellular surfaces of the exchanger. No other monovalent cation, such as Li^+, can substitute for Na^+ in the Na^+–Ca^{2+} exchange reaction. This contrasts with the Na^+–Ca^{2+} exchanger of the mitochondrial inner membrane, which transports Li^+ as well as Na^+.

Na^+ and Ca^{2+} compete for transport sites on the exchanger. Na^+, on the same side of the cardiac sarcolemmal membrane as Ca^{2+}, inhibits Ca^{2+} flux through the exchanger with a K_i of about 15 m*M* and a Hill coefficient of 2 at high Na^+ levels (Reeves and Sutko, 1983). Thus, when analyzing exchanger Na^+ or Ca^{2+} dependencies, it is important to be cognizant of the concentration of each ion on both sides of the membrane.

3. K^+ Dependence

It is the ability of the ROS Na^+–(Ca^{2+},K^+) exchanger to transport K^+ (Cervetto *et al.*, 1989; Schnetkamp *et al.*, 1989) that distinguishes that transporter from Na^+–Ca^{2+} exchangers which have been examined in other issues. The stoichiometry of the ROS exchanger is 4 Na^+/ 1 Ca^{2+} + 1 K^+, with an apparent $K_m(K^+)$ of about 1 m*M* (Schnetkamp *et al.*, 1989; Nicoll *et al.*, 1991). An interesting feature, however, is that coupling between Ca^{2+} and K^+ transport in ROS is not absolute (Schnetkamp, 1989). Schnetkamp has observed Ca^{2+}–Ca^{2+} exchange in the absence of K^+. Apparently, Ca^{2+} can be translocated without simultaneous translocation of K^+. It appears that the modes of the ROS exchanger and the possible interactions of ion binding sites are more complicated than those for the cardiac exchanger.

Although the cardiac Na^+–Ca^{2+} exchanger has generally been found to be unaffected by the presence or absence of K^+, Gadsby *et al.* (1991) finds that extracellular K^+ or Li^+ produces a modest activation of Na^+–Ca^{2+}

exchange of voltage clamped myocytes. In the squid giant axon, external monovalent cations induce a substantial activation of Na_i^+–Ca_o^{2+} exchange (DiPolo and Rojas, 1984; Allen and Baker, 1986). However, with both the cardiac and the squid axon exchangers, there is no evidence that transport of K^+ accompanies Na^+–Ca^{2+} exchange.

E. Inhibitors

Many attempts to identify pharmacological agents specific for the Na^+–Ca^{2+} exchanger have been unsuccessful. The inhibitor that has been used most frequently is dichlorobenzamil, an amiloride derivative. Dichlorobenzamil has numerous effects on other membrane components, which limits its use in physiological experiments. Dichlorobenzamil has been useful for investigating the interactions between ion binding sites on the exchanger, as reviewed by Kaczorowski *et al.* (1989).

The cloning of the cardiac Na^+–Ca^{2+} exchanger has led to the identification of a peptide inhibitor of the Na^+–Ca^{2+} exchanger (Li *et al.*, 1991; see below). The exchanger inhibitory peptide (XIP) is much more potent and specific than dichlorobenzamil but still is not completely specific. In addition, the site of action for XIP is at the intracellular surface of the exchanger and thus inaccessible for most physiological experiments.

F. Temperature Dependence

The Na^+–Ca^{2+} exchanger in mammals shows the typical dependence of membrane transporters on temperature (Bartschat and Lindenmayer, 1980). Activity declines about two-fold per 10° above 25°C and more sharply below 20°C. Three recent studies have compared the temperature dependencies of Na^+–Ca^{2+} exchange in mammals with those in poikilothermic species. Bersohn *et al.* (1991) compared exchange in mammalian and amphibian hearts, Tibbits *et al.* (1992) compared mammalian and teleost hearts, and Tessari and Rahamimoff (1991) compared exchange in rat brain and *Torpedo* electric organ. Although exact details varied, the general conclusion is that the Na^+–Ca^{2+} exchanger of poikilotherms maintains substantial activity at low temperatures, which inactivate mammalian exchangers. For example, Na^+–Ca^{2+} exchange activity in sarcolemmal vesicles from trout heart is about 65% of maximal activity at 7°C, whereas the activity in dog heart is only about 3% of maximal at this temperature (Tibbits *et al.*, 1992).

The lower temperature sensitivity of Na^+–Ca^{2+} exchange in poikilotherms seems to be a feature inherent to the Na^+–Ca^{2+} exchange protein itself. The alternative explanation, that the different membrane lipid environment of poikilotherms determines the temperature dependence of Na^+–Ca^{2+} exchange, is apparently ruled out by reconstitution studies (Bersohn *et al.*, 1991; Tessari and Rahamimoff, 1991; Tibbits *et al.*, 1992). For example, if sarcolemmal vesicles from trout or frog heart are solubilized and then reconstituted in asolectin, the temperature dependence of exchange in the reconstituted vesicles is unchanged from that in the native sarcolemmal vesicles (Bersohn *et al.*, 1991; Tibbits *et al.*, 1992). The implication is that the exchanger proteins of poikilotherms and homeotherms have evolved to cope with a different range of temperatures independent of the lipid environment.

IV. Regulation

Most information on the regulation of Na^+–Ca^{2+} exchange comes from experiments with the following preparations: cardiac sarcolemmal vesicles, squid giant axons, whole-cell voltage clamped myocytes, and giant excised patches of sarcolemma. Little regulatory information has emerged on the ROS exchanger.

A. Ca_i^{2+} Regulation

In addition to having a transport site for Ca^{2+}, the cardiac-type Na^+–Ca^{2+} exchanger has a high-affinity Ca^{2+} regulatory site at the intracellular surface. This secondary regulation by internal Ca^{2+} is seen in studies of Na_i^+–Ca_o^{2+} exchange and was first described in the squid axon by Baker and McNaughton (1976). Later work by DiPolo and Beauge characterized the regulatory role of intracellular Ca^{2+} more directly (DiPolo, 1979; DiPolo and Beauge, 1986, 1987b). The activation by Ca^{2+} has now also been described in barnacle muscle (Rasgado-Flores and Blaustein, 1987), dialyzed myocytes (Kimura *et al.*, 1986), and excised sarcolemmal patches (Hilgemann, 1990). The phenomenon of regulation by intracellular Ca^{2+} is generally the same in each case. When internal Ca^{2+} drops, Na_i^+–Ca_o^{2+} exchange is inactivated. The K_d for Ca^{2+} at the regulatory site varies in the different systems and also varies with conditions. For example, in the squid axon, the K_d of the regulatory site for Ca^{2+} is about 2 μM in the presence of ATP and is about 12 μM in the absence of ATP (DiPolo and Beauge, 1988). In dialyzed myocytes, the K_d is about 50 nM (Kimura and

Miura, 1988; Noda *et al.*, 1988). In giant excised sarcolemmal patches, the K_d is about 1 μM but varies considerably with conditions. Hilgemann *et al.* (1991) have observed that these effects take a few seconds each for completion upon changing internal Ca^{2+} in the excised patch, suggesting that the kinetics of association and dissociation of Ca^{2+} are slow, or alternatively that conformational changes of the exchanger protein which accompany Ca^{2+}-binding and unbinding are slow.

The significance of the Ca^{2+} regulatory site is unclear. In the cardiac system, Ca^{2+} entry through the Ca^{2+} channel upon depolarization could increase Ca^{2+} binding to the exchanger regulatory site and activate Ca^{2+} influx through the exchanger during the early phase of the action potential. Alternatively, during diastole the Ca^{2+} regulatory site might act as a safety valve to prevent the exchanger from lowering cytoplasmic Ca^{2+} too far. That is, once Ca^{2+} was reduced below about 100 nM, the exchanger would inactivate to prevent further Ca^{2+} efflux. Interestingly, the ROS exchanger is unable to lower free internal Ca^{2+} to a level such that the exchanger is at equilibrium (Schnetkamp *et al.*, 1992a). Although an internal Ca^{2+} regulatory site has not been described in ROS, the system is apparently somehow regulated to prevent the exchanger from lowering Ca^{2+} below physiologic levels.

The presence of a high-affinity Ca^{2+} regulatory site on the exchanger has never been demonstrated in isolated plasma membrane vesicles. Reeves and Poronnik (1987) described a stimulation of Na_i^+-dependent Ca^{2+} influx in sarcolemmal vesicles by intravesicular Ca^{2+}. The stimulatory effect, however, was due to a very low-affinity (0.1–0.5 mM range) Ca^{2+} binding site. The secondary Ca^{2+} regulation observed in vesicles is apparently due to a mechanism independent of that seen in more intact preparations.

B. ATP and Phosphorylation

Until recently, the only system in which unequivocal effects of ATP on Na^+–Ca^{2+} exchange could be demonstrated was the squid giant axon. In the axon, stimulation of exchange by ATP requires internal Ca^{2+} and hydrolyzable ATP analogs (DiPolo and Beauge, 1987a). The data suggest the involvement of a Ca^{2+}-dependent protein kinase. The effect is potentiated by using ATPγS consistent with this hypothesis (DiPolo and Beauge, 1987b). The effects of ATP are primarily to increase the affinity of the exchanger for its substrates, Na^+ and Ca^{2+}.

In cardiac tissues, effects of ATP on Na^+–Ca^{2+} exchange had been difficult to substantiate until the development of the giant excised sarcolemmal patch technique. Outward Na^+–Ca^{2+} exchange current in sarcolemmal patches is stimulated by Mg-ATP. Unlike the exchanger of squid axon, however, the stimulation of the cardiac exchanger by ATP does not require internal Ca^{2+} and is not mimicked by ATPγS. Hilgemann (1990) was unable to find evidence for the involvement of a kinase or phosphatase. Thus, the mechanisms by which ATP affects the exchanger in squid axon and sarcolemma appear to be different.

One hypothesis is that the effects of ATP on the cardiac exchanger are due to an indirect effect on the lipid asymmetry of the sarcolemmal membrane. Excised patches used to measure exchanger activity are derived from myocytes treated overnight at 4°C and are presumably ATP depleted. At low ATP concentrations, the normal asymmetric arrangement of sarcolemmal phospholipids is expected to break down. Phosphatidylserine, a negatively charged phospholipid normally constrained to the inner leaflet of the bilayer (Post *et al.*, 1988), becomes randomly distributed. Exchange is inactivated due to the loss of anionic lipids at the intracellular surface. When Mg-ATP is added to sarcolemmal patches, a sarcolemmal "flippase" can recreate normal phospholipid asymmetry of Na^+–Ca^{2+} exchange activity (see below). Extensive data supporting this interesting model are found in Collins *et al.* (1992) and Hilgemann and Collins (1992).

There are currently no data indicating that phosphorylation or ATP directly affect the cardiac Na^+–Ca^{2+} exchanger. The marked ATP action on Na^+–Ca^{2+} exchange in excised patches may be mediated by phospholipids. The physiological significance of this effect is unclear. Presumably ATP levels in myocardium are always sufficient to maintain phospholipid asymmetry. Perhaps during myocardial ischemia or other pathological situations, phospholipid distribution is disturbed and exchange activity is affected.

In other systems, there have been sporadic reports of effects of ATP on Na^+–Ca^{2+} exchange. Caroni and Carafoli (1983) reported that the exchange activity of cardiac sarcolemmal vesicles was modulated by a calmodulin-dependent kinase and phosphatase. These data, however, have proven difficult to reproduce. Vigne *et al.* (1988) and Furukawa *et al.* (1991) found that phorbol esters stimulated the Na^+–Ca^{2+} exchange of cultured smooth muscle cells, suggesting the involvement of protein kinases. Perhaps the exchangers of smooth and cardiac muscle sarcolemma can be regulated by different mechanisms.

C. Stimulation in Vesicles

There are a number of ways to stimulate Na^+–Ca^{2+} exchange, which have been first noted using isolated sarcolemmal vesicles. This topic has recently been described in detail (Reeves and Philipson, 1989) and will only be briefly reviewed here.

1. Proteinase Treatment

Mild treatment of sarcolemmal vesicles with a variety of proteinases markedly activates Na^{+}–Ca^{2+} exchange activity (Philipson and Nishimoto, 1982b). Concurrent with the stimulation is a conversion of the 120-kDa Na^{+}–Ca^{2+} exchange protein to a 70-kDa form (see below). Although initially described as a way to stimulate exchange in vesicles, chymotrypsin also activates Na^{+}–Ca^{2+} exchange in giant excised patches when applied to the intracellular surface (Hilgemann, 1990). After treatment of the patches with chymotrypsin, outward exchange currents are no longer modulated by either ATP or internal Ca^{2+}, and XIP is a less effective inhibitor.

2. Lipid Environment

Na^{+}–Ca^{2+} exchange activity in sarcolemmal vesicles is quite sensitive to membrane lipid environment. In general, many (but not all) anionic lipid components stimulate exchange activity. A specific interaction appears to develop between the negatively charged group of membrane lipid components and the exchanger. The stimulation is affected by both the hydrophobic portion of amphiphiles and the headgroup. Specifically, anionic amphiphiles activate exchange most potently if the amphiphile also causes some disordering within the lipid bilayer. The stimulation of exchange by anionic lipid components is not due to effects on surface charge or the diffuse double layer. These studies are summarized in a recent review (Philipson, 1990b). Experiments on the dependence of the exchange on membrane environment have used phospholipases (Philipson *et al.*, 1983; Philipson and Nishimoto, 1984), a variety of amphipathic molecules (Philipson, 1984; Philipson and Ward, 1985, 1987), and reconstitution techniques (Soldati *et al.*, 1985; Vemuri and Philipson, 1989, 1990).

Again, perturbations to study the Na^{+}–Ca^{2+} exchanger had been applied only to sarcolemmal vesicles until the development of the giant excised patch technique. As described above, anionic phospholipid asymmetry has been implicated in the effect of ATP on sarcolemmal patches. Direct evidence that anionic lipids at the cytoplasmic surface activate Na^{+}–Ca^{2+} exchange comes from experiments applying phospholipase D or dodecyl sulfate to the cytoplasmic surface of excised patches (Hilgemann and Collins, 1992). In retrospect, the stimulation of vesicular Na^{+}–Ca^{2+} exchange by increasing anionic lipids is likely due to an increase in anionic lipids in the outer leaflet of inside-out sarcolemmal vesicles (i.e., at the intracellular surface). At the time, however, it was not clear that only inside-out vesicles were taking part in the exchange reactions (see Section III,B). Thus, the vesicle and excised patch data are completely consistent.

The membrane sterol component can also have a profound effect on Na^+–Ca^{2+} exchange activity. Vemuri and Philipson (1989) did reconstitution experiments in which Na^+–Ca^{2+} exchange activity was assayed in phosphatidylcholine:phosphatidylserine proteoliposomes containing several different sterols. High cholesterol levels (20% by weight) were required for maximal exchange activity. Surprisingly, several other sterols with only minor structural changes from cholesterol were unable to support Na^+–Ca^{2+} exchange activity. This implies a specific interaction between the exchanger and cholesterol. Kutryk and Pierce (1988) have also suggested that cholesterol can stimulate sarcolemmal Na^+–Ca^{2+} exchange activity.

3. Redox Modification

Reeves *et al.* (1986) have shown that a variety of redox reagents (e.g., Fe^{2+} + H_2O_2) stimulate the exchange activity of sarcolemmal vesicles. The stimulation may involve a sulfhydryl–disulfide interchange. Redox modification has also been seen to stimulate outward exchange currents in excised sarcolemmal patches (D. Hilgemann, personal communication).

For over 5 years, regulatory interventions such as proteolysis, lipid perturbation, and redox modification had been studied only in sarcolemmal vesicles. There was an uncertainty whether those results had relevance to other preparations or whether they were artifacts of the vesicular preparation. The physiological relevance of such modes of regulation is still unclear, but it is somewhat comforting that vesicular regulatory mechanisms have now also been observed in the giant excised sarcolemmal patch.

4. Ca^{2+} Chelators

The presence of EGTA, EDTA, or CDTA can stimulate the Na^+–Ca^{2+} exchanger activity of cardiac sarcolemmal vesicles (Trosper and Philipson, 1984). That is, when free Ca^{2+} is adjusted to equal levels in the presence and absence of a Ca^{2+} chelator, higher exchange activity is observed in the presence of the chelator. The system behaves as though some of the chelator-bound Ca^{2+} is available for transport. Although the exact mechanism of this effect of Ca^{2+} chelators is unknown, the effect has important consequences for the design and interpretation of experiments. To work at physiologic intracellular Ca^{2+} levels, it is often required that a Ca^{2+} chelator be present. However, the Ca^{2+} chelator itself may be affecting the measurement.

Each of the four mechanisms just described for stimulating the Na^+–Ca^{2+} exchange of sarcolemmal vesicles acts primarily by increasing the apparent affinity of the transport mechanism for Ca^{2+}. It is not yet deciphered whether each of the mechanisms is working via a common pathway or not. Possibly, some of the mechanisms induce the same conformational change to stimulate exchange activity.

V. Modeling and Reaction Mechanism

During the cardiac action potential and excitation–contraction coupling, intracellular Ca^{2+} undergoes rapid concentration fluctuations. There have been several attempts to model the role of Na^+–Ca^{2+} exchange during these events. The exact role of exchange on a rapid time scale is a subtle function of membrane potential, Ca^{2+} concentration, competing Ca^{2+} transport processes, and spatial localization of the Ca^{2+} transients. The complexity has made it difficult to provide an unequivocal model of exchanger function under all the varying conditions. The mathematical approaches to these important physiological questions will not be reviewed here but are elegantly presented in several references (see, e.g., Hilgemann and Noble, 1987; Hilgemann, 1988; Kline *et al.*, 1990).

More pertinent to the present review is modeling relevant to the reaction mechanism of the Na^+–Ca^{2+} exchanger. For a countertransporter like the exchanger, two general kinetic schemes are most likely: simultaneous or consecutive mechanisms. In a simultaneous reaction mechanism, the exchanger is viewed as having two sets of binding sites that are simultaneously exposed on opposite surfaces of the exchanger. Na^+ and Ca^{2+} bind at the same time and exchange then occurs. In contrast, for a consecutive reaction mechanism (Fig. 1), there is only one set of binding sites that can bind either 1 Ca^{2+} or 3 Na^+ and that is exposed only at one surface of the membrane. Na^+ or Ca^{2+} binds, moves across the membrane with the binding site, and is then released. The exchanger binding sites are then available to again bind Na^+ or Ca^{2+} to return across the membrane and complete the reaction cycle.

Despite the drastic differences in these two reaction schemes, the appropriate experimental data to distinguish the correct scheme had not been available. In general, the simultaneous model was favored without strong experimental evidence (Hilgemann, 1988). Recent data, however, strongly suggest that the consecutive model correctly describes the behavior of the cardiac exchanger. First, Khananshvili (1990) studied Na^+-dependent

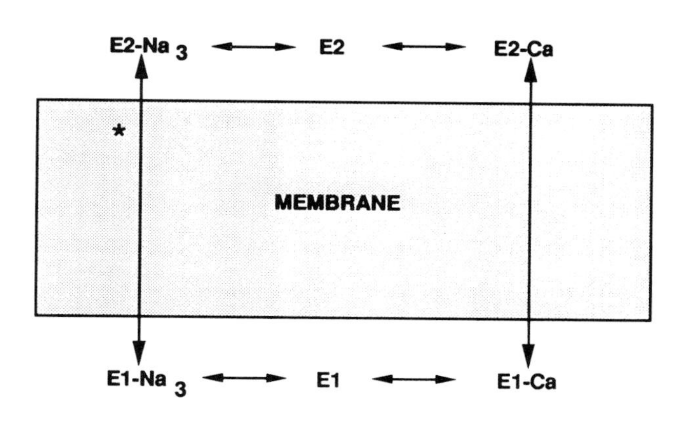

FIG. 1 The cardiac Na^+–Ca^{2+} exchanger has a consecutive reaction mechanism. In one state of the exchanger (E1), the ion binding site is exposed to the cytoplasm and can bind one Ca^{2+} or three Na^+ ions. Following a conformational change, the binding site becomes exposed to the extracellular surface and the ions are released. The exchanger (E2) can again bind either Na^+ or Ca^{2+} and complete the reaction cycle. There is no indication that the transition between exposure to the extra- and intracellular surfaces can occur when the binding site is empty.

Ca^{2+} uptake in reconstituted proteoliposomes loaded with a Ca^{2+} chelator. This permitted exchange rates to be measured under zero-*trans* conditions. By studying the kinetic characteristics of Ca^{2+} uptake as function of intravesicular Na^+ concentration, it was deduced that the exchanger operated in a consecutive mechanism. In a follow-up study, Khananshvili (1991) suggested that only one step in the exchange reaction mechanism (either Na^+ or Ca^{2+} translocation) was voltage dependent. Niggli and Lederer (1991) examined current transients induced by the photorelease of intracellular Ca^{2+} in myocytes. They concluded that the transients were due to an initial translocation of Ca^{2+} across the sarcolemma by the exchanger with a consecutive exchanger reaction mechanism. Alternative explanations to these latter data exist, however, and it is unclear whether the current transients are related to Na^+–Ca^{2+} exchange.

Strong support for a consecutive reaction mechanism for the cardiac Na^+–Ca^{2+} exchanger comes from the use of the giant excised patch technique for studying exchange currents. Hilgemann *et al.* (1991) noted that the apparent affinity of one ion (i.e., Na^+ or Ca^{2+}) for transport was a function of the concentration of the other ion on the opposite side of the membrane (Fig. 2). This characteristic is the hallmark of a consecutive reaction mechanism for a countertransporter.

In this study (Hilgemann *et al.*, 1991), half-cycle reactions were also examined by ion pulse experiments. With the rapid application of Na^+ to the intracellular surface, Na^+ becomes bound and is translocated. If this step of the reaction cycle involves the net movement of charge, a transient current should be observed (Fig. 3). Transient currents with these characteristics were observed both in sarcolemmal patches and in membrane patches from oocytes expressing the cloned Na^+–Ca^{2+} exchanger. Transient currents were absent in patches from control, water-injected, oocytes providing unequivocal evidence that the transients were associated with Na^+–Ca^{2+} exchange. The overall data were consistent with a consecutive reaction mechanism in which net charge moves across the membrane electric field during Na^+ translocation but in which Ca^{2+} translocation is largely electrically silent.

Although data are accumulating that the cardiac exchanger operates with a consecutive reaction mechanism, it is possible that the actual mechanism is more complex and involves multiple steps. A contrasting view is put forward by Milanick (1991), who has argued that the exchanger mechanism in ferret red blood cells is simultaneous. Perhaps different mechanisms are possible with different exchanger types or under different conditions.

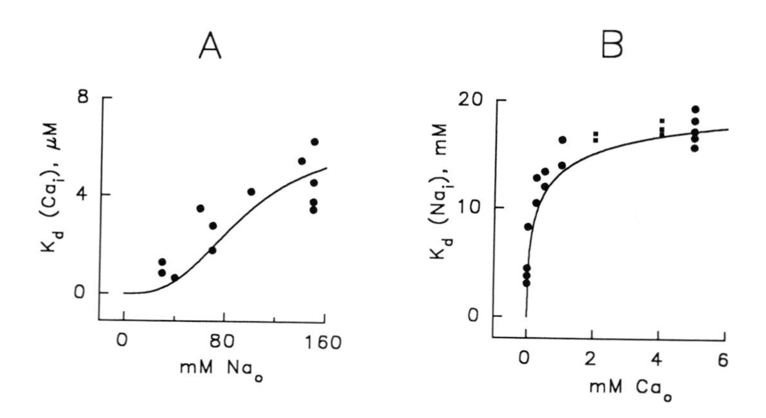

FIG. 2 Exchanger ion affinities depend on the concentration of counterion. Na^+–Ca^{2+} exchange currents were measured using giant excised patches from myocytes. (A) Dependence of the affinity for Ca^{2+} at the internal surface on Na^+ concentration at the extracellular surface. (B) Dependence of the affinity for Na^+ at the internal surface on Ca^{2+} concentration at the extracellular surface. The dependence of K_d on the counterion concentration is characteristic of a consecutive reaction. (After Hilgemann *et al.*, 1991, Fig. 3, p. 717.)

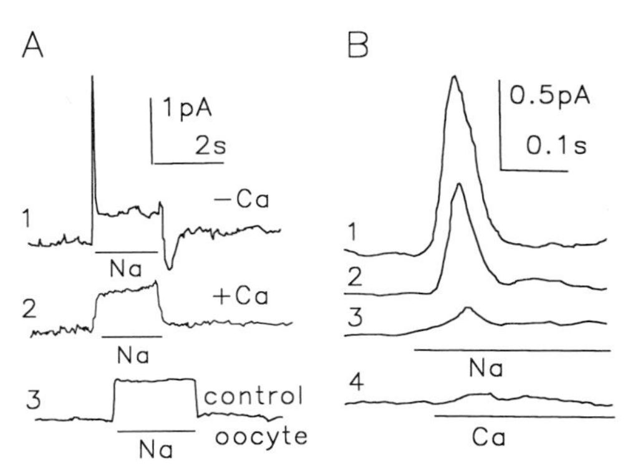

FIG. 3 Charge movements due to half-reaction cycles of the Na^+–Ca^{2+} exchanger. Measurements were made using giant excised patches from oocytes expressing the cardiac exchanger. (A) Trace 1: Transient outward current observed when Na^+ is rapidly applied at the intracellular surface. The current is due to Na^+ translocation but is transient because there is no Ca^{2+} and only very low Na^+ at the extracellular surface so that the reaction cycle cannot be completed. The steady-state current is due to a greater permeability of the membrane to Na^+ than to Li^+. Trace 2: The transient current is abolished if the patch is preincubated with Ca^{2+} at the intracellular surface. This forces exchanger binding sites to the external surface making the sites unavailable to Na^+ (see Fig. 1). Trace 3: The transient current is not present in patches from control oocytes not expressing exchange activity. This provides unequivocal evidence that the transient is associated with the Na^+–Ca^{2+} exchanger. (B) Transient currents, as in (A), at faster time resolution. Trace 1: Transient current due to an exchanger half-reaction cycle. Traces 2 and 3: The magnitude of the transient current is reduced by preincubation with 2 or 5 m*M* Na^+, respectively, prior to the 100 m*M* Na^+ pulse. The preincubation with Na^+ causes the binding sites to become externally oriented and unavailable to the Na^+ pulse. Trace 4: Transient currents are not induced by a Ca^{2+} pulse. This indicates that charge movement is associated with Na^+ translocation, but not with Ca^{2+} translocation. (After Hilgemann *et al.*, 1991, Fig. 3, p. 716.)

VI. Purification of the Exchangers

A. Heart

The cardiac Na^+–Ca^{2+} exchanger has proven to be a difficult protein to isolate. Sarcolemmal vesicles are first solubilized, and the solubilized proteins fractionated. Since the only marker for the exchanger is transport, each fraction is then reconstituted to assay for exchange activity. The low abundance and lability of the cardiac exchanger have made this task difficult. There have been several reports misidentifying the cardiac exchanger protein (for reviews see Philipson, 1990a; Reeves, 1990).

The first successful identification of the cardiac Na^+–Ca^{2+} exchange proteins used a combination of alkaline extraction, solubilization with the detergent decylmaltoside, ion-exchange chromatography on DEAE–Sepharose, and affinity chromatography with wheat-germ agglutinin (Philipson *et al.,* 1988). Analysis of the purified exchanger by SDS–PAGE under reducing conditions shows three major polypeptides migrating at 160, 120, and 70 kDa. Under nonreducing conditions only one prominent polypeptide at 160 kDa is observed. More recently, Durkin *et al.* (1991) have also described the isolation of the cardiac exchanger principally using anion exchange chromatography.

The relationship among the three polypeptides was investigated by proteolytic analysis (Philipson *et al.,* 1988). Treatment of the purified exchanger with chymotrypsin under mild conditions resulted in a decrease in the amount of the 120-kDa and an increase in the amount of 70-kDa polypeptide. This indicates that the 70-kDa polypeptide probably arises from proteolysis of the 120-kDa polypeptide. Additionally, affinity-purified antibodies specific for the 160-, the 120-, or the 70-kDa polypeptide each reacted with all three of the 160-, 120-, and 70-kDa polypeptides (Philipson *et al.,* 1988), proving that the three polypeptides were immunologically related. The 120- and 160-kDa proteins have the same amino-terminal amino acid sequence, and the relative intensities of the two bands can be altered by the procedures used for preparing the samples for electrophoresis (Durkin *et al.,* 1991).

These data suggest that the exchanger is a single polypeptide. The nonreduced exchanger appears to migrate as a 160-kDa species. Reduction shifts the protein primarily to a 120-kDa species and proteolysis reduces the protein to a smaller 70-kDa form.

The identity of the exchanger polypeptides was further confirmed following the successful cloning, sequencing, and expression of the exchanger (see below). Based on the deduced amino acid sequence of the exchanger a short peptide was synthesized and antibodies to the peptide were produced. On Western blots of sarcolemmal membrane proteins, the antibody to the peptide gave the same reactions as an antibody raised against the purified exchanger, thereby verifying the molecular identity of the exchanger (Nicoll *et al.,* 1990).

B. Rod Outer Segments

The bovine ROS exchanger was independently purified by two groups (Cook and Kaupp, 1988; Nicoll and Applebury, 1989). Cook and Kaupp (1988) employed a three-step purification protocol. Starting with solubilized ROS membranes, the first step was ion-exchange chromatography through a DEAE–HPLC column to yield a 40-fold enrichment of exchanger activity. The exchanger is a very anionic protein and binds to the anion exchange resin at relatively high salt concentrations, allowing for a significant separation from more neutrally charged proteins.

Following elution of the exchanger from the DEAE column, the exchanger-containing solution was applied to an AF Red HPLC column. The exchanger did not bind to the resin of this column. One of the major contaminants, the cGMP-gated channel, was bound and the exchanger activity was collected in the flow-through fraction. This step provided an additional 1.6-fold enrichment of exchanger activity.

The final step in the purification scheme was to bind the exchanger to a concanavalin A (Con A)–Sepharose column and elute the exchanger with the competing sugar, α-methyl-D-mannoside. Since Con A is a lectin and binds specific sugar residues, the binding of the exchanger to Con A verifies that the exchanger is glycosylated.

Purification of the exchanger from solubilized membranes through the three steps resulted in a 108-fold enrichment of exchanger activity and a single polypeptide on Coomassie blue-stained gels of about 220 kDa. That the 220-kDa polypeptide was the exchanger was verified by gel filtration chromatography, which separates proteins on the basis of their size. Exchanger activity corresponded directly with the 220-kDa polypeptide.

Nicoll and Applebury (1989) also purified the ROS exchanger by selective solubilization of ROS membrane proteins, DEAE–Sepharose chromatography, and a different lectin–resin, wheat germ agglutinin–agarose, to yield a single polypeptide of approximately 215 kDa.

C. Brain

Although the heart and ROS exchangers were purified by traditional chromatographic procedures, the brain exchanger has been partially purified by the use of a novel technique called transport specificity fractionation (Papazian *et al.*, 1979). This technique utilizes the ability of a transport protein, such as the exchanger, to transport Ca^{2+}. Membrane proteins are reconstituted into liposomes under conditions that result in one protein per liposome and in the presence of a sodium salt containing a counterion that can form a precipitant with Ca^{2+} (e.g., phosphate). Those liposomes containing a Ca^{2+} transporter take up Ca^{2+} and form a dense precipitant within the liposome. The more dense liposomes are then separated by density gradient centrifugation.

Barzilai *et al.* (1984) used transport specificity fractionation to obtain 128-fold enrichment of exchanger activity from brain synaptic plasma membranes. Analysis of the resultant polypeptides by Coomassie blue staining after SDS–PAGE revealed several major polypeptides and significant enrichment of a 70-kDa polypeptide.

To identify which polypeptide(s) in the enriched fraction might be the exchanger, Barzilai *et al.* (1987) immunized rabbits with each of the polypeptides. Antibodies to 70- and 33-kDa polypeptides were obtained and analyzed by Western blots and immunoprecipitation of exchanger activity. Antibodies raised against each of the polypeptides were able to immunoprecipitate activity and react with both polypeptides. It was concluded that the 33- and 70-kDa proteins were immunologically related and associated with the brain exchanger.

Yip *et al.* (1992) have recently compared the brain and heart exchangers on Western blots. Proteins from rat synaptic plasma membranes and canine cardiac sarcolemma were probed with the sarcolemmal exchanger polyclonal antibody. In the synaptic plasma membrane, the antibody recognized polypeptides at 150, 120, and 70 kDa, and in the cardiac sarcolemma, the antibody recognized polypeptides at 160, 120, and 70 kDa. Hence, the exchanger of brain synaptosomes is likely to be a polypeptide similar to that which is expressed in the cardiac sarcolemma. The 70- and 33-kDa polypeptides that were obtained by Barzilai *et al.* (1984) by transport specificity fractionation may be the result of proteolytic degradation. Their fractionation procedure is time intensive and brain tissue is high in proteolytic activity, giving ample opportunity for degradation of the exchanger.

In summary, the purified sarcolemmal exchanger is 120 kDa, the ROS exchanger is 220 kDa, and the brain exchanger also appears to be 120 kDa. This suggests that there may be two families of exchangers, an SL-type which is 120-kDa and has a stoichiometry of 3 Na^+:1 Ca^{2+}, and a ROS-type with a molecular mass of 220 kDa and a stoichiometry of 4 Na^+:1 Ca^{2+} + 1 K^+. The extent of similarity between the sarcolemmal and brain exchangers remains to be determined.

VII. Cloning the Exchangers

A. Heart

A cDNA clone for the exchanger was obtained by screening a λgt11 expression library with an antibody to the exchanger (Nicoll *et al.*, 1990). A 6.5-kb clone, which hybridized with mRNA of 7 kb on Northern blots, was isolated. The coding region of the clone was sequenced and found to specify a protein of 970 amino acids with a predicted molecular mass of 110 kDa. Ribonucleic acid was synthesized from the exchanger clone and injected into *Xenopus* oocytes to induce exchange activity. Cells injected with exchanger cRNA had high levels of exchange activity compared to cells injected with water.

The exchanger contains six potential asparagine-linked glycosylation sites. *In vitro* translation analysis of the exchanger indicates that the exchanger is glycosylated only at the first of these sites and that the exchanger has a cleaved leader peptide (D. A. Nicoll and K. D. Philipson, unpublished observations). Glycosylation of the exchanger does not appear to be necessary for exchanger function. A mutant with asparagine 9 replaced with a tyrosine is still capable of expressing exchanger activity in oocytes at levels equivalent to that of the wild-type exchanger (D. A. Nicoll and K. D. Philipson, unpublished observations).

The leader peptide was predicted by the presence of a consensus leader peptide cleavage site (von Heijne, 1986; Nicoll *et al.*, 1990). Durkin *et al.* (1991) have confirmed that the amino terminus of the exchanger is cleaved posttranslationally by amino acid sequence analysis of the purified protein. They have determined that the amino-terminal sequence of the exchanger is ETEMEG, which is 32 amino acids from the initiating methionine. Following cleavage of the leader peptide, the predicted molecular mass of the polypeptide portion of the exchanger is 105 kDa.

A proposed model for the exchanger is illustrated in Fig. 4. The model is based on the following:

(*a*) Hydropathy analysis of the sequence indicates that the exchanger contains 12 membrane-spanning segments. The first corresponds to the leader sequence, which is cleaved so that the mature protein is predicted to have 11 membrane spanning segments.

(*b*) The asparagine at position 9 is glycosylated. Since the vast majority of glycosylation sites in other membrane proteins are located at the extracellular surface, the amino terminus of the exchanger is modeled to be extracellular.

(*c*) The region spanning membrane segments 4 and 5 is 48% identical to a region in the Na/K-ATPase (Fig. 4), which has been modeled to be extracellular (Schull *et al.*, 1985). A similar region in the sarcoplasmic reticular Ca^{2+} pump is involved in Ca^{2+} binding (Clarke *et al.*, 1989).

(*d*) The large hydrophilic region (520 amino acids) between membrane-spanning segments 5 and 6 contains the sequence from which XIP (see below), a peptide inhibitor of the exchanger was derived; XIP inhibits the exchanger at the intracellular surface. The region between membrane-spanning segments 5 and 6 is therefore likely to be intracellular. However, hydropathy analysis is far from foolproof, and the topology of this hydrophilic region may be more complicated than predicted.

The role of the hydrophilic region between membrane-spanning segments 5 and 6 is being investigated in different ways. First, a potential autoinhibitory region of the exchanger has been identified. Residues

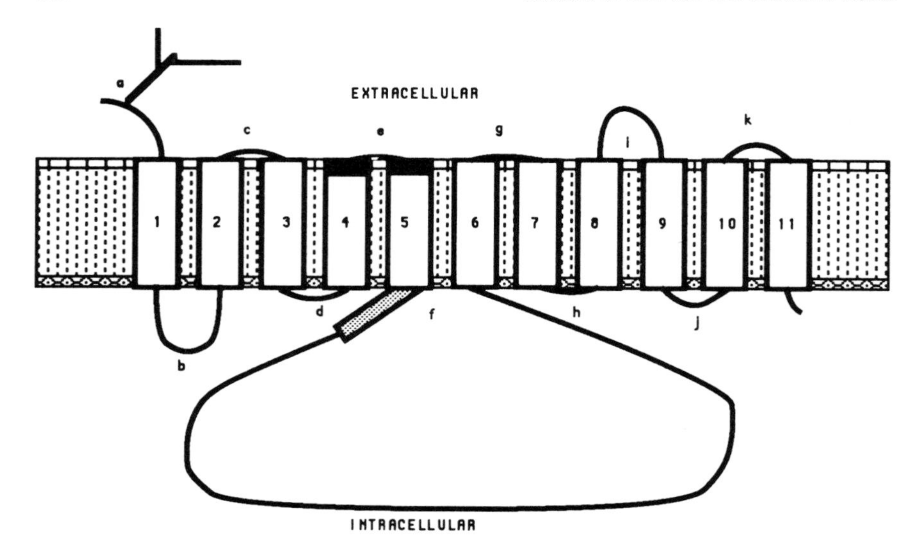

FIG. 4 Model for the cardiac exchanger protein. The exchanger protein is modeled to have the amino terminus, with residue 9 glycosylated, on the extracellular surface and the carboxy terminus on the intracellular surface. The exchanger contains 11 putative membrane-spanning segments (numbered rectangles). Regions of interest that are highlighted are the region of sequence similarity to the Na/K-ATPase (in black) and the region from which the XIP peptide sequence was derived (stippled).

219–238, in the hydrophilic region between membrane-spanning segments 5 and 6, is similar to the calmodulin binding domains of a number of proteins. For some proteins, such domains have been found to be autoinhibitory. That is, a peptide corresponding to the calmodulin binding domain can inhibit protein function (see, e.g., Enyedi *et al.*, 1989). Similarly, the peptide XIP, corresponding to residues 219–238 of the exchanger, inhibits the exchanger at the intracellular surface (Li *et al.*, 1991) and is more potent and more specific than any other known inhibitor of the exchanger. The region with which XIP interacts has not yet been defined, and whether the "endogenous" XIP region has any regulatory function has not yet been determined.

The role of the large hydrophilic domain has also been examined by deletion mutagenesis. A mutant with a deletion encompassing residues 239–680 was constructed. It is striking that this protein, with over 45% of the amino acids deleted, was still capable of catalyzing Na^+ gradient-dependent Ca^{2+} uptake (K. D. Philipson *et al.*, unpublished observations). Therefore, the deleted region is not essential for ion translocation. Roles for the deleted region may include regulatory functions and/or attachment sites for cytoskeletal elements.

Additional regions that may be of functional significance include the membrane-spanning segments and the region that is similar to the Na/K-ATPase. A number of the membrane-spanning segments contain charged or hydroxyl-containing residues and could form amphipathic helices. These helices could be arranged in the membrane to surround an ion translocation pathway (Nicoll and Philipson, 1991).

B. Rod Outer Segments

The ROS exchanger was cloned by screening a λgt11 expression library with poly- and monoclonal antibodies and also by screening a λgt10 library with a degenerate oligonucleotide derived from the amino acid sequence of a cyanogen bromide fragment of the ROS exchanger (Achilles *et al.*, 1992). The predicted molecular mass of the protein encoded by the open reading frame of the ROS exchanger clone is 130 kDa, considerably less than the 220 kDa observed for the purified protein. The difference between the predicted and the actual size of the ROS exchanger is probably due, in part, to the high level of glycosylation of the ROS exchanger (Reid *et al.*, 1990).

The similarities between the ROS and the sarcolemmal exchangers are limited. Both proteins appear to have cleaved leader sequences and extracellular, glycosylated amino termini. Both proteins have a region containing five potential membrane-spanning segments followed by a long hydrophilic region and then another region forming six potential membrane-spanning segments. The long hydrophilic regions of both proteins are apparently located at the intracellular surface; in the ROS exchanger the localization was determined by using a monoclonal antibody that binds the exchanger at the intracellular surface and that also reacts with a fusion protein containing part of the hydrophilic region (Achilles *et al.*, 1992). The long hydrophilic regions of both proteins are very acidic. The ROS and sarcolemmal exchangers share two short segments (about 60 amino acids) in the hydrophobic regions with 30 to 40% identity. Thus, the similarity between the two exchangers is primarily at the overall topological level rather than at the level of the amino acid sequence.

More notable are the differences in the sequences of the two proteins. Except for the two short segments in the hydrophobic regions, the exchangers have no sequence similarity. Also, the amino terminus is considerably longer in the ROS than the sarcolemmal exchanger. Although the long hydrophilic regions of both exchangers are very acidic, that of the ROS exchanger is more strikingly so. The ROS exchanger contains 8 repeats of a 17-amino acid sequence with half of the residues in the repeat being acidic. Following the region of repeats, the ROS exchanger has a string of 26 consecutive acidic residues. In retrospect, considering the different stoichiometries and ion specificities of the two exchangers, it is perhaps not surprising that the amino acid sequences are so different.

A note of caution is necessary regarding the topological models of the two exchangers. The topological similarities are based on hydropathy analysis. Very little information on the validity of this analysis is available. The correct number of transmembrane segments and their orientations may be very different from the initial models. Experiments should be interpreted with this caveat in mind.

VIII. Concluding Comments

Na^+–Ca^{2+} exchange research has reached a productive new phase. The availability of molecular probes is expected to accelerate the pace of research. New information on structure and function and also on the presence of other isoforms and gene products should be forthcoming. Immunolocalization and gene regulation studies are in progress and initial results are anticipated in the near future.

References

Achilles, A., Fiedel, U., Haase, W., Reilander, H., and Cook, N. J. (1991). *Ann. N.Y. Acad. Sci.* **634,** 234–244.

Allen, T. J. A., and Baker, P. F. (1985). *Nature (London)* **315,** 755–756.

Allen, T. J. A., and Baker, P. F. (1986). *J. Physiol. (London)* **378,** 53–76.

Allen, T. J. A., Noble, D., and Reuter, H., eds. (1989). "Sodium–Calcium Exchange." Oxford Univ. Press, Oxford.

Baker, P. F., and McNaughton, P. A. (1976). *J. Physiol. (London)* **259,** 104–114.

Baker, P. F., Blaustein, M. P., Hodgkin, A. L., and Steinhardt, R. A. (1969). *J. Physiol. (London)* **200,** 431–458.

Barcenas-Ruiz, L., Beuckelmann, D. J., and Wier, W. G. (1987). *Science* **328,** 1720–1722.

Bartschat, K. K., and Lindenmayer, G. E. (1980). *J. Biol. Chem.* **255,** 9626–9634.

Barzilai, A., Spanier, R., and Rahamimoff, H. (1984). *Proc. Natl. Acad. Sci. U.S.A.* **81,** 6521–6525.

Barzilai, A., Spanier, R., and Rahamimoff, H. (1987). *J. Biol. Chem.* **262,** 10315–10320.

Berlin, J. R., Hume, J. R., and Lederer, W. J. (1988). *J. Physiol. (London)* **407,** 128P.

Bers, D. M., and Bridge, J. H. B. (1989). *Circ. Res.* **65,** 334–342.

Bersohn, M. M., Vemuri, R., Schuil, D. W., Weiss, R. S., and Philipson, K. D. (1991). *Biochim. Biophys. Acta* **1062,** 19–23.

Blaustein, M. P. (1984). *In* "Electrogenic Transport: Fundamental Principles and Physiologic Implications" (M. P. Blaustein and M. Lieberman, eds.), pp. 129–147. Raven, New York.
Blaustein, M. P., Ashida, T., Goldman, W. F., Weir, W. G., and Hamlyn, J. M. (1986). *Ann. N.Y. Acad. Sci.* **488,** 199–216.
Blaustein, M. P., DiPolo, R., and Reeves, J. P., eds. (1991). *Ann. N.Y. Acad. Sci.* **639.**
Bridge, J. H. B., Smolley, J. R., and Spitzer, K. W. (1990). *Science* **248,** 376–378.
Caroni, P., and Carafoli, E. (1983). *Eur J. Biochem.* **132,** 451–460.
Cervetto, L., Lagnado, L., Perry, R. J., Robinson, D. W., and McNaughton, P. A. (1989). *Nature (London)* **337,** 740–743.
Clarke, D. M., Loo, T. W., Inesi, G., and MacLennan, D. H. (1989). *Nature (London)* **339,** 476–478.
Collins, A., Somlyo, A. V., and Hilgemann, D. W. (1992). *J. Physiol.* **454,** 27–57.
Cook, N. J., and Kaupp, U. B. (1988). *J. Biol. Chem.* **263,** 11382–11388.
Crespo, L. N., Grantham, C. J., and Cannell, M. B. (1990). *Nature (London)* **345,** 619–621.
Crompton, M. (1990). "Calcium and the Heart" (G. A. Langer, ed.), pp. 167–198. Raven, New York.
DiPolo, R. (1979). *J. Gen. Physiol.* **73,** 91–113.
DiPolo, R., and Beauge, L. (1986). *Biochim. Biophys. Acta* **854,** 298–306.
DiPolo, R., and Beauge, L. (1987a). *Biochim. Biophys. Acta* **897,** 347–354.
DiPolo, R., and Beauge, L. (1987b). *J. Gen. Physiol.* **90,** 505–525.
DiPolo, R., and Beauge, L. (1988). *Biochim. Biophys. Acta* **947,** 549–569.
DiPolo, R., and Rojas, H. (1984). *Biochim. Biophys. Acta* **776,** 313–316.
Durkin, J. T., Ahrens, D. A., Pan, Y.-C. E., and Reeves, J. P. (1991). *Arch. Biochem. Biophys.* **290,** 369–375.
Enyedi, A., Vorherr, T., James, P. McCormick, D. J., Filoteo, A. G., Carafoli, E., and Penniston, J. T. (1989). *J. Biol. Chem.* **264,** 12313–12321.
Furukawa, K.-I., Ohshima, N., Tawada-Iwata, Y., and Shigekawa, M. (1991). *J. Biol. Chem.* **266,** 12337–12341.
Gadsby, D. C. (1991). *Ann. N.Y. Acad. Sci.* **639,** 140–146.
Hamlyn, J. M., Harris, D. W., and Ludens, J. H. (1989). *J. Biol. Chem.* **264,** 7395–7404.
Hilgemann, D. W. (1988). *Prog. Biophys. Mol. Biol.* **51,** 1–45.
Hilgemann, D. W. (1989). *Pfluegers Arch.* **415,** 247–249.
Hilgemann, D. W. (1990). *Nature (London)* **344,** 242–245.
Hilgemann, D. W., and Collins, A. (1992). *J. Physiol. (London)* **454,** 59–82.
Hilgemann, D. W., and Noble, D. (1987). *Proc. R. Soc. London, Ser. B* **230,** 163–205.
Hilgemann, D. W., Nicoll, D. A., and Philipson, K. D. (1991). *Nature (London)* **352,** 715–718.
Hilgemann, D. W., Collins, A., Cash, D. P., and Nagel, G. A. (1991). *Ann. N.Y. Acad. Sci.* **639,** 126–139.
Hodgkin, A. L., McNaughton, P. A., and Nunn, B. J. (1985). *J. Physiol. (London)* **358,** 447–468.
Kaczorowski, G. J., Slaughter, R. S., King, V. F., and Garcia, M. L. (1989). *Biochim. Biophys. Acta* **988,** 287–302.
Khananshvili, D. (1990). *Biochemistry* **29,** 2437–2442.
Khananshvili, D. (1991). *J. Biol. Chem.* **266,** 13764–13769.
Kimura, J., and Miura, Y. (1988). *J. Mol. Cell. Cardiol.* **20,** S19.
Kimura, J., Noma, A., and Irisawa, H. (1986). *Nature (London)* **319,** 596–597.
Kimura, J., Miyamae, S. and Noma, A. (1987). *J. Physiol. (London)* **384,** 199–222.
Kline, R. P., Zablow, L., and Cohen, I. S. (1990). *J. Gen. Physiol.* **95,** 499–522.
Kutryk, M. J. B., and Pierce, G. N. (1988). *J. Biol. Chem.* **263,** 13167–13172.
Lagnado, L., Cervetto, L., and McNaughton, P. A. (1988). *Proc. Natl. Acad. Sci. U.S.A.* **85,** 4548–4552.

Leblanc, N., and Hume, J. R. (1990). *Science* **248,** 372–375.
Li, Z., Nicoll, D. A., Collins, A., Hilgemann, D. W., Filoteo, A. G., Penniston, J. T., Weiss, J. N., Tomich, J. M., and Philipson, K. D. (1991). *J. Biol. Chem.* **266,** 1014–1020.
Milanick, M. (1991). *Am. J. Physiol.* **261,** C185–C193.
Miura, Y., and Kimura, J. (1989). *J. Gen. Physiol.* **93,** 1129–1145.
Nicoll, D. A., and Applebury, M. L. (1989). *J. Biol. Chem.* **264,** 16207–16213.
Nicoll, D. A., and Philipson, K. D. (1991). *Ann. N.Y. Acad. Sci.* **639,** 181–188.
Nicoll, D. A., Longoni, S., and Philipson, K. D. (1990). *Science* **250,** 562–565.
Nicoll, D. A., Barrios, B. R., and Philipson, K. D. (1991). *Am. J. Physiol.* **260,** C1212–C1216.
Niggli, E., and Lederer, W. J. (1991). *Nature (London)* **349,** 621–624.
Noda, M., Shepherd, R. N., and Gadsby, D. C. (1988). *Biophys. J.* **53,** 342a.
Papazian, D, Rahamimoff, H., and Goldin, S. M. (1979). *Proc. Natl. Acad. Sci. U.S.A.* **76,** 3708–3712.
Philipson, K. D. (1984). *J. Biol. Chem.* **259,** 13999–14002.
Philipson, K. D. (1990a). *In* "Calcium and the Heart" (G. A. Langer, ed.), pp. 85–108. Raven, New York.
Philipson, K. D. (1990b). *Cell Biol. Int. Rep.* **14,** 305–309.
Philipson, K. D., and Nishimoto, A. Y. (1982a). *J. Biol. Chem.* **257,** 5111–5117.
Philipson, K. D., and Nishimoto, A. Y. (1982b). *Am. J. Physiol.* **243,** C191–C195.
Philipson, K. D., and Nishimoto, A. Y. (1984). *J. Biol. Chem.* **259,** 16–19.
Philipson, K. D., and Ward, R. (1985). *J. Biol. Chem.* **260,** 9666–9671.
Philipson, K. D., and Ward, R. (1987). *Biochim. Biophys. Acta* **897,** 152–158.
Philipson, K. D., Frank, J. S., and Nishimoto, A. Y. (1983). *J. Biol. Chem.* **258,** 5905–5910.
Philipson, K. D., Longoni, S., and Ward, R. (1988). *Biochim. Biophys. Acta* **945,** 298–306.
Post, J. A., Langer, G. A., Op den Kamp, J. A. F., and Verkleij, A. J. (1988). *Biochim. Biophys. Acta* **943,** 256–266.
Rasgado-Flores, H., and Blaustein, M. P. (1987). *Am. J. Physiol.* **252,** C499–C504.
Reeves, J. P. (1990). *In* "Intracellular Calcium Regulation" (F. Bonner, ed.), pp. 306–347. Alan R. Liss, New York.
Reeves, J. P., and Hale, C. C. (1984). *J. Biol. Chem.* **259,** 7733–7739.
Reeves, J. P., and Philipson, K. D. (1989). *In* "Sodium–Calcium Exchange" (T. J. A. Allen, D. Noble, and H. Reuter, eds.), pp. 27–53. Oxford Univ. Press, Oxford.
Reeves, J. P., and Poronnik, P. (1987). *Am. J. Physiol.* **252,** C16–C23.
Reeves, J. P., and Sutko, J. L. (1979). *Proc. Natl. Acad. Sci. U.S.A.* **76,** 590–594.
Reeves, J. P., and Sutko, J. L. (1983). *J. Biol. Chem.* **258,** 3178–3182.
Reeves, J. P., Bailey, C. A., and Hale, C. (1986). *J. Biol. Chem.* **261,** 4948–4955.
Reid, D. M., Friedel, U., Molday, R. S., and Cook, N. J. (1990). *Biochemistry* **29,** 1601–1607.
Reuter, H., and Seitz, N. (1968). *J. Physiol. (London)* **195,** 451–470.
Sanchez–Armass, S., and Blaustein, M. P. (1987). *Am. J. Physiol.* **252,** C595–C603.
Schnetkamp, P. P. M. (1989). *Prog. Biophys. Mol. Biol.* **54,** 1–29.
Schnetkamp, P. P. M., Azerencsei, R. T., and Basu, D. K. (1988). *Biophys. J.* **53,** 389a.
Schnetkamp, P. P. M., Basu, D. K., and Szerencsei, R. T. (1989). *Am. J. Physiol.* **257,** C153–C157.
Schnetkamp, P. P. M., Szerencsei, R. T., and Basu, D. K. (1991). *J. Biol. Chem.* **266,** 198–206.
Schnetkamp, P. P. M., Basu, D. K., Li, X.-B., and Szerencsei, R. T. (1992a). *J. Biol. Chem.* **266,** 22983–22990.
Schnetkamp, P. P. M., Li, X.-B., Basu, D. K., and Szerencsei, R. T. (1992b). *J. Biol. Chem.* **266,** 22975–22982.
Schull, G. E., Schwartz, A., and Lingrel, J. B. (1985). *Nature (London)* **316,** 691.
Schwartz, E. A. (1985). *Annu. Rev. Neurosci.* **8,** 339–367.

Soldati, L., Longoni, S., and Carafoli, E. (1985). *J. Biol. Chem.* **260,** 13321–13327.
Tessari, M., and Rahamimoff, H. (1991). *Biochim. Biophys. Acta* **1066,** 208–218.
Tibbits, G. F., Philipson, K. D., and Kashihara, H. (1992). *Am. J. Physiol.* **262,** C411–C417.
Trosper, T. L., and Philipson, K. D. (1984). *Cell Calcium* **5,** 211–222.
Vemuri, R., and Philipson, K. D. (1989). *J. Biol. Chem.* **264,** 8680–8685.
Vemuri, R., and Philipson, K. D. (1990). *Biochem. Biophys. Res. Commun.* **168,** 917–922.
Vemuri, R., Longoni, S., and Philipson, K. D. (1989). *Am. J. Physiol.* **256,** C1273–C1276.
Vigne, P., Breittmayer, J.-P., Duval, D., Frelin, C., and Lazdunski, M. (1988). *J. Biol. Chem.* **263,** 8078–8083.
von Heijne, G. (1986). *Nucleic Acids Res.* **14,** 4683–4690.
Wakabayashi, S., and Goshima, K. (1981). *Biochim. Biophys. Acta* **645,** 311–317.
Yau, K.-W., and Baylor, D. A. (1989). *Annu. Rev. Neurosci.* **12,** 289–327.
Yip, R. K., Blaustein, M. P., and Philipson, K. D. (1992). *Neurosci. Lett.* **136,** 123–126.

Molecular Analysis of the Role of Na^+/H^+ Antiporters in Bacterial Cell Physiology

Shimon Schuldiner and Etana Padan
Division of Microbial and Molecular Ecology, The Alexander Silberman Institute of Life Sciences, The Hebrew University of Jerusalem, 91904 Jerusalem, Israel

I. Introduction

All living cells maintain a Na^+ cycle across the cytoplasmic membrane. This cycle is driven by Na^+ extruding systems that excrete the ion and maintain a Na^+ concentration gradient ($\Delta\mu_{Na^+}$) directed inward (reviews in Rosen, 1986; Dimroth, 1987, 1992a,b; Skulachev, 1987, 1988; Skou *et al.,* 1988; Tokuda, 1989, 1992; Glynn and Karlish, 1990; Anraku, 1992; Krulwich, 1992; Muller and Gottschalk, 1992; Padan and Schuldiner, 1992; Schonheit, 1992). Na^+ reenters the cells via Na^+ gradient consumers: Na^+-coupled cotransport systems (reviews in Maloy, 1990; Anraku, 1992), Na^+ motive flagella (Hirota *et al.,* 1981; Sugiyama *et al.,* 1985; Dibrov *et al.,* 1986), and Na^+ gradient-driven ATP synthases (Hoffmann *et al.,* 1990; Schonheit, 1992). Hence the $\Delta\mu_{Na^+}$ like $\Delta\mu_{H^+}$ is a convertible energy currency that is transduced to osmotic, mechanical, or chemical work. No ion, other than H^+ and Na^+, is known to play such a central role in bioenergetics.

In addition to its importance in bioenergetics, the Na^+ cycle has additional roles in signal transduction and cell homeostasis including regulation of intracellular pH, cell Na^+ content, and cell volume (Padan *et al.,* 1981; Pouyssegur *et al.,* 1984; Booth, 1985; Grinstein *et al.,* 1989).

In animal cells the Na^+/K^+-ATPase uses ATP to energize the cytoplasmic membrane (Fig. 1). It pumps Na^+ out, in exchange for K^+, generating a $\Delta\mu_{Na^+}$ that serves as the primary energy source at the membrane. This Na^+ cycle we designate primary, since it is driven by a primary Na^+ pump.

The cytoplasmic membrane of plants, fungi, and many bacteria is energized by primary proton pumps (Fig. 1): H^+-ATPase of the P-type in plants and fungi and the F-type in most bacteria excluding the archaebacteria,

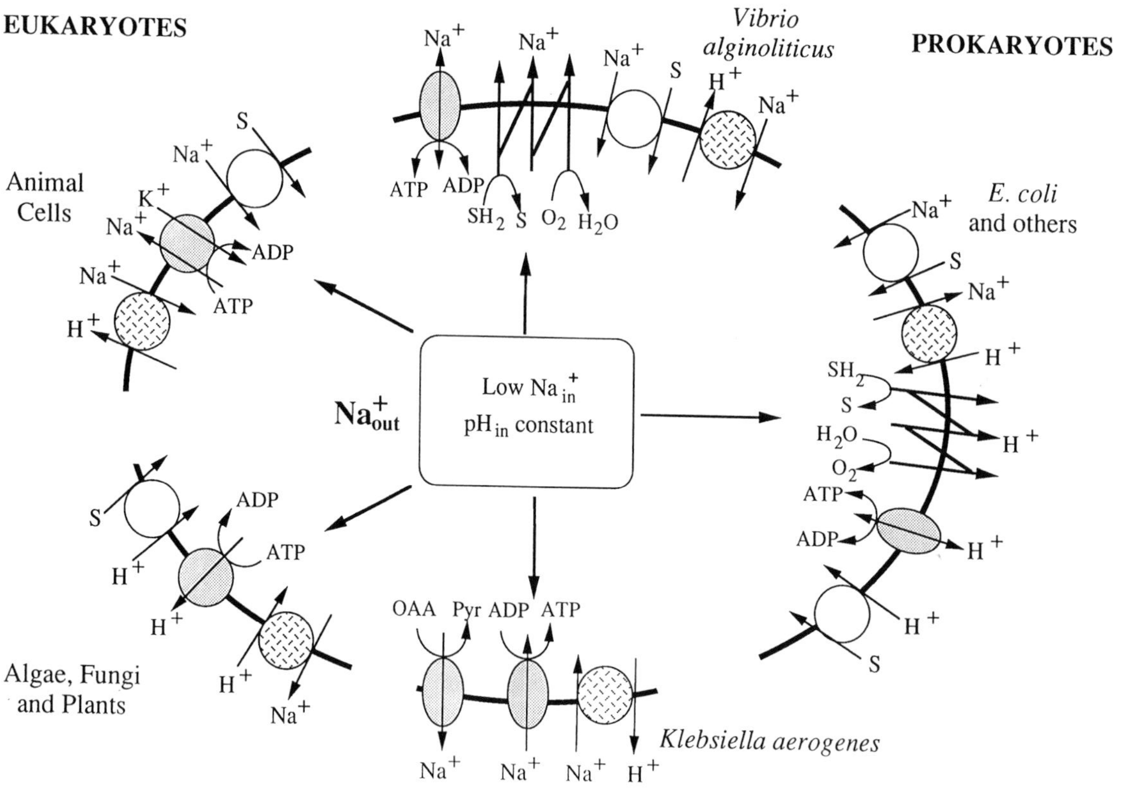

FIG. 1 Circulation of hydrogen and sodium ions. Multiple strategies for generation of Na^+ gradients in the living world. The Na^+/H^+ antiporters are ubiquitous and found in organisms that possess either H^+ or Na^+ primary cycles.

which possess the V-type H^+-ATPase (Nelson and Taiz, 1989). The prokaryotes also possess primary proton pumps linked to electron transport (mitochondrial, chloroplast type) or photoreaction (bacteriorhodopsin). All these systems pumps protons out generating a $\Delta\mu_{H^+}$, which serves as the primary energy source at the cytoplasmic membrane. Utilizing the $\Delta\mu_{H^+}$ as a driving force, Na^+ extrusion in many plants, fungi, and bacteria is conducted by Na^+/H^+ antiporters (Padan and Schuldiner, 1992; Schuldiner and Padan, 1992), initiating a Na^+ cycle that we designate secondary since it is dependent on the primary proton cycle.

Irrespective of the specific Na^+ cycles, sodium proton antiporters are ubiquitous membrane proteins that are found in the cytoplasmic and organellar membranes of cells of many different origins, including plants and animals, and in microorganisms such as bacteria, algae, and fungi (Krulwich, 1983, 1986; Blumwald and Poole, 1985; Katz *et al.*, 1986; Grinstein, 1988; Haigh and Phillips, 1989; Krulwich and Guffanti, 1989) including organisms that possess primary Na^+ cycle (Krulwich, 1983; Muller *et al.*, 1987; Krulwich and Ivey, 1990; Dimroth, 1992; Muller and Gottschalk, 1992; Schonheit, 1992; Tokuda, 1992). This ubiquitousness is in line with the important roles ascribed to these antiporters, additional to their creation of the secondary Na^+ cycle needed for cell energetics. Since they exchange $\Delta\mu_{H^+}$ for $\Delta\mu_{Na^+}$ they serve as energy buffers (Oesterhelt *et al.*, 1978; Schuldiner and Fishkes, 1978). In methanogens they have even been suggested to produce $\Delta\mu_{H^+}$ at the expense of the $\Delta\mu_{Na^+}$ (Muller and Gottschalk, 1992; Schonheit, 1992). They play primary roles in signal transduction, regulation of intracellular pH (pH_i), cell Na^+ content, and cell volume.

The existence of cation/H^+ antiporters was first postulated by Mitchell (1961) and demonstrated in mitochondria by Mitchell and Moyle (1967). In bacteria, antiporter activity was first reported in *Streptococcus faecalis* (Harold and Papineau, 1972).

In *Escherichia coli* the antiporter activity has been studied using a wide variety of techniques at multiple levels from the intact cell (West and Mitchell, 1974; Borbolla and Rosen, 1984; Castle *et al.*, 1986a,b) to the pure protein reconstituted in proteoliposomes (Taglicht *et al.*, 1991) and including a large body of work in isolated membrane vesicles (see below for references on specific topics).

In intact cells, coupling between the movement of Na^+ and H^+ was first studied by West and Mitchell (1974), who observed proton extrusion following addition of Na^+ to an anaerobic cell suspension. Coupling between $\Delta\mu_{Na^+}$ and $\Delta\mu_{H^+}$ was described in a series of studies by Macnab and collaborators (Castle *et al.*, 1986a,b; Macnab and Castle, 1987; Pan and Macnab, 1990). Borbolla and Rosen (1984) also demonstrated in intact cells that extrusion of Na^+ against its electrochemical gradient is coupled to $\Delta\mu_{H^+}$ and ATP is not required for this process.

Coupling between the ions in subcellular preparations has been demonstrated in a variety of modes: (*a*) an imposed Na^+ gradient drives the generation of pH gradients in right-side-out membrane vesicles (Schuldiner and Fishkes, 1978) and of $\Delta\mu_{H^+}$ in proteoliposomes reconstituted with pure NhaA (Taglicht *et al.*, 1991); (*b*) $\Delta\mu_{H^+}$ generated by respiration (Reenstra *et al.*, 1980) or ΔpH generated by an ammonium gradient (Nakamura *et al.*, 1986; Goldberg *et al.*, 1987; Karpel *et al.*, 1988; Taglicht *et al.*, 1991) drives Na^+ uptake against its concentration gradient in inverted membrane vesicles and in reconstituted systems with membrane extract or pure antiporter; (*c*) addition of Na^+ ions to inverted membrane vesicles in which a pH gradient was generated by respiration causes a decrease in the magnitude of the preexisting pH gradient (Brey *et al.*, 1978; Schuldiner and Fishkes, 1978; Beck and Rosen, 1979; but see also Reenstra *et al.*, 1980); and (*d*) the rate of Na^+ extrusion from right-side-out membrane vesicles (Schuldiner and Fishkes, 1978; Bassilana *et al.*, 1984a,b) and from proteoliposomes reconstituted with pure NhaA (Taglicht *et al.*, 1991) is dependent on the $\Delta\mu_{H^+}$ across the membrane.

It is only during recent years (Padan *et al.*, 1989) that it has become evident that in wild-type *E. coli* cells there are two distinct systems that catalyze Na^+/H^+ exchange, namely NhaA and NhaB. Since these findings are central for our understanding of the activities we will discuss the genetic evidence that proves this contention before we discuss other properties of the antiporters.

II. Molecular Genetic Approach to Study of the Bacterial Na^+/H^+ Antiporter

A genetic approach has been undertaken in several laboratories to study the Na^+/H^+ antiporter system in *E. coli*. Different mutants have been isolated that lost the Na^+/H^+ activity as well as the capacity to grow at alkaline pH, implying the role of the antiporter in pH homeostasis (Zilberstein *et al.*, 1980; Ishikawa *et al.*, 1987; McMorrow *et al.*, 1989). The selections were based on the inability of the mutants to grow on carbon sources that require a sodium gradient for their uptake: mutation in locus *phs* was obtained based on an inability to grow on both glutamate and melibiose (Zilberstein *et al.*, 1980); Hit1 could not grow on serine (Ishikawa *et al.*, 1987) and HS3051 was isolated as a strain that failed to grow on melibiose in the presence of 100 m*M* NaCl (McMorrow *et al.*, 1989). Thus far only the *phs* mutation (Zilberstein *et al.*, 1980) has been characterized in some detail. Mapping of the *phs* mutation showed it to be an allele of *rpo*A, the gene encoding the α-subunit of the RNA polymerase

(Rowland *et al.*, 1984). The mutation causes a relatively selective transcription defect which affects several genes that seem to have a common regulatory mechanism (Giffard *et al.*, 1985; Thomas and Glass, 1991). In conclusion, even though the classical genetic approach has yielded several mutants, none of them has yet been shown to be defective in a structural gene coding for the antiporter.

A different type of mutant, with increased rather than decreased antiporter activity, has been isolated by Tsuchiya, Wilson, and collaborators based on its resistance to Li^+ (Niiya *et al.*, 1982). Li^+ ions are toxic to *E. coli* cells due mostly to their effect on the cell pyruvate kinase (Umeda *et al.*, 1984). When grown on melibiose as a carbon source the toxicity is augmented due to their effect on the melibiose transporter which is inhibited by Li^+ ions. Tsuchiya, Wilson, and colleagues isolated a mutant in *E. coli* that tolerates Li^+ concentrations that are otherwise toxic to the wild-type cells (Niiya *et al.*, 1982). This mutant harbors at least two mutations responsible for the acquirement of the resistance: one in the *mel*B allele (*mel*BLiD), in which replacement of proline at position 122 with serine brings about a modification in the melibiose transporter such that it can now cotransport the sugar with Li^+ (Niiya *et al.*, 1982; Yazyu *et al.*, 1985). The second mutation is in an additional locus, which brings about an enhanced antiporter activity capable of an increased excretion of the toxic ion (which is also a substrate of the antiporter). We have separated the two mutations and showed that the one that increased antiporter activity (*nha*Aup, previously called *ant*up), is necessary to confer resistance to toxic levels of Li^+ ions and maps at about 0.3 min on the *E. coli* chromosome (Goldberg *et al.*, 1987).

We have taken advantage of the toxicity of Li^+ ions and the resistance associated with high activity of the antiporter in order to clone the wild-type *nha*A gene (Goldberg *et al.*, 1987; Karpel *et al.*, 1988). We assumed that when in high copy number (plasmidic + chromosomal) the wild-type *nha*A will increase Na^+/H^+ antiporter activity and thereby confer Li^+ resistance to cells, i.e., an $NhaA^{up}$ phenotype. We surveyed a library of plasmids containing inserts covering 15 kbp from *car* to *dna*J (Mackie, 1980, 1986), which should include the wild-type locus affecting the antiporter, and succeeded cloning an insert bearing *nha*A. This achievement initiated the molecular biology of the Na^+/H^+ exchange.

To study the role of the antiporter, the strategy was to inactivate the chromosomal gene by replacing most or all of it with a selectable marker. The appropriate constructs were engineered in plasmids taking care to include enough flanking sequences to allow for homologous recombination to take place. After selection and transduction into our isogenic strains, recombination in the desired locus was tested by southern hybridizations (Padan *et al.*, 1989). We constructed Δ*nha*A strains in which either two-thirds (Padan *et al.*, 1989) or the whole gene (Karpel *et al.*, 1991) was replaced with *kan* without disrupting any of the neighboring genes. The Δ*nha*A strains obtained grow normally in low sodium medium indicating that, at least under these conditions, *nha*A is not an essential gene. Δ*nha*A cannot adapt to high sodium concentrations that do not affect the wild type (0.7 *M* NaCl at pH 6.8). The Na^+ sensitivity of Δ*nha*A is pH dependent, increasing at alkaline pH (0.1 *M* NaCl at pH 8.5). The Δ*nha*A strains also cannot challenge the toxic effects of Li^+ ions (0.01 *M*), a substrate of the Na^+/H^+ antiporter system. It is concluded that *nha*A is indispensable for adaptation to high salinity, for challenging Li^+ toxicity and for growth at alkaline pH (in the presence of Na^+). Although a high dose of *nha*A confers resistance to Li^+, it does not increase the limits of pH or salt that *E. coli* can cope with, suggesting that factors other than *nha*A are limiting in setting the upper limits of tolerance. In addition, the fact that the Δ*nha*A strain can grow at alkaline pH at Na^+ concentrations below 0.1 *M* suggests the presence of additional or alternative mechanisms for pH homeostasis.

Indeed, growth of the Δ*nha*A strain is normal on carbon sources that require Na^+ ions for transport, suggesting that it can generate a sodium gradient large enough to support growth even in the absence of NhaA activity. Moreover, antiporter activity, as measured in everted membrane vesicles, decreases only to 50% of the wild-type level. A detailed analysis of the remaining antiporter activity in the Δ*nha*A strain reveals that the kinetic properties of the residual activity differ from those displayed by the NhaA protein: (*a*) K_m for transport of Li^+ ions is about 15 times higher and (*b*) the activity is practically independent of intracellular pH. Hence our results demonstrate the presence of a novel alternative Na^+/H^+ antiporter in *E. coli* (designated *nha*B) additional to the *nha*A system (Padan *et al.*, 1989).

In addition to learning about the role of *nha*A in cell physiology and to the unveiling of the alternative system, construction of the Δ*nha*A strain provided us with a system for cloning of genes coding for other antiporters by functional complementation. The Δ*nha*A strain is sensitive to Na^+ and its sensitivity to Li^+ ions is highly increased; transformation of a Δ*nha*A strain by multicopy plasmid carrying *nha*A renders the transformants resistant to the ions. It was therefore anticipated that homologous and even heterologous antiporter genes will be able to complement the deletion. Using this paradigm we have succeeded in cloning three other genes: one from *Salmonella enteritidis,* homologous to *nha*A (Pinner *et al.*, 1992a); the *nha*B gene from *E. coli* (Pinner *et al.*, 1992b); and a novel gene from the alkaliphilic *Bacillus firmus* OF4 (Ivey *et al.*, 1991) that codes for a putative antiporter. For the cloning of *nha*B it was also strictly necessary to prepare the library from the Δ*nha*A strain in order to prevent recloning of *nha*A (Pinner *et al.*, 1992b).

Cloning of *nha*B made possible the generation of Δ*nha*B and Δ*nha*A Δ*nha*B strains that already supplied further invaluable information about the components involved in the metabolism of Na^+ and H^+ in *E. coli*. The Δ*nha*B strain showed no impairment in its ability to adapt to high salt or alkaline pH or its resistance to Li^+ (Pinner *et al.*, 1993). These findings suggest that NhaA alone can cope with the salt and pH stress and it has a capacity high enough for these functions. Also, and as will be discussed below, expression of *nha*A is highly regulated and increases significantly at the conditions in which it is essential: high salt, alkaline pH (in the presence of Na^+ ions), and presence of toxic Li^+ ions (Karpel *et al.*, 1991).

The double mutant, Δ*nha*AΔ*nha*B, grows very poorly in the presence of Na^+ concentrations as low as 10 m*M*, which are the contaminating levels of LBK broth (Padan *et al.*, 1989). At concentrations of 100 m*M* Na^+ (pH 7.5), growth is completely arrested. Analysis of the antiporter activity in membranes prepared from the Δ*nha*AΔ*nha*B strain shows no residual activity of Na^+/H^+ antiporter (Pinner *et al.*, 1993).

We can tentatively conclude that in relation to *nha*B, *nha*A is more flexible and capable of handling a higher load of Na^+ at the entire pH range. Thus, it can support growth in the absence of *nha*B even at high salinity and pH; *nha*B, on the other hand, can support growth only at relatively low sodium concentrations. In addition, NhaB shows a higher affinity to Na^+ than NhaA (see below), suggesting that the recurrent theme described for many other transport systems possibly holds also for the systems handling Na^+ and H^+: a low-affinity, high-capacity system (in our case *nha*A) and another high-affinity, low-capacity system (*nha*B) are required to cope with adaptation to a large range of concentrations. Interestingly, and unlike most other chromosomally encoded mineral transport systems (Silver and Walderhaug, 1992), in the case of the Na^+/H^+ antiporters the high-capacity one is regulated (Karpel *et al.*, 1991). Very little is known thus far about the regulation of *nha*B or the kinetic properties of the system. Also, we still do not know what is the actual contribution of each protein in the wild-type strain under various conditions.

The nucleotide sequence of the *nha*A and *nha*B genes have been determined. They encode membrane proteins of M_r 41,316 (Karpel *et al.*, 1988; Talglicht *et al.*, 1991) and 55,543 (Pinner *et al.*, 1992b), respectively. A hydropathic evaluation of the aminoacid sequence of both proteins reveals the presence, respectively of 11 and 12 putative transmembrane spanning segments linked by hydrophilic segments of variable length. The proteins were specifically labeled using the T7 polymerase system as described by Tabor and Richardson (1985), a system designed to label, in the intact cell, solely the product of genes cloned downstream of T7 promoters. In both cases a protein is labeled that displays an apparent M_r in SDS–PAGE lower than that expected from the analysis of the sequence (33 and 45 kDa, respectively; Taglicht *et al.*, 1991; Pinner *et al.*, 1992b), a phenomenon that has been described for many hydrophobic proteins. The label is associated with the membrane fraction even after washes with 5 *M* urea, a finding which corroborates that both proteins are intrinsic membrane components. NhaA has been purified in a functional state and shown to catalyze Na^+/H^+ exchange (Taglicht *et al.*, 1991, and see also below).

Sequencing of the genes allowed for an accurate mapping in the *E. coli* chromosome: *nha*A is located at 0.35 min distal to *dna*J and between two insertion sequences detected in various strains: IS186 and IS1. Between IS186 and *nha*A we find *gef,* a member of a gene family encoding small, toxic proteins of approximately 50 amino acids (Poulsen *et al.*, 1989). Downstream of *nha*A, there is an additional open reading frame that had already been determined (Mackie, 1986) and recently found to be involved in regulation of expression of *nha*A and designated *nha*R (Rahav-Manor *et al.*, 1992, and see also below). *rps*20 is located further downstream of IS1 (Mackie, 1986). *nha*B is located at 25.5 min between *fad*R and the *umu*CD operon. Downstream of *nha*B there is a small potential open reading frame, whose function is unknown yet.

A mutant, Hit1, that cannot grow on serine as a carbon source in a Na^+-dependent fashion was isolated (Ishikawa *et al.*, 1987). Since this mutant showed impaired Na^+ extrusion capacity and lack of growth at alkaline pH (with glycerol as a carbon source) it was concluded that the mutation affects the Na^+/H^+ antiporter activity. Recently (Thelen *et al.*, 1991) the mutation was mapped at 25.6 min on the *E. coli* chromosome, suggesting that *hit1* resides in the *nha*B locus. However, we could not complement Hit1 by the cloned *nha*B, neither for growth on serine at neutral pH nor for growth on glycerol at alkaline pH. The phenotype of Hit1 also differs from Δ*nha*A (see above). Molecular characterization of the *hit1* mutation is required.

The sequences of the two antiporters NhaA and NhaB show very little similarity. The overall identity, as determined with the algorithm of Smith and Waterman (Devereux *et al.*, 1984), is 20% (but see also Section IX).

III. Biochemical Approach

Molecular genetics provides important tools for overexpression of the transport protein, a crucial step for purification. As yet few of the Na^+ transporters have been purified to homogeneity and reconstituted in liposomes in a functional form. Using a site-specifically cleavable fusion protein, the proline/Na^+ cotransporter of *E. coli* has been purified (Hanada *et al.*, 1988). The redox-linked Na^+ pump of *Vibrio alginolyticus* (Hayashi and Unemoto, 1986; Tokuda, 1989, 1992), the decarboxylation-linked Na^+ pump (Dimroth, 1992), and the Na^+/ATPase of *Propionigenium modestum* (Laubinger and Dimroth, 1988a,b; Dimroth, 1992a) have been purified by conventional biochemical methods. Purification of a functional protein is the only way to prove the involvement of a gene product in the transport reaction. Furthermore, it is crucial for the study of molecular properties in the transport systems, specifically those that cannot easily and/or unequivocally be determined in membrane preparations or in the intact cell (Taglicht *et al.,* 1991).

There are at least three reasons for the scarcity of proteins that have been purified in a functional state: (*a*) the biochemistry of membrane proteins is more difficult than that of soluble proteins and necessitates the choice of proper detergents; (*b*) techniques for reconstitution are not always trivial and may be somewhat cumbersome and laborious for quick assessment of purifications, and (*c*) the proteins of interest are not always abundant, causing the need for processing of large amounts of material.

In the case of the Na^+/H^+ antiporter of *E. coli,* successful reconstitution has been already reported (Tsuchyia *et al.,* 1982; Nakamura *et al.,* 1986). In both reports octyl glucoside was used as a detergent; the driving force being a ΔpH generated either by the reconstituted respiratory chain (Tsuchyia *et al.,* 1982) or by an ammonium gradient (Nakamura *et al.,* 1986). In addition, since the transporter could be specifically labeled using the T7 polymerase system (Karpel *et al.,* 1988), we were able to follow its fate rapidly and efficiently during the various purification steps (Taglicht *et al.,* 1991). Probably the most important aid in the purification was the overproduction of the antiporter 100 to 200-fold. In order to overproduce the *nha*A gene product, a plasmid was constructed (pEP3T) in which the promoterless gene was cloned downstream of the strong inducible *tac* promoter. When cells carrying pEP3T are induced with IPTG, a diffuse protein band of about 33 kDa becomes evident after 30 min and induction is maximal after 90–120 min. The band is undetectable before induction, as expected from our estimate that under these conditions it represents only 0.1–0.2% of the membrane protein. At the end of the induction period it is the most abundant protein in the membrane. Essentially all NhaA appears to be membrane associated. Cell growth declines with kinetics similar to that for NhaA induction, demonstrating the detrimental effect on growth upon overproduction of NhaA. In dodecyl maltoside extracts of membranes prepared from induced cells the antiporter is about 20% of the total protein, about 100 to 200-fold more abundant than in extracts from wild-type cells. When membranes produced from such induced cells are solubilized and reconstituted the Na^+/H^+ antiport activity observed is about 100-fold higher than the activity measured in proteoliposomes prepared from membranes derived from wild-type cells, suggesting that most, if not all, of the overproduced protein is catalytically active. An additional purification of only 4-fold on DEAE and hydroxylapatite columns was sufficient to yield a highly purified fraction (Taglicht *et al.*, 1991).

When the activity of the purified protein (600 nmol/min/mg protein) is compared to the activity of reconstituted total membrane extract (120 nmol/min/mg protein) the rise in specific activity (5-fold) is comparable to the purification fold calculated from the ^{35}S-labeling of the protein. Hence, the overall purification obtained, including the overproduction step, is about 400 to 500-fold.

This protein, which is the only antiporter purified thus far, when reconstituted into liposomes, catalyzes all the modes of action that were previously documented in intact cells and in membrane preparations. Thus, proteoliposomes reconstituted with NhaA accumulate Na^+ ions against their concentration gradient upon imposition of a pH gradient generated by ammonium diffusion (Fig. 2A) (Taglicht *et al.*, 1991). The activity displays an apparent K_m of 110 μM (pH 8.6).

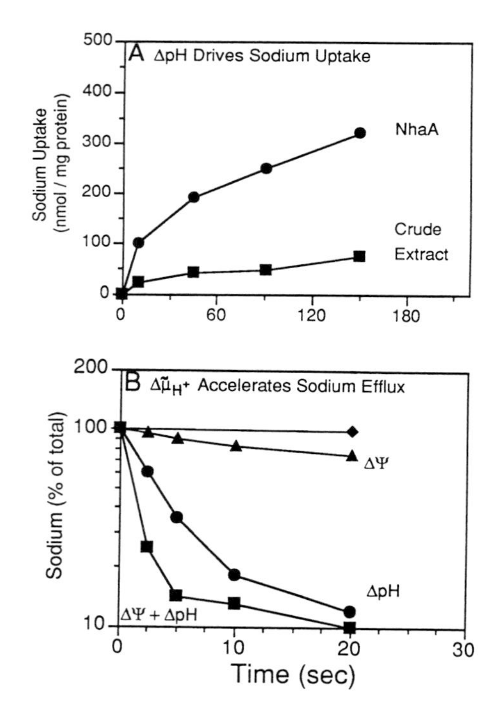

FIG. 2 Modes of catalysis of NhaA. Proteoliposomes reconstituted with pure NhaA catalyze uphill uptake of ^{22}Na driven by a pH gradient (A) or downhill efflux $^{22}Na^+$ (B), which is influenced by ΔpH, $\Delta\psi$.

Another mode of catalysis that has been measured in the purified proteoliposomes is downhill transport of $^{22}Na^+$ in the presence or in the absence of imposed $\Delta\mu_{H^+}$ (Fig. 2B). The most striking property of this reaction is that the antiporter is virtually turned off when the reconstituted proteoliposomes are loaded with K-acetate and $^{22}Na^+$ as pH 6.5 and diluted into a medium of identical composition containing valinomycin and devoid of Na^+. The rate of efflux increases upon imposition of either a proton gradient or an electrical potential across the membrane. Thus, in proteoliposomes diluted into media containing choline acetate and valinomycin the membrane potential, generated by the outwardly directed K^+ gradient, accelerates sodium efflux. Upon dilution of the proteoliposomes into media containing potassium-gluconate, the proton gradient formed by the outwardly directed acetate gradient accelerates efflux many-fold (Taglicht *et al.*, 1991).

The downhill movement of Na^+ ions is coupled to the movement of H^+ ions against their concentration gradient. This has been tested in a series of experiments in which proteoliposomes reconstituted with NhaA were loaded with 10 m*M* NaCl and with the pH indicator pyranine. Upon dilution of the proteliposomes into a medium devoid of Na^+, a rapid acidification of the internal milieu is observed, as indicated by the changes in the fluorescence of the trapped pyranine (Fig. 3, traces C and D) (Taglicht, 1992). This acidification, as expected, is prevented by, and reversed upon, addition of ammonium salts or nigericin.

A purified functional antiporter allows the discernment of a direct effect of various inhibitors. Amiloride has been reported as an inhibitor of most but not all antiporters (Raley-Susman *et al.*, 1991) in eukaryotes (Benos, 1988) and in some bacteria as well (Ivey *et al.*, 1991). There is some controversy in the literature over the effect of amiloride-like compounds on the *E. coli* antiporter activity. Amiloride was claimed to inhibit Na^+-dependent changes in pH gradients in everted vesicles (K_i=40 μM,) (Mochizuki-Oda and Oosawa, 1985). These data were interpreted as an effect on the antiporter. However, results from both our laboratory (D. Taglicht, S. Schuldiner, and E. Padan, unpublished observations) and others (Leblanc *et al.*, 1988) detected a potent uncoupler activity of amiloride. We were unable to detect any significant inhibition of downhill sodium efflux catalyzed by purified NhaA, with amiloride, MK-685 or benzamil derivatives (Taglicht *et al.*, 1991).

Of a long list of chemical modifiers tested (reviewed in Leblanc *et al.*, 1988) the only one that was found to inhibit the antiporter is the histidyl reagent diethylpyrocarbonate (DEPC). Diethylpyrocarbonate was found to inhibit Na^+ efflux in a specific way, since hydroxylamine reverses the inactivation; DEPC also inhibits Na^+ efflux by purified NhaA (Taglicht, 1992).

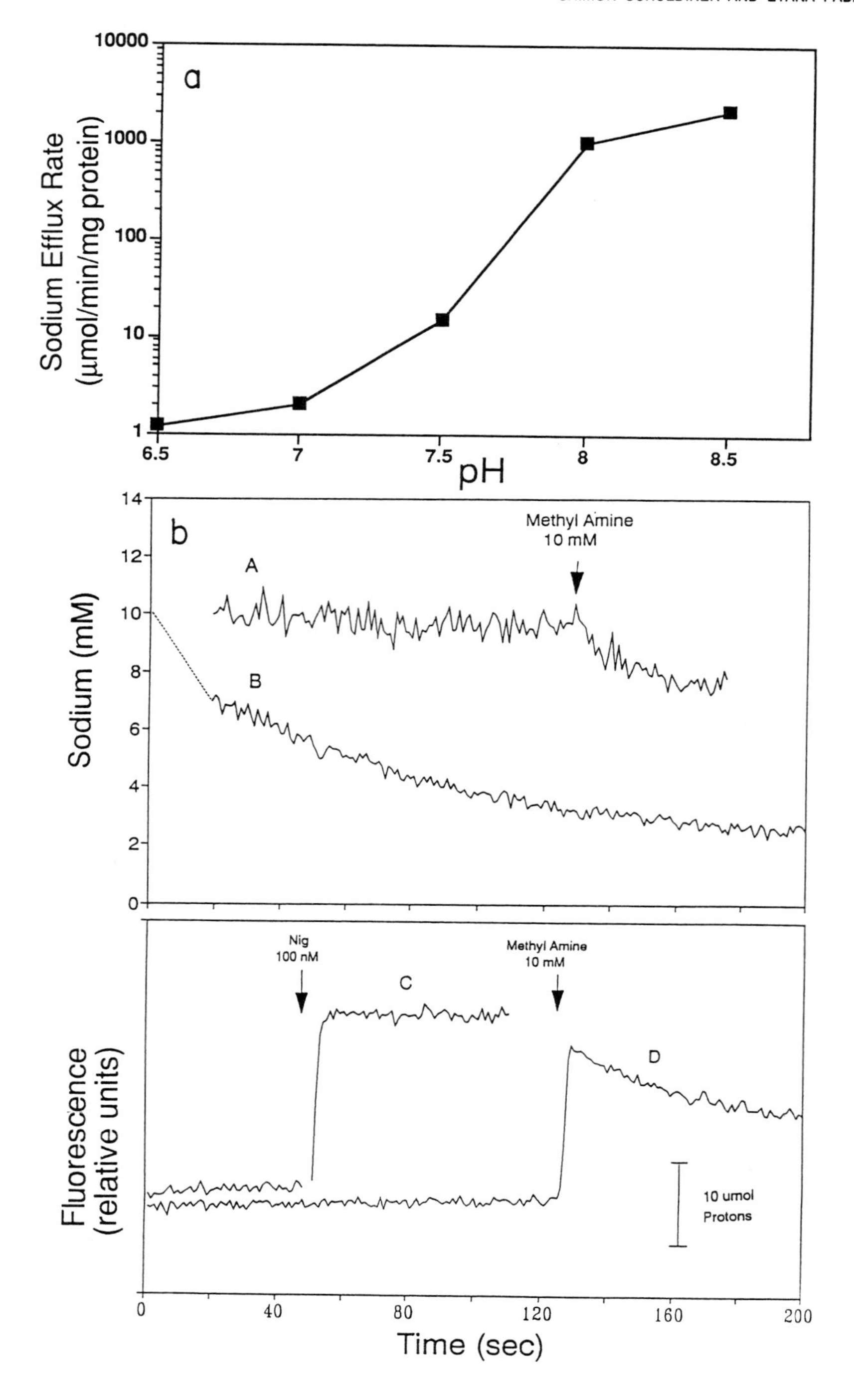
a
Sodium Efflux Rate (μmol/min/mg protein)
10000
1000
100
10
1
6.5
7
7.5
8
8.5
pH
b
Sodium (mM)
14
12
10
8
6
4
2
0
A
B
Methyl Amine 10 mM
Nig 100 nM
C
Methyl Amine 10 mM
D
Fluorescence (relative units)
10 umol Protons
0
40
80
120
160
200
Time (sec)

IV. Kinetic Studies

Most of the kinetic studies on the antiporter published thus far have been done in either whole cells or membrane vesicles from wild-type strains. As we now know these membranes contain at least three proteins that exchange Na^+ and H^+ ions: NhaA, NhaB (Padan *et al.*, 1989; Pinner *et al.*, 1992c), and the nonspecific Kha system (Brey *et al.*, 1978). Since Kha has a higher affinity for K^+ than for Na^+, in a high K^+ medium its contribution is usually minimal. Still, we have two systems whose relative contributions under most conditions are not yet well characterized. However, since there is a strong dependence of the antiporter activity on pH (Bassilana *et al.*, 1984a,b; Taglicht *et al.*, 1991), and since NhaB shows practically no dependence on pH (Padan *et al.*, 1989; Pinner *et al.*, 1992c), we can assume, until further information becomes available, that the dominant activity measured at pH 8.0 is that of NhaA.

Another complication of many of the kinetic measurements is the use in many of the experiments of a very indirect and qualitative measurement of the antiporter activity, namely the ability of Na^+ or Li^+ ions to collapse a pH gradient as measured by fluorescent acridine derivatives. The ability of various concentrations of the ions to change the steady-state level of fluorescence has been measured (Schuldiner and Fishkes, 1978; Beck and Rosen, 1979; Niiya *et al.*, 1982; Padan *et al.*, 1989), a parameter that is not necessarily linearly related to the rate of the antiporter activity. The range of apparent K_m values obtained with this technique in wild-type strains (pH 8.0) is 3.0 to 16 m*M* (Schuldiner and Fishkes, 1978; Beck and

FIG. 3 Downhill sodium efflux coupled to proton influx and effect of pH. (a) Proteoliposomes reconstituted with pure NhaA catalyze downhill efflux of ^{22}Na which is regulated by pH. The experiment is done under conditions in which generation of $\Delta\psi$ is prevented by valinomycin (5 μM). ΔpH is short-circuited with a weak acid ($[\text{acetic acid}]_{in}$ = $[\text{acetic acid}]_{out}$ = 100 m*M*). For more details see Taglicht *et al.* (1991). (b) Proteoliposomes were reconstituted with purified NhaA (4 μg protein/ml) in the presence of NaCl (15 m*M*) and SBFI (0.5 m*M*, traces A and B), or pyranine (0.05 m*M*, traces C and D). The buffers used for loading the liposomes also contained 10 m*M* potassium phosphate, pH 7.5 (10 m*M*), and either 140 m*M* KCl (traces A, C, and D) or 100 m*M* potassium acetate (trace B). The external probe was removed by gel filtration through a 2-ml sephadex G-50 column preswollen in the same buffer in which the proteoliposomes were loaded. The proteoliposomes were diluted 50-fold into the same buffer containing valinomycin (20 n*M*) but devoid of NaCl. Where indicated by arrows, nigericin (100 n*M*) or methylamine (10 m*M*) were added. SBFI fluorescence was measured at 510 nm (with a 12-nm slit) using an excitation light at 335 nm with a 3-nm slit). Pyranine fluorescence was measured at 510 nm (3-nm slit) using an excitation light at 465 nm (3-nm slit). Additions of aliquots of sodium or acid in the presence of nigericin were used to calibrate the fluorescence response.

Rosen, 1979; Niiya *et al.*, 1982); in $NhaA^{up}$ strains or strains containing multiple doses of *nha*A, the range is between 0.68 and 4.5 m*M* (Niiya *et al.*, 1982; Padan *et al.*, 1989). Δ*nha*A strains yield an apparent K_m value of 0.25 m*M* for NhaB (Padan *et al.*, 1989) and strains containing multiple dose of *nha*B display an even lower K_m of 40–70 μM (Pinner *et al.*, 1992b). Therefore, from the data existant in the literature, taken as a whole it would seem that the range of measurements of apparent K_m of NhaA fall, at pH 8, in the low millimolar range, whereas NhaB is in the submillimolar range, suggesting that NhaA is the low-affinity system and NhaB has a higher affinity to Na^+ ions.

The apparent K_m values obtained when measuring $\Delta\mu_{H^+}$-driven $^{22}Na^+$ uptake in the vesicles from equivalent strains cannot be directly compared, since the pH of the reaction is different: at pH 7.3 the system did not seem to saturate even at 50 m*M* (Reenstra *et al.*, 1980), at pH 8.0 it was 2.4 m*M* (Nakamura *et al.*, 1986), and at pH 8.8 it was 0.5 m*M* (Goldberg *et al.*, 1987). The apparent K_m of the pure NhaA at pH 8.6 is 0.1 m*M* (Taglicht *et al.*, 1991). The published data, taken at face, would suggest a dependence of the apparent K_m on pH similar to what has been claimed for the efflux reaction (see below). However, we should stress that the three K_m values compared were obtained in three different preparations (inverted membrane vesicles, proteoliposomes reconstituted with crude extract, or with NhaA).

In experiments in which $^{22}Na^+$ efflux from right-side-out vesicles from wild-type strains was measured, the apparent K_m was dependent on the internal pH of the vesicle: 40 m*M* at pH 6.8 and 3.5 m*M* at pH 7.7 with intermediary values of 20 and 7 m*M* when the pH is 7 and 7.5, respectively (Bassilana *et al.*, 1984a,b). The authors interpreted this effect as competition between internal H^+ and Na^+ ions.

An apparent gating of $^{22}Na^+$ efflux from right-side-out vesicles by ΔpH was reported by Bassilana *et al.* (1984a,b). The authors considered the possibility that this is not an intrinsic gating phenomenon as suggested for the antiporter of *Halobacterium halobium* (Lanyi, 1979), but rather an effect of low pH_{in} on the protein. Similarly, an apparent gating for uptake of $^{22}Na^+$ was reported in inverted membrane vesicles by Nakamura *et al.* (1986). Interestingly, in inverted membrane vesicles, the antiporter activity, as measured by the rate of $^{22}Na^+$ uptake driven by an ammonium gradient (Nakamura *et al.*, 1986) or by the acridine orange technique (Padan *et al.*, 1989), seems to show a dependence on the external pH (which corresponds to the internal pH of cells or right-side-out membranes). The existence of a gating phenomenon or a need for energy even for downhill transport has also been claimed in intact cells: when $^{22}Na^+$-loaded cells were diluted into sodium-free medium most of the radioactivity remained in the cells unless an energy source was added (Borbolla and

Rosen, 1984). It would seem, however, that the lack of activity in the absence of $\Delta\mu_H+$ is not necessarily an intrinsic energetic requirement but is most likely due to a kinetic block. In proteoliposomes reconstituted with NhaA, very rapid downhill efflux could be detected in the absence of $\Delta\mu_{H^+}$, provided the reaction was carried out at the appropriate pH (Fig. 3a) (Taglicht *et al.*, 1991).

Homologous sodium/sodium exchange or counterflow has not been observed until now in any of the systems used. External sodium has no effect on efflux of $^{22}Na^+$ from intact cells (Borbolla and Rosen, 1984), or right-side-out membrane vesicles (Bassilana *et al.*, 1984b), a phenomenon that suggests that the translocation of sodium across the membrane may be the rate-limiting step in the overall reaction. Another possibility to explain the lack of effect of sodium in *trans* is that the antiporter is very assymetric and does not recognize the sodium ions in the external face of the cell or the membrane. Although it is very appealing to postulate an assymetry of this type, since it would indeed facilitate the function of a protein whose role is to extrude Na^+, there is not yet solid experimental data to support it.

V. Na^+/H^+ Antiport Activity of NhaA Is Electrogenic

The issue whether the antiporter is electrogenic has been the matter of some controversy over the years. Although it was first suggested that the exchange is electroneutral (West and Mitchell, 1974), further evidence has indicated an electrogenic exchange ($H^+:Na^+ > 1$) (Schuldiner and Fishkes, 1978; Beck and Rosen, 1979; Bassilana *et al.*, 1984a; Castle *et al.*, 1986a,b; Pan and Macnab, 1990). It has also been proposed that below a certain pH_{out}, the antiporter is electroneutral, whereas above it, electrogenic (Schuldiner and Fishkes, 1978). This question was addressed in an extensive and careful study by Macnab and colleagues (Castle *et al.*, 1986a,b; Macnab and Castle, 1987; Pan and Macnab, 1990). In this study, steady-state values of $\Delta\mu_{H^+}$ and $\Delta\mu_{Na^+}$ were measured under various conditions in endogenously respiring *E. coli* using $^{23}Na^+$ and ^{31}P NMR spectroscopy. Na^+ extrusion and maintenance of a low intracellular Na^+ concentration were found to correlate with the development and maintenance of $\Delta\mu_{H^+}$. At pH 6.7 a concentration ratio ($[Na^+]_{out}/[Na^+]_{in}$ of about 25 was observed and this was independent of extracellular Na^+ concentrations over the measured range 4–285 m*M*, indicating that intracellular Na^+ concentration is not regulated. When the gradients were measured at various pH values, it was found that in the acidic-to-neutral pH range the Na^+ chemical potential followed the proton chemical potential quite closely, always exceeding it slightly. Above pH 7.4, there was a progressive divergence between the two values. Thus, whereas the ΔpH continued to decrease, reached zero at pH 7.5, and changed signs (pH_{in} becoming more acidic than pH_{out}), ΔpNa $\{\Delta pNa = -\log([Na^+]_{in}/[Na^+]_{out})\}$ practically leveled off at a value of 25 to 40 mV, corresponding to a Na^+ concentration gradient of 2.5 to 5-fold at the alkaline pH values. As a consequence, the apparent overall stoichiometry changes from 1.1 at pH_{out} 6.5 to 1.4 at pH_{out} 8.5 (Pan and Macnab, 1990).

It was suggested that this change in apparent overall stoichiometry may reflect a change in the relative rates of two antiporters with different stoichiometries rather than a change in stoichiometry of a single protein (Macnab and Castle, 1987). Our studies with the purified NhaA support the notion that it is electrogenic: as described above downhill movement of ^{22}Na is stimulated by the imposition of a membrane potential (negative inside) even at pH's as low as 6.5 (Figs. 2 and 3) (Taglicht *et al.*, 1991). The downhill movement of Na^+ via NhaA generates a membrane potential, as suggested by the fact that in the absence of valinomycin the efflux rate of $^{22}Na^+$ is several-fold lower than in its presence (Taglicht *et al.*, 1991). Direct measurement of membrane potential with Oxonol VI further supports the rehogenic nature of the antiporter (Taglicht, 1992; and see below).

Direct measurement of the stoichiometry as evaluated from rate measurements is now approachable with pure NhaA: Na^+ movements are being followed by monitoring either $^{22}Na^+$ or changes in fluorescence of a novel sodium indicator SBFI (sodium-binding benzofuran isophtalate) (Taglicht, 1992). Some of the problems that we face in these experiments beautifully illustrate important properties of the antiporter: massive (measurable) movements of Na^+ can be detected only under conditions in which the formation of $\Delta\mu_{H^+}$ by the antiporter is prevented or when preformed gradients are collapsed. Representative results are shown in Fig. 3b: proteoliposomes loaded with Na^+ lost very small amounts of the ion (trace A), whereas a pH gradient reaches its maximal value as soon as it can be measured (trace C). Addition of 10 m*M* methylamine transiently alkalinizes the internal milieu and allows for exit of some of the Na^+ ions. Addition of three identical aliquots of methylamine is necessary to release most of the internal Na^+. If the generations of $\Delta\mu_{H^+}$ is prevented by performing the experiment in the presence of K-acetate and valinomycin, half of the internal Na^+ is lost after about 60 sec (trace B). In conclusion, it is enough to move a very small amount of Na^+ ions to allow for the rapid generation of $\Delta\mu_{H^+}$, so that perturbations in the intracellular pH can be corrected quickly and efficiently without massive movements of Na^+. In these experiments a rough estimate of the stoichiometry H^+/Na^+ can be obtained from the rate of the buildup of the pH gradient and the rate of release of Na^+ after the addition of methylamine: the values are slightly higher than two H^+ per Na^+.

A stoichiometry of $2H^+/Na^+$ has also been estimated in similar experiments using a thermodynamic rather than a kinetic approach. The size of the $\Delta\mu_{H^+}$ generated by a Na^+ gradient can be predicted from the fact that at equilibrium $\Delta\mu_{Na^+} = n\Delta\mu_{H^+}$, which means that $\Delta pNa^+ = n\Delta pH + (n-1)\,\Delta\psi$. We have measured with Oxonol VI the size of the $\Delta\psi$ generated at various ΔpNa^+, in the presence of nigericin, which allows for an electroneutral exchange of K^+ and H^+ and thereby discharges the ΔpH (Fig. 4). The magnitude of the $\Delta\psi$ generated at various pH values (7.1–8.2) was consistent with a stoichiometry of 2 (Taglicht, 1992).

Our results indicate that the apparent changes in stoichiometry measured in the intact cell at alkaline pH are not due to a change in stoichiometry of NhaA but rather in its relative contribution to the Na^+ cycle.

VI. pH Sensor in NhaA

The proposed role of the antiporter in pH_{in} regulation, i.e., acidification of the cytoplasm at alkaline extracellular pH, implied that the activity should be dependent on and/or regulated by pH so that the higher the pH the higher the activity (Padan *et al.*, 1976). Studies of Na^+ effects upon lactose-dependent H^+ circulation suggested that the Na^+/H^+ antiporter was more active at alkaline than at neutral pH (Zilberstein *et al.*, 1979). Studies in right-side-out membrane vesicles by Leblanc and collaborators (Bassilana *et al.*, 1984a,b) showed that the activity of the antiporter is

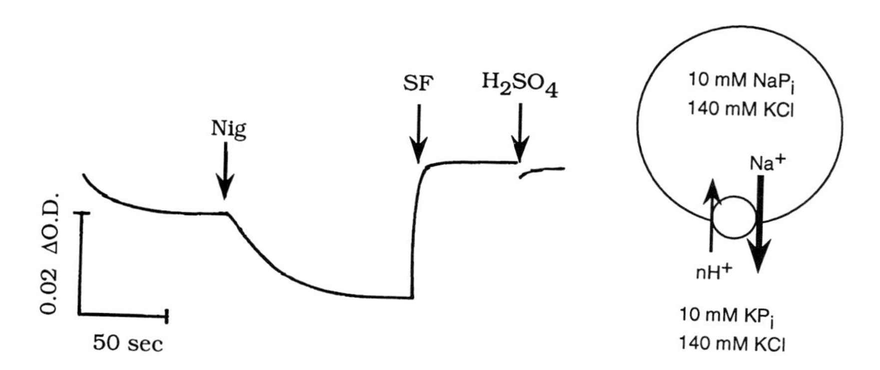

FIG. 4 Oxonol VI absorbance. Generation of a membrane potential by the Na^+/H^+ antiporter. Determination of the H^+/Na^+ stoichiometry at various pH values.

extremely dependent on pH_{in}. In these studies right-side-out membrane vesicles were loaded with $^{22}Na^+$ at various pH values and the rate of downhill efflux (V_{Na^+}) was monitored. The authors found that imposition of $\Delta\mu_{H^+}$ stimulated Na^+ efflux at all external pH values between 5.5 and 7.5. The contributions of the electrical ($\Delta\Psi$) and chemical (ΔpH) potential to the acceleration mechanism were studied by their selective dissipation with valinomycin and nigericin in respiring vesicles and by imposition of artificial gradients that stimulated efflux even at acid pH (5.5), provided the internal pH was increased by imposition of a pH gradient, suggesting that also at this pH the antiporter is electrogenic. The effect of ΔpH on V_{Na^+} was dependent on the external pH value: at low pH the relation is nonlinear and indicates the existence of an apparent threshold. This threshold progressively decreases as the pH rises, and at pH7 and above disappears completely. These variations in behavior can be accounted for by variations in pH_{in}. The authors propose that the high internal H^+ concentration inhibits Na^+ efflux by competition, as suggested by the change in the apparent K_m for Na^+ from 40 m*M* at pH_{in} 6.8 to 3.5 m*M* at pH_{in} 7.7. These important conclusions are somewhat complicated by several factors, which could not be controlled at the time: (*a*) "nonstimulated" passive leaks increased also with pH and, above pH 7.5, were too high to ignore or to correct for; (*b*) the existence of more than one transport system for Na^+ was not known and the relative contributions of each therefore could not be analyzed; and (*c*) changes in pH_{in} always necessitated changes in ΔpH as well and therefore the two factors could not be isolated from one another. The first two problems do not exist in proteoliposomes reconstituted with pure NhaA.

As described above (see also Taglicht *et al.*, 1991) when efflux of $^{22}Na^+$ from proteoliposomes is monitored, it is stimulated upon imposition of ΔpH, a stimulation that can be ascribed, at least in part, to the change in pH_{in}. Efflux has also been measured under conditions where the only driving force is the gradient of Na^+ generated upon dilution. In these experiments, generation of $\Delta\Psi$ or ΔpH by the action of the antiporter is prevented by the presence of valinomycin and high concentrations of K^+ inside and outside the proteliposomes and with the penetrating weak acid acetate, which rapidly short circuits any pH gradient generated. Under these conditions we have measured a stimulation of up to 2000-fold in the efflux rate from proteoliposomes reconstituted with NhaA upon increase of the pH from 6.5 to 8.5 (Fig. 3). This stimulation of the purified NhaA is solely a kinetic effect, since no H^+ ion gradient is generated under the conditions tested. Moreover, the stimulation reflects the effect solely on NhaA, without background activity of the other Na^+/H^+ antiporter(s) (NhaB). In the experiments described above, both internal and external pH were modified and we do not know yet whether a change in the *cis* side alone is sufficient to bring about the stimulation. Also, we still have to determine how the kinetic parameters are affected by pH. The downhill efflux rates are very high, increasing up to 2.2 mmol/min/mg at pH 8.5 (Na^+_{in} = 15 m*M*) a value that corresponds to a turnover number of $10^3 \cdot sec^{-1}$. This turnover number is one of the highest reported thus far for an ion-coupled transport system, and is only 10 times lower than the turnover number of the erythrocyte anion exchanger (Cabantchik, 1990).

We suggest that the steep pH dependence of NhaA defines a "set point" for the activity such that NhaA is practically inactive in pH values below the intracellular homeostatic one (7.6–7.8) (Fig. 5). When the pH increases the antiporter is activated so that it can acidify the cytoplasm back to the "resting pH_{in}" in a self-regulated mechanism. This idea of a molecular pH meter and titrator in the same molecule seems to be quite a successful one, since it was chosen also by completely different molecules: the animal Na^+/H^+ antiporter (Aronson, 1985) and the nonerythroid Cl/HCO_3 exchanger (Olsnes *et al.*, 1986, 1987). The set point of the human protein seems to be regulated by various hormones through phosphorylation of the cytoplasmic domain of the protein (Sardet *et al.*, 1990). We do not know yet whether the setpoint of NhaA or the Cl/HCO_3 exchanger are regulated or modulated by physiological factors. Also, it remains to be tested whether the H^+-sensing and the H^+-transporting sites are identical or overlap somehow.

VII. Regulation of Expression of *nha*A

The existence of a multicomponent system responsible for homeostasis of Na^+ and H^+ ions requires the cell to carefully regulate each individual transporter. Thus far, we have only information on the regulation of expression of *nha*A. In the *nha*A gene we have mapped two promoters by primer extension in the 5′ upstream region (Karpel *et al.*, 1991). In addition, a quite extensive putative secondary structure in the RNA can be predicted in the 5′ end of the gene (Karpel *et al.*, 1988) and the first codon is GTG rather than ATG (Taglicht *et al.*, 1991) (Fig. 6). GTG has been found to mediate initiation of translation in about 8% of the *E. coli* documented proteins (Gold and Stromo, 1987) and it has been suggested that it may be used in mRNA's that are poorly translated. Also, the codon usage in *nha*A is classical of poorly expressed proteins (Pinner *et al.*, 1992a). We estimate that at the growth conditions that are standard in our laboratory (L broth adjusted to pH 7.5 in which the Na^+ is replaced with K^+ and the contamination levels of Na^+ are around 10 m*M* or minimal salt medium to which sodium is not added) NhaA is a minor component

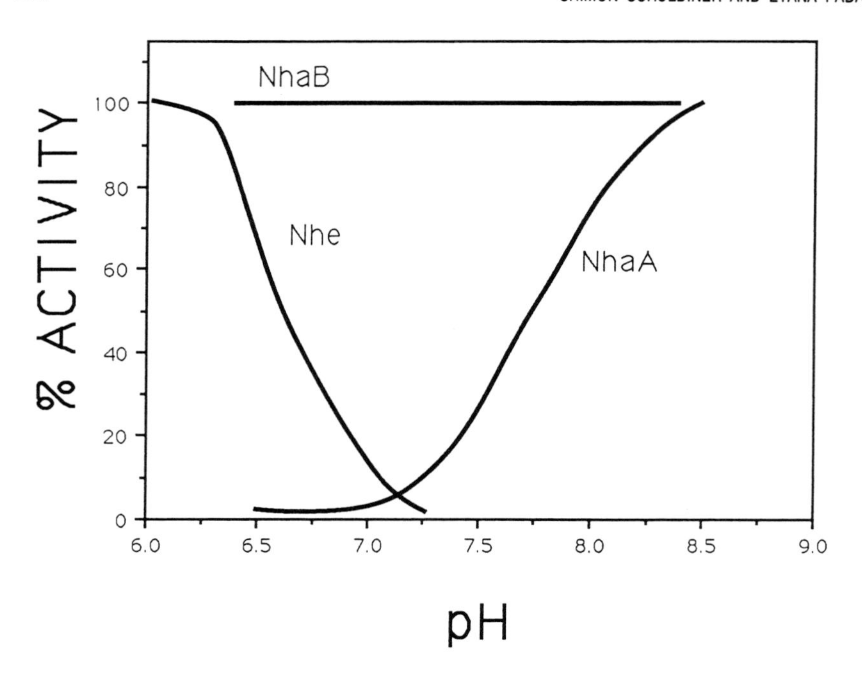

	Nhe (Human)	NhaA	NhaB
STOICHIOMETRY (H^+/Na^+)	1:1	2:1	N.D.
AMILORIDE SENSITIVITY	+	−	N.D.
AFFINITY	LOW 20–60 mM	MEDIUM 0.1–20 mM (varies with pH)	HIGH (0.05 mM)

FIG. 5 Classes of Na^+/H^+ antiporters.

of the membrane (less than 0.2%, or an equivalent of less than 500 copies per cell; (Taglicht *et al.,* 1991). Expression with an exogenous promoter (*tac*) is much higher when the regulatory sequences of *nha*A are deleted (Taglicht *et al.,* 1991), implying that, at least under some conditions, the upstream region has an inhibitory effect on expression of *nha*A.

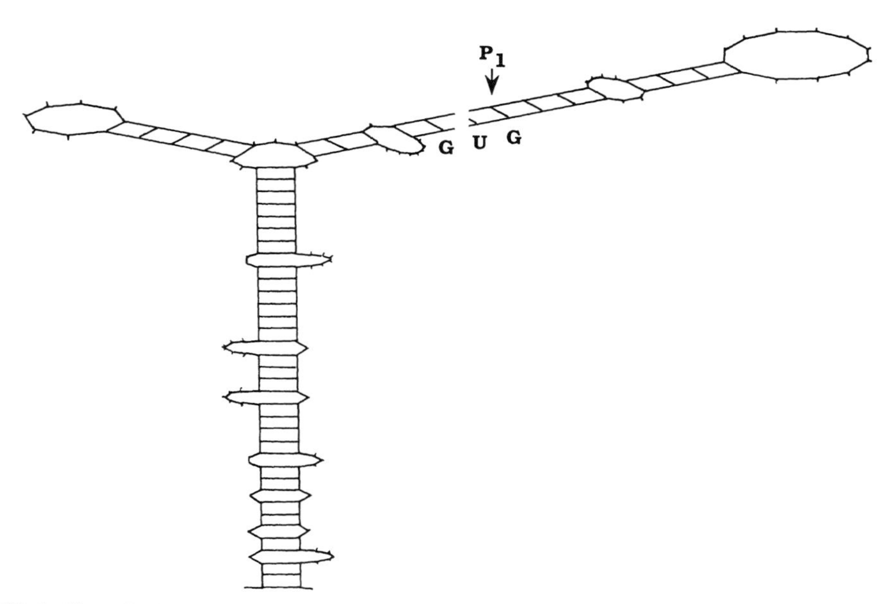

FIG. 6 Hypothetical secondary structure of *nha*A mRNA. The initiation codon GTG is depicted.

We have constructed a chromosomal translation fusion between *nha*A and *lac*Z (*nha*A′–′*lac*Z) and found that the levels of expression are very low unless Na^+ or Li^+ are added. Na^+ and Li^+ ions increase expression in a time- and concentration-dependent manner (Karpel *et al.*, 1991): maximal increase is detected when the cells are exposed to 50–100 m*M* of either ion for a period of 2 hr. The effect is specific to the nature of the cation and is not related to a change in osmolarity. When a functional NhaA or NhaB is coexpressed to high levels from multicopy plasmids the effect of 100 m*M* Na^+ is undetectable (Rahav-Manor *et al.*, 1992). Since the antiporter genes do not share significant homology we conclude that it is their acitivity, rather than their DNA or proteins, that is responsible for eliminating the induction by Na^+. Since the antiporter activity reduces intracellular Na^+, the findings suggest that intracellular, rather than extracellular, Na^+ is the signal for this process. In the same vein, alkaline pH potentiates the effect of the ions: whereas 10 m*M* Na^+ has no effect at pH 7.5, its effect is maximal at pH 8.5. This synergistic effect with pH could reflect, at least in part, the fact that the Na^+ gradient that the cell can maintain, decreases, and therefore $[Na]_{in}$ increases with an increase in pH of the medium (Castle *et al.*, 1986a; Pan and Macnab, 1990).

The pattern of regulation of *nha*A reflects therefore its role in adaptation to high salinity and alkaline pH in *E. coli,* as revealed by the analysis of the Δ*nha*A strains. This pattern also suggests the possibility of involvement of novel regulatory genes in addition to *nha*A and *nha*B.

A gene downstream of *nha*A, *nha*R [previously known as *ant*O (Henikoff *et al.*, 1988), or 28-kDa protein (Mackie, 1986)] is proposed to play a role in regulation of *nha*A. In addition to its proximity to *nha*A, and the fact that there are no conspicuous consensus sequences of either terminators or promoters between the two genes, expression with foreign promoters cloned upstream of *nha*A brings about expression of *nha*R as well (Karpel *et al.*, 1988). Multiple doses of *nha*R enhance the Na^+-dependent induction of the *nha*A′–′*lac*Z fusion (Rahav-Manor *et al.*, 1992). The fact that the dose level affects the induction by Na^+, but not the basic level of expression, suggests that the induction involves *nha*R either directly or indirectly. NhaR exerts its effect in *trans* as shown in the latter experiments. Furthermore, extracts derived from cells overexpressing *nha*R exhibit DNA binding capacity specific to the upstream sequences of *nha*A, as observed by gel retardation assays (Rahav-Manor *et al.*, 1992). Inactivation of chromosomal *nha*R by insertion unveils a phenotype of sensitivity to Li^+ higher than that displayed by the wild type (Rahav-Manor *et al.*, 1992). A change of tolerance toward Na^+ in these cells becomes apparent only at pH 8.5, under conditions in which the load seems to be more pronounced, as suggested by the phenotype of the Δ*nha*A strain and by the pattern of regulation of *nha*A. Both phenotypes are corrected by *nha*R in *trans* (Rahav-Manor *et al.*, 1992).

On the basis of the latter results, it is proposed that NhaR is a positive regulator of *nha*A. This suggestion accords the fact that NhaR belongs to the OxyR–LysR family of positive regulators first described by Henikoff *et al.* (1988) and studied also by Christman *et al.* (1989). All the proteins in this group have in their N-terminus a conserved helix–turn–helix domain that is supposed to bind to DNA. Several of these proteins are involved in the response of the organism to stress, such as, for example, OxyR, which is essential for the resistance of the organism to oxidative stress (Storz *et al.*, 1990). NhaR may represent a component of another type of stress response essential for tolerance to Na^+ and Li^+.

The regulation of gene expression of anion and cation transport systems is gaining more and more attention. Some of them, such as Pst, Kdp, Fur, magnesium, and sulfate, are quite well understood (reviewed in Silver and Walderhaug, 1992). Regulation of these systems seems to fall into two categories: two component pairs of regulatory proteins or a single one that both senses and regulates. Also, a common theme of the chromosomally encoded mineral transport systems studied thus far is the presence of multiple transport systems for each ion: a low-affinity, high-capacity transport system that is expressed constitutively, and another high-affinity, low-capacity transport system that is carefully regulated in response to ion starvation. In iron-regulated systems, the sensor and effector activities are carried out by the same protein, Fur. The protein responds to the intracellular, rather than extracellular, iron levels as opposed to the systems that sense Pi and Mg^{2+}, which respond to the extracellular levels. One possible reason is that intracellular Fe^{2+} levels are generally low, whereas intracellular Pi and Mg^{2+} always remain high (above m*M* levels) to support metabolism even when the extracellular concentration drops toward zero (Silver and Walderhaug, 1992). Hence regulation of iron transport can employ an intracellular signal. The regulation of the Na^+ transport systems seems to respond to intracellular Na^+ sharing some of the properties displayed by the Fur systems rather than with the other ions.

However, Fur is a repressor of transcription, whereas PhoB is an activator. NhaR belongs to the family of positive activators of transcription as well, similar to the case of plasmid-based toxic ion resistance systems in which ion binding seems to lead to the stimulation of mRNA synthesis, instead of its repression (Silver and Walderhaug, 1992).

VIII. Physiological Roles of the Na^+ Cycle

A. Extreme Environments

The dependence of the Na^+/H^+ antiporters on $\Delta\mu_H+$ for Na^+ extrusion suggests that this form of Na^+ export is not efficient when $\Delta\mu_H+$ is limiting and/or when the Na^+ concentration is high and imposes a heavy load on the $\Delta\mu_H+$. Accordingly, alternative systems initiating primary Na^+ cycles exist in anaerobic and marine bacteria. Decarboxylases utilizing energy from decarboxylation reaction serve as primary Na^+ pumps in many anaerobes (Dimroth, 1992). Anaerobes also possess Na^+/ATPase (Heefner and Harold, 1980; Unemoto *et al.*, 1990; Dimroth, 1992). In methanogens several metabolic reactions are coupled to the formation and utilization of the Na^+ gradient (Muller and Gottschalk, 1992; Schonheit, 1992). In marine bacteria (Ken-Dror *et al.*, 1986; Dimroth, 1987; Unemoto *et al.*, 1990) and other halotolerant and alkaline-tolerant bacteria (for classification of alkaliphiles and alkaline-tolerant bacteria see Krulwich, 1989), primary Na^+ pumps are linked to electron transport (Skulachev, 1988). It has been suggested that stress caused by low $\Delta\mu_H+$ induces the primary Na^+ pump even in organisms such as *E. coli* (Dibrov, 1991).

In contrast to this general theme, primary Na^+ pumps have not yet been discovered in either extreme halophiles or extreme alkaliphiles (nonmarine nonhalotolerant). The primary proton cycles of these organisms operate against the heaviest load of extracellular Na^+ (>3.5*M* NaCl) (Lanyi, 1979; Reed, 1986) or with the lowest $\Delta\mu_H+$ (-25 to -50 mV) (Krulwich and Guffanti, 1989; see, however, Hoffmann and Dimroth, 1991), respectively, yet to extrude Na^+ they use the Na^+/H^+ antiporter-dependent secondary Na^+ cycle (Lanyi, 1979; Krulwich, 1986, 1992; Krulwich and Ivey, 1990; Krulwich *et al.*, 1990).

The extreme halophiles seem to tolerate high cytoplasmic concentration of Na^+ (up to 1 *M*) (Kushner and Kamekura, 1988). It appears therefore that they compromise with a Na^+ gradient lower than 10, which can be easily produced by the Na^+/H^+ antiporter despite of the heavy Na^+ load imposed on cell energetics. The properties of this antiporter appear to be unique; it requires a gating potential of 100 mV and has properties that restrict Na^+ backflow even when $\Delta\mu_H+$ is reversed (Murakami and Konishi, 1990).

In the extreme alkaliphiles the low $\Delta\mu_H+$ is due to the Na^+/H^+ antiporter, which generates a reversed ΔpH (3 units). This secondary Na^+ cycle has a crucial role in homeostasis of cytoplasmic pH of the extreme alkaliphiles, facing values of extracellular pH up to 11 (Krulwich, 1992).

It may be concluded that prokaryotes have most versatile Na^+ extrusion systems. This versatility can be related to the modes of energization across the cytoplasmic membranes, and to the constraints imposed by the multifarious nature of the environments microorganisms face.

B. Utilization of Na^+ Gradients in Bacteria

Na^+ cotransport systems use the energy of the Na^+ electrochemical gradient to drive substrate accumulation. These cotransporters are the rule in animal cells. In accordance with the demonstration of both primary and secondary Na^+ cycles in bacteria, Na^+ cotransport systems are now known to be common in bacteria that grow under a wide variety of conditions (Anraku, 1992; Maloy, 1990). In many bacteria living at moderate conditions of salinity (<1 *M*) and pH (neutral), transport is coupled mainly to H^+ rather than to Na^+. In *E. coli* only four Na^+/cotransport systems are known (MacDonald *et al.*, 1977; Tsuchiya *et al.*, 1977; Stewart and Booth, 1983; Ishikawa *et al.*, 1987). Most transport systems in obligate halophiles and extreme alkaliphiles are Na^+ symporters (Krulwich, 1992; Lanyi, 1979). In the case of the latter an obvious selective advantage is conferred, since the ΔpH of opposite polarity lowers the available $\Delta\mu_H+$ (Krulwich, 1992; Krulwich and Guffanti, 1989).

With the change of primary Na^+ cycle in the anaerobes (Dimroth, 1992; Muller and Gottschalk, 1992; Schonheit, 1992), marine bacteria (Tokuda, 1992), and similar halotolerant alkaline-tolerant organisms (Skulachev, 1987, 1988), Na^+ becomes the main coupling ion for transport. In marine bacteria (Dibrov *et al.*, 1986; Tokuda, 1992) and extreme alkaliphilic bacteria (Hirota *et al.*, 1981) flagellar motility is also Na^+ dependent. Interestingly, protein translocation across the membrane in *V. alginolyticus* has also been suggested to be dependent on $\Delta\mu_{Na^+}$ (Tokuda *et al.*, 1990).

A major consumer of energy is the reaction of ATP synthesis, which in all bacteria growing at moderate Na^+ and H^+ concentrations is driven, in a chemiosmotic mechanism, by $\Delta\mu_{H^+}$ via F-type H^+-ATPase. This is also the case in extreme halophiles (Lanyi, 1979) and in the extreme, nonmarine nonhalotolerant alkaliphiles (Krulwich, 1992). The very low available $\Delta\mu_{H^+}$ of the latter led Krulwich to propose a non-Mitchellian mechanism, still involving the F-type H^+-ATPase (Krulwich, 1992; but see also Hoffmann and Dimroth, 1991; Dimroth, 1992). Na^+-ATPases accompany the primary Na^+ pumps in various anaerobes (Heefner and Harold, 1980, 1982; Dimroth, 1992b). Their existence has also been suggested in marine bacteria (Dibrov *et al.*, 1986; Sakai *et al.*, 1989; Sakai-Tomita *et al.*, 1991). However, recently the *unc* operon has been cloned from *V. alginolyticus* and expressed in *E. coli* (Krumholz *et al.*, 1990). Since the purified and reconstituted enzyme exhibits only H^+-pumping activity, the authors conclude that if *V. alginolyticus* retains Na^+/ATPase it is not likely to be the F-type enzyme.

In allowing the reuptake of Na^+, all $\Delta\mu_{Na^+}$ consumers complete the Na^+ cycle. If no other substantial Na^+ leaks exist, they may also determine the rate of the Na^+ cycle.

It is apparent that the various Na^+ cycles observed seem to be an adaptation to a unique environment. The complete primary Na^+ cycle allows the ability to cope with limited or absence of $\Delta\mu_{H^+}$ in anaerobic niches. The possibility of alternating between primary and secondary Na^+ cycles affords marine bacteria a very wide spectrum of adaptation in terms of pH and salinity (Tokuda *et al.*, 1988; Sakai *et al.*, 1989; Tokuda, 1992). Whereas the secondary Na^+ cycle of extreme halophiles differs from that of ordinary bacteria in its accommodation to higher intracellular Na^+, that of the extreme alkaliphiles is unique and essential to the adaptation to extreme alkaline pH (moderate salinity, nonmarine) niche (Krulwich, 1992).

C. Na^+ Channels

It has been implied that the rate of Na^+ entry into the cells determines the rate of operation of the Na^+ cycle (Booth, 1985). Indeed, the presence of substrates of Na^+-coupled cotransport systems markedly enhances acidification of the cytoplasm via the Na^+/H^+ antiporter during an alkaline shift of extreme alkaliphile (Krulwich *et al.*, 1982). Nevertheless both alkaliphiles and other bacteria may possess a symport-independent, pH-regulated Na^+ entry route as suggested for *Exiguobacterium aurantiacum* (McLaggan *et al.*, 1984) and *E. coli* (Booth, 1985). This raises the questions of the passive pathways of Na^+ leaks and Na^+ channels in prokaryotes.

The outer membrane of Gram-negative bacteria presents a barrier to the passage of macromolecules but is relatively permeable to hydrophilic solutes due to the presence of porins. Porins are abundant proteins that form nonspecific channels across the outer membrane *E. coli* (Rosenbusch, 1974). Using salt taxis as a sensitive *in vivo* assay for outer membrane permeability it has been shown that porins hardly affect salt permeability via the outer membrane (Ingham *et al.*, 1990). Relative to amino acid taxis, salt taxis is surprisingly little impaired by the loss of porins. Like the amino acids tested these salts are sensed by one or more of the methyl-accepting chemotaxis proteins and must therefore penetrate the outer membrane. These results imply that an additional mechanism allows salt penetration through the outer membrane. Accordingly, the study of *E. coli* cell membrane with the patch clamp technique reveals new ion channels located within the outer membrane (Buechner *et al.*, 1990).

Although there is evidence for K^+ channels in the cytoplasmic membrane of *E. coli* (Booth *et al.*, 1992; Zoratti, 1992), there is no such evidence for Na^+ channels. Using NMR spectroscopy, Castle *et al.* (1986b) showed that the downhill Na^+ movement from deenergized cells (under anaerobic conditions or in the presence of uncoupler) is very slow, in agreement with the results of other studies (measuring fluxes of $^{22}Na^+$ (Tsuchiya and Takeda, 1979; Bassilana *et al.*, 1984a,b; Borbolla and Rosen, 1984). Although this slow Na^+ movement is probably mediated by Na^+/H^+ antiporter, the contribution of Na^+ leaks or channels was not ruled out.

D. Toxicity of Na^+

There is a universal asymmetry in the distribution of Na^+ and K^+ ions across the plasma membrane: intracellular K^+ is high, whereas intracellular Na^+ is low (Harold, 1986). The reason can be the more ubiquitous occurrence of Na^+ compared to K^+ and therefore first using the former for ion coupling. Alternatively it is possible that Na^+ is toxic compared to K^+.

Nevertheless, all organisms require salts and there is an optimum concentration of salts for each; accordingly too little and too much are both avoided. *Escherichia coli* has been shown to exhibit salt taxis. Although the threshold is high (0.1–1 m*M*), this is a powerful attraction, in some cases nearly as strong as for the best *E. coli* attractants known, the amino acids L-aspartate and L-serine. The optimum concentration for NaCl taxis is between 10 and 100 m*M* (Qi and Adler, 1989). Higher concentrations of salt, for example 500–400 m*M* act as repellent (Li *et al.*, 1988). Lower concentrations are also less attractive and pure water is avoided. The nature of the cation is critical, the order of taxis is being $NH_4^+ > Li^+ > Na^+ > K^+ > Rb^+ > Cs^+$. Mg^{2+} is the best divalent cation. The nature of the anion has a lesser but definite effect; Cl^- is the best halide. It appears that both methly-accepting chemotaxis proteins, MCPI and MCPII, participate in salt taxis, the former being more important. Salt taxis is different from osmotaxis, which gives less than one-tenth of the response and does not operate via MCP.

The repellent effect of NaCl suggests that Na^+, above certain concentration (500–400 m*M* at pH 7) (Li *et al.*, 1988), has a deleterious effect on *E. coli* cells. This effect is not due to increased osmolarity nor to increased ionic strength, since osmotaxis is different from salt taxis (see above) and various salts have different specific effects.

Determination of NaCl concentrations inhibiting growth of the wild type may reflect the concentrations toxic to cell components exposed to the environment. However, it is also possible that intracellular Na^+ may be responsible for the inhibition, suggesting toxic effects of the ion on cytoplasmic components. In this case, and given the existence of Na^+ extrusion systems in all bacteria, it is predicted that impairment of Na^+ excretion will increase the Na^+ sensitivity of the cells.

Comparison of the salt tolerance of a Δ*nha*A mutant to the wild type revealed a specific toxic effect of Na^+. Thus at pH of 6.8 the mutant does not grow in the presence of NaCl concentration of above 600 m*M*, whereas the wild type grows up to 900 m*M*. This effect is independent of ionic strength or osmolarity, since the mutant grows at 900 m*M* KCl like the wild type. Furthermore the toxicity is pH dependent, increasing upon alkalinization. At pH 8.6, whereas 100 m*M* NaCl has no effect on the wild type, it completely inhibits the growth of Δ*nha*A. Deleting both *nha*A and *nha*B in mutant EP432 further increased the Na^+ sensitivity of the cells (Pinner *et al.*, 1992b). These cells are already inhibited at pH 7.5 by 100 m*M* NaCl. Since mutant EP432 does not possess any measurable Na^+/H^+ antiporter activity under the conditions used (Schuldiner and Padan, 1992), it is suggested that it does not extrude Na^+ and its intracellular Na^+ reaches toxic concentrations at much lower extracellular concentrations than the wild type. Therefore, it is tempting to conclude that intracellular Na^+ has a specific toxic effect on a cytoplasmic component(s) in *E. coli*. It will be most interesting to directly measure these toxic intracellular Na^+

concentrations and compare them to cytoplasmic concentrations causing the Na^+-repellent response of *E. coli* (see above).

The increased sensitivity of the mutants to Na^+ caused by alkaline pH is intriguing. It may be related to the interchangeability of Na^+ and H^+ in some systems. At alkaline pH Na^+ will compete better with H^+ and therefore may inhibit proton-requiring processes. It is also plausible that with alkalinization intracellular Na^+ reaches toxic levels at lower extracellular Na^+ concentrations. Indeed the Na^+ gradient that the cells can maintain decreases when the medium pH is increased (Pan and Macnab, 1990). Another alternative is that homeostasis of intracellular pH is impaired in the mutants and Na^+ aggravates this condition.

Which is the most sensitive reaction to Na^+? If this reaction is essential, a straightforward answer is difficult to obtain by inhibiting or deleting it. Using peptidyl-puromycin synthesis on polyribosomes as an *in vitro* model for protein synthesis in *E. coli,* Pestka (1972) showed that 100–200 m*M* NaCl inhibits the reaction. Interestingly this *in vitro*-inhibiting concentration of Na^+ is similar to the concentration inhibiting growth of the Δ*nha*AΔ-*nha*B mutant (EP432) (Schuldiner and Padan, 1992).

In marked contrast to Na^+, K^+ activates a number of cell enzymes from all living cells (review in Walderhaug *et al.,* 1987). Some of these effects are specific to the ion and independent of osmolarity or ionic strength. Activation of protein synthesis by K^+ was demonstrated by Ennis and Lubin (1961) in studies of a K^+ transport mutant of *E. coli.* As cell K^+ fell and Na^+ rose, protein synthesis was progressively inhibited, whereas DNA and RNA synthesis continued. The concentration of K^+ needed to activate enzymes (about 10 m*M*) is far below the intracellular concentration (above 150 m*M*) found in most bacteria.

There is no clear answer why K^+ is preferred to Na^+. It appears that ions of similar size, such as NH_4^+ and Rb^+, usually can replace K^+, whereas smaller monovalent cations with larger hydration shells, such as Na^+ and Li^+, often antagonize activation by K^+ (Walderhaug *et al.,* 1987).

E. Lack of Homeostasis of Cytoplasmic Na^+

The demonstration of Na^+ toxicity and repellent activity on the one hand and Na^+ taxis on the other hand raises the question of homeostasis of intracellular Na^+ concentrations. By the use of a perfusion system that stably maintains the external pH of dense cultures of endogenously respiring *E. coli,* Pan and Macnab (1990) extended studies by ^{31}P and ^{23}Na NMR spectroscopy of Na^+ and H^+ bioenergetics (Castle *et al.,* 1986a,b) to a wide pH range (pH 6.4–8.4). These studies substantiated previous results (Padan *et al.,* 1976, 1981; Slonczewski *et al.,* 1981; Padan and Schuldiner,

1986, 1987) showing that intracellular pH in neutrophilic bacteria is regulated to 7.6–7.8 over the entire range of pH permitting growth.

In marked contrast to the extensive homeostasis of pH, intracellular Na^+ was not found constant (Pan and Macnab, 1990). Measuring $[Na^+]_{in}$ as a function of $[Na^+]_{out}$ in the range 4 to 285 m*M*, at 25°C, extracellular pH of 6.6–6.7, and constant osmolarity (375 mOsm) shows that ΔpNa $\{= -\log([Na^+]_{in}/[Na^+]_{out})\}$ is constant (1.24) and intracellular Na^+ is not regulated under these conditions varying from 0.23 to 16.4 m*M*, respectively. Lack of regulation of intracellular concentration of Na^+ was observed at various pH values between 6.25 and 7.6; (Castle *et al.*, 1986b) ΔpNa changes with pH (see below), but at each pH value it is constant and independent of extracellular Na^+ concentrations (80 and 180 m*M* at constant osmolarity of 375 mOsm).

These results show that extracellular pH affects intracellular Na^+ concentration. At 80 m*M* $[Na^+]_{out}$, increasing extracellular pH to 7.4 causes a parallel decrease in ΔpH and ΔpNa, although the latter is always somewhat larger. Above pH 7.4 ΔpNa increases with external pH but beyond pH 8 it again slowly decreases. These variations with pH suggest different modes of Na^+ export (Pan and Macnab, 1990). It also shows that intracellular Na^+ increases with pH at a given extracellular Na^+ concentration. In the presence of 80 m*M* Na^+, intracellular Na^+ at pH 6.7 is 4.6 m*M*, whereas at pH 8.4 it is 25.3 m*M*.

On the basis of the published data (Pan and Macnab, 1990) it can be estimated that at neutral pH and at 400–500 m*M* extracellular Na^+, the intracellular Na^+ is 40–50 m*M*. As this extracellular concentration is the repellent concentration (see above), we suggest that 40–50 m*M* intracellular Na^+ becomes toxic to the cytoplasm.

External concentrations in the same range induce β-galactosidase activity of a *nha*A′–′*lac*Z fusion (*in vivo*), and alkaline pH increases the sensitivity of the expression system to the ions (Karpel *et al.*, 1991). As described above our results (Rahav-Manor *et al.*, 1992) show that *nha*R, is a positive regulator of *nha*A. NhaR works in *trans* and its effect is Na^+ dependent. This Na^+ dependency is related to intracellular rather than extracellular Na^+. Most interestingly, on the basis of protein similarity, NhaR has been shown to belong to the LysR family of positive regulators (Henikoff *et al.*, 1988), which share homologous sequences in the N-terminal region of the protein where a helix–turn–helix domain is found. Some of these proteins are involved in the response of bacteria to various environmental stresses. An example is OxyR (Storz *et al.*, 1990), which is crucial for adaptation to oxidative damage. We postulate that *nha*R is involved in a novel signal transduction pathway responding to Na^+ stress. Taken together it is suggested that *E. coli* cells react to a high Na^+ load, when the intracellular concentration approaches the stress concentrations, by swimming away from the toxic ion and at the same time by initiating the Na^+-dependent *nha*R response, which leads to induction of *nha*A for efficient excretion of the ion.

On the other hand, it appears that as long as toxic concentrations are not reached, the cells can accommodate quite high variations in intracellular Na^+. Furthermore as long as $[K^+]_{in}$ is high, which is the usual case, there is no reason to maintain constant $[Na^+]_{in}$ either for osmotic balance or for ionic strength. Hence, there is a wide flexibility in intracellular Na^+ below its toxic concentration. This is consistent with the function of Na^+/H^+ antiporters in pH homeostasis, manipulating the concentration of Na^+ to keep intracellular H^+ concentration constant.

F. Role of Na^+/H^+ Antiporters in Regulation of Intracellular pH

One of the central roles assigned to the antiporter is regulation of intracellular pH (pH_{in}) mainly at alkaline extracellular pH. In *E. coli,* pH_{in} has been shown to be clamped at around 7.8 despite huge changes in the extracellular medium pH (Padan *et al.,* 1976, 1981; Booth, 1985; Padan and Schuldiner, 1986). Since then many bacteria as well as eukaryotic cells have been shown to maintain strictly their cytoplasmic pH constant at around neutrality (Padan *et al.,* 1976, 1981; Slonczewski *et al.,* 1981; Pouyssegur *et al.,* 1984; Booth, 1985; Krulwich, 1986; Häussinger, 1988; Grinstein *et al.,* 1989; Krulwich *et al.,* 1990; Pan and Macnab, 1990). Relatively small increases of pH_{in} stop cell division and activate expression of specific genes (Bingham *et al.,* 1989) and of regulons (Schuldiner *et al.,* 1986; Padan and Schuldiner, 1987). It is therefore not surprising that both eukaryotic and prokaryotic cells have evolved several pH_{in}-regulative mechanisms to eliminate metabolic induced changes in pH_{in} or to counter extreme environmental conditions (Booth, 1985; Grinstein, 1988; Grinstein *et al.,* 1989; Sardet *et al.,* 1989, 1990).

We proposed that Na^+/H^+ antiporters in conjunction with the primary H^+ pumps are responsible for homeostasis of intracellular pH in *E. coli* (Padan *et al.,* 1976, 1981). This suggestion had its most dramatic experimental validation in akaliphiles in which it was shown that Na^+ ions are required for acidification of the cytoplasm and for growth (Krulwich *et al.,* 1982, 1985; McLaggan *et al.,* 1984). In neutrophiles, such as *E. coli,* there is no direct evidence that supports this contention, since it is not clearly established that Na^+ is required for growth at alkaline pH. In some alkaliphiles the requirement for Na^+ was not easy to demonstrate either, presumably due to a very high affinity for Na^+ (as low as 0.5 m*M*), such that the contaminations present in most media suffice to support growth

(Sugiyama *et al.*, 1985). McMorrow *et al.* (1989) have taken special precautions to reduce Na^+ to very low levels (5–15 μM) and reported a strict requirement for Na^+ (saturable at 100 μM) for growth of *E. coli* at pH 8.5. This range of concentrations of Na^+ required for growth is well within the range of the K_m of the NhaB system (40–70 μM) (Pinner *et al.*, 1992b).

The Na^+/H^+ antiporter mutants generated in our laboratory (Padan *et al.*, 1989; Pinner *et al.*, 1993) show that, in the absence of added Na^+, cells can grow at alkaline pH, provided one of the antiporter genes is functional. When Na^+ is added (>0.1 M), *nha*A, but not *nha*B alone, can cope with the load. As described above, *nha*A seems to be more flexible and capable of handling higher loads, as suggested not only by the phenotypes obtained but also because this system is highly regulated, both at the level of the protein (Taglicht *et al.*, 1991) and at the level of gene expression (Karpel *et al.*, 1991; Rahav-Manor *et al.*, 1992). In all of the above documented cases, we are assuming that growing cells regulate their pH, but no measurements of internal pH are available yet with any of the mutants described.

Although the findings described strongly support the involvement of the antiporters in regulation of pH, at least in the presence of sodium ions, there are still many unknown components. The primary means of H^+ extrusion under aerobic conditions is the electron transport chain which, in the absence of permeable ions, would develop a negligible ΔpH and a large $\Delta\psi$. There is now evidence suggesting that Na^+ and K^+ are involved in the modulation of $\Delta\mu_{H^+}$ components: electrogenic uptake of K^+, with or without H^+, would compensate for charge extrusion and thus permit the development of ΔpH; an electrogenic antiport of Na^+ and H^+ would be a means of generating an inverted ΔpH. The reported change in apparent overall stoichiometry may reflect the change in the relative rates of NhaA and NhaB, two antiporters that may have different stoichiometries. We know that NhaA is electrogenic, with a stoichiometry of close to 2 (Taglicht, 1992). We do not know whether NhaB is electrogenic. However, we have measured rates of downhill Na^+ transport in membrane vesicles of Δ*nha*A and wild-type strains at various pH values assuming that they reflect the rate of NhaB and NhaB + NhaA, respectively. From the maximal rate of NhaA activity in its purified state and from its estimated abundance, we have calculated its contribution to the overall activity of the cell under various conditions. It seems that NhaB is the main activity at acid pH, whereas NhaA is dominant at alkaline one.

NhaA is a highly active system: the maximal value of downhill Na^+ efflux at pH 8.5 is around 400 nmol/min/mg cell protein and it can increase up to 4000 when *nha*A is fully induced. Since the rate of H^+ extrusion through the respiratory chain is around 1000 nmol/min/mg cell protein, and the rate of K^+ transport through the Trk systems is around 500

(Walderhaug *et al.*, 1987), our measurements show that indeed NhaA can quantitatively account for a rapid response to changes in ion content.

As discussed above the rate of the Na^+ leak into the cell is most probably the rate-limiting step for the operation of the Na^+ cycle. Na^+ solute cotransport is apparently the main route of Na^+ entry in *B. firmus* since without the solute pH homeostasis is impaired (Krulwich, 1992). In *E. coli,* however, regulation of intracellular pH has been shown in bacteria respiring on solutes that are taken up without Na^+ (Padan *et al.*, 1976) or on endogenous substrates (Pan and Macnab, 1990). As discussed above (Booth, 1985) the question of reentry of Na^+ into the cells to keep the Na^+ cycle rolling in *E. coli* is most important but is still unanswered.

Since many biomolecules (including all proteins and nucleic acids) are acid–base species, whose conformations are pH dependent, the regulation of cytoplasmic pH serves a centrally important physiological function by providing a suitable environment for maintenance of stable structure and enzymatic activity. Furthermore it appears that intracellular pH itself can serve as a specific signal for certain regulons (Padan and Schuldiner, 1987). The SOS regulon has been shown to be induced upon alkalinization of intracellular pH (Schuldiner *et al.*, 1986). Remarkably, many aspects of the central regulatory roles played by the Na^+ cycle in the prokaryotes are shared by the eukaryotes. This includes the role of Na^+/H^+ antiporters in pH regulation (Grinstein *et al.*, 1989) and in pH-related control mechanisms affecting cell proliferation (Pouyssegur *et al.*, 1988).

IX. Concluding Remarks: Multiple Proteins Catalyze Na^+/H^+ Exchange

It is becoming evident that the catalysis of Na^+/H^+ exchange is performed by various distinct proteins that differ significantly at least in their amino acid sequence and in some of the details of the exchange reaction. Genes coding for Na^+/H^+ antiporters have been cloned from *E. coli* (Karpel *et al.*, 1988; Pinner *et al.*, 1992a), *S. enteritidis* (Pinner *et al.*, 1992a), *B. firmus* (Ivey *et al.*, 1991), and *Enterococcus hirae* (Waser *et al.*, 1992) as well as from several eukaryotic sources: human (Sardet *et al.*, 1989), rabbit heart (Fliegel *et al.*, 1991), and rabbit ileal villus cell from basolateral membrane (Tse *et al.*, 1991). There seems to be very little conservation of sequences between the eukaryotic and the prokarytic sequences and also within the prokaryotic sequences themselves. In one 16-amino acid stretch, however, the homology between NhaA and NhaB is quite striking: 35% identity and 52% similarity (Fig. 7). In addition, the homologous stretch is in approximately the same area of the protein, starting at amino

Homologous domain
in
three antiporters

```
102. P F E I S L W I L L A C L M K I G   Nhe1
       | : : | | | : . : | |   : | : :   (41, 82)
300. P L G I S L F C W L A L R L K L A   NhaA
         : | : . | . .   | | | : |   .   (35, 52)
295. I I G V W L V T A L A L H L A E V   NhaB
```

FIG. 7 Homologous domain in three antiporters. Homologous stretch of 16 amino acids is highly conserved in NhaA, NhaB, and Nhe1. Residues G, L, and A spaced as in the three antiporters are found also in the rat brain GABA transporter (Guastella *et al.*, 1990); the rabbit Na^+/glucose symporter (Hediger *et al.*, 1987); the *E. coli* proline, pantothenate, and melibiose symporters (Nakao *et al., 1987; Reizer et al.*, 1990; Yazyu *et al.*, 1984); and the Na^+/H^+ antiporters from *B. firmus* F4 (Ivey *et al.*, 1991) and *E. hirae* (Waser *et al.*, 1992).

acids 295 and 300 in NhaB and NhaA, respectively. Also, Nhel is 41% identical and 83% similar to NhaA in this segment. Interestingly, this domain overlaps with a small part of the diffuse SOB motif previously identified by Anraku and collaborators (Anraku, 1992; Deguchi *et al.*, 1990, see also Reizer *et al.*, 1990) in various transporters that move Na^+ ions across membranes: three residues, glycine, leucine, and alanine, are found in a fixed spacing in all the sequences available to us except for the one coding for the human exchanger (Nhe1), in which the glycine is replaced by a glutamic acid residue. In all cases none of the residues between the glycine and the alanine are charged.

The catalytic properties of the various antiporters seem to differ as well (Fig. 5): whereas Nhe1 is electroneutral and inhibitable by amiloride-like compounds, NhaA is electrogenic and not sensitive to various amiloride derivatives tested in our laboratory (Taglicht *et al.*, 1991). Both antiporters have a "pH sensor," but the direction of the switch is opposite: Nhe1 is activated by acid pH, whereas NhaA is activated by alkaline. Obviously, the properties of the various antiporters seem to have developed to better fit the functions of each of them, namely Nhe1 is the main acid-extruding mechanism in cells of higher organisms, whereas NhaA is a major acidifier of *E. coli* cytoplasm at alkaline extracellular pH. NhaB seems to have an activity almost insensitive to pH, in line with its putative housekeeping role. We will probably see in the near future an increase in the number of types known, mainly from those localized in subcellular organelles, such as mitochondria, storage organelles, and plant vacuoles, as well as from bacteria dwelling in specialized and/or extreme environments, such as methanogens and extreme halophiles.

Acknowledgment

Research was supported by research grants from the U.S./Israel Binational Science Foundation (BSF).

References

Anraku, Y. (1992). *In* "Alkali Cation Transport Systems in Procaryotes" (E. Bakker, ed.). CRC Press, Boca Raton, Florida. In press.

Aronson, P. S. (1985). *Annu. Rev. Physiol.* **47,** 545–560.

Bassilana, M., Damiano, E., and Leblanc, G. (1984a). *Biochemistry* **23,** 5288–5294.

Bassilana, M., Damiano, E., and Leblanc, G. (1984b). *Biochemistry* **23,** 1015–1022.

Beck, J. C., and Rosen, B. P. (1979). *Arch. Biochem. Biophys.* **194,** 208–214.

Benos, D. J. (1988). *In* "Na^+/H^+ Exchange " (S. Grinstein, ed.), pp. 121–136. CRC Press, Boca Raton, Florida.

Bingham, R. J., Hall, K. S., and Slonczewski, J. L. (1989). *J. Bacteriol.* **172,** 2184–2186.

Blulmwald, E., and Poole, R. J. (1985). *Plant Physiol.* **78,** 163–167.

Booth, I. R. (1985). *Microbiol. Rev.* **49,** 359–378.

Booth, I. R., Douglas, R. M., Ferguson, G. P., Lamb, A. J., Munro, A. W., and Ritchie, G. Y. (1992). *In* "Alkali Cation Transport Systems in Prokaryotes" (E. P. Bakker, ed.). CRC Press, Boca Raton, Florida.

Borbolla, M. G., and Rosen, B. P. (1984). *Arch. Biochem. Biophys.* **22,** 98–103.

Brey, R. N., Beck, J. C., and Rosen, B. P. (1978). *Biochem. Biophys. Res. Commun.* **83,** 1588–1594.

Buechner, M., Delcour, A. H., Martinac, B., Adler, J., and Kung, C. (1990). *Biochim. Biophys. Acta* **1024,** 111–121.

Cabantchik, Z. (1990). *In* "Blood Cell Biochemistry, 1: Erythroid Cells" (J. R. Harris, ed.), pp. 359–364. Plenum New York.

Castle, A. M., Macnab, R. M., and Schulman, R. G. (1986a). *J. Biol. Chem.* **261,** 3288–3294.

Castle, A. M., Macnab, R. M., and Schulman, R. G. (1986b). *J. Biol. Chem.* **261,** 7797–7806.

Christman, M. F., Storz, G., and Ames, B. N. (1989). *Proc. Natl. Acad. Sci. U.S.A.* **86,** 3484–3488.

Deguchi, Y., Yamato, I., and Anraku, Y. (1990). *J. Biol. Chem.* **265,** 21704–21708.

Devereux, J., Haeberli, P., and Smithies, O. (1984). *Nucleic Acids Res.* **12,** 387–395.

Dibrov, P. A. (1991). *Biochim. Biophys. Acta* **1056,** 209.

Dibrov, P. A., Lazarova, R. L., Skulachev, V. P., and Verkhovskaya, M. L. (1986). *Biochim. Biophys. Acta* **850,** 458–465.

Dimroth, P. (1987). *Microbiol. Rev.* **51,** 320–340.

Dimroth, P. (1992). *In* "Alkali Cation Transport Systems in Procaryotes" (E. Bakker, ed.). CRC Press, Boca Raton, Florida. In press.

Ennis, H. L., and Lubin, M. (1961). *Biochim. Biophys. Acta* **50,** 399–402.

Fliegel, L., Sardet, C., Pouyssegur, J., and Barr, A. (1991). *FEBS Lett.* **279,** 25–29.

Giffard, P. M., Rowland, G. C., Kroll, R. G., Stewart, L. M. D., Bakker, E. P., and Booth, I. R. (1985). *J. Bacteriol.* **164,** 904–910.

Glynn, I. M., and Karlish, S. J. K. (1990). *Annu. Rev. Biochem.* **59,** 171–205.

Gold, L., and Stromo, G. (1987). *Am. Soc. Microbiol.* Washington, D. C. pp. 1302–1307.

Goldberg, B. G., Arbel, T., Chen, J., Karpel, R., Mackie, G. A., Schuldiner, S., and Padan, E. (1987). *Proc. Natl. Acad. Sci. U.S.A.* **84,** 2615–2619.

Grinstein, S., ed. (1988). "Na^+/H^+ Exchange" CRC Press, Boca Raton, Florida.

Grinstein, S., Rotin, D., and Mason, M. J. (1989). *Biochim. Biophys. Acta* **988,** 73–97.
Guastella, J., Nelson, N., Nelson, H., Czyzyk, L., Keynan, S., Miedel, M. C., Davidson, N., Lester, H. A., and Kanner, B. I. (1990). *Science,* **249,** 1303–1306.
Häussinger, D. (1988). "pH Homeostasis: Mechanisms and Control." Academic Press, San Diego.
Haigh, J. R., and Phillips, J. H. (1989). *Biochem. J.* **257,** 499–507.
Hanada, K., Yamato, I., and Anraku, Y. (1988). *J. Biol. Chem.* **263,** 7181–7185.
Harold, F. M. (1986). "The Vital Force: A Study of Bioenergetics." Freeman, New York.
Harold, F. M., and Papineau, D. (1972). *J. Membr. Biol.* **8,** 45–62.
Hayashi, M., and Unemoto, T. (1986). *FEBS Lett.* **202,** 327–330.
Hediger, M. A., Coady, M. J., Ikeda, T. S., and Wright, E. M. (1987). *Nature (London)* **330,** 379–381.
Heefner, D. L., and Harold, F. M. (1980). *J. Biol. Chem.* **255,** 11396–11402.
Heefner, D. L., and Harold, F. M. (1982). *Proc. Natl. Acad. Sci. U.S.A.* **79,** 2798–2802.
Henikoff, S., Haughn, G. W., Calvo, J. M., and Wallace, J. C. (1988). *Proc. Natl. Acad. Sci. U.S.A.* **85,** 6602–6606.
Hirota, N., Kitada, M., and Imae, Y. (1981). *FEBS Lett.* **132,** 278–280.
Hoffmann, A., and Dimroth, P. (1991). *Eur. J. Biochem.* **201,** 467–473.
Hoffmann, A., Laubinger, W., and Dimroth, P. (1990). *Biochim. Biophys. Acta* **1018,** 206–210.
Ingham, C., Buechner, M., and Adler, J. (1990). **172,** 3577–3583.
Ishikawa, T., Hama, H., Tsuda, M., and Tsuchiya, T. (1987). *J. Biol. Chem.* **262,** 7443–7446.
Ivey, D. M., Guffanti, A. A., Bossewitch, J. S., Padan, E., and Krulwich, T. A. (1991). *J. Biol. Chem.* **266,** 23483–23489.
Karpel, R., Olami, Y., Taglicht, D., Schuldiner, S., and Padan, E. (1988). *J. Biol. Chem.* **263,** 10408–10414.
Karpel, R., Alon, T., Glaser, G., Schuldiner, S., and Padan, E. (1991). *J. Biol. Chem.* **266,** 21753–21759.
Katz, A., Kaback, R., and Avron, M. (1986). *FEBS Lett.* **202,** 141–144.
Ken-Dror, S., Lanyi, J. K., Schobert, B., Silver, B., and Avi-Dor, Y. (1986). *Arch. Biochem. Biophys.* **244,** 766–772.
Krulwich, T. A. (1983). *Biochim. Biophys. Acta* **726,** 245–264.
Krulwich, T. A. (1986). *J. Membr. Biol.* **89,** 113–125.
Krulwich, T. A. (1992). *In* "Alkali Cation Transport Systems In Procaryotes" (E. Bakker, ed.). CRC Press, Boca Raton, Florida. In press.
Krulwich, T. A., and Guffanti, A. A. (1989). *Ann. Rev. Microbiol.* **43,** 435–463.
Krulwich, T. A., and Ivey, D. M. (1990). *In* "Bacterial Energetics" (T. A. Krulwich, ed.), pp. 417–447. Academic Press, San Diego.
Krulwich, T. A., Guffanti, A. A., Bornstein, R. F., and Hoffstein, T. (1982). *J. Biol. Chem.* **257,** 1885–1889.
Krulwich, T. A., Federbush, J. G., and Guffanti, A. A. (1985). *J. Biol. Chem.* **260,** 4055–4058.
Krulwich, T. A., Guffanti, A. A., and Seto-Young, D. (1990). *FEMS Microbiol. Rev.* **75,** 271–278.
Krumholz, L. R., Esser, U., and Simoni, R. D. (1990). *J. Bacteriol.* **172,** 6809–6817.
Kushner, D. J., and Kamekura, M. (1988). *In* "Halophilic Bacteria" (F. Rodriguez-Valira, ed.), Vol. 1, pp. 109–138. CRC Press, Boca Raton, Florida.
Lanyi, J. K. (1979). *Biochim. Biophys. Acta* **559,** 377–398.
Laubinger, W., and Dimroth, P. (1988a). *Biochemistry* **28,** 7194–7198.
Laubinger, W., and Dimroth, P. (1988b). *Biochemistry* **27,** 7531–7537.
Leblanc, G., Bassilana, M., and Damiano, E. (1988). *In* "Na^+/H^+ Exchange" (S. Grinstein, ed.), pp. 103–117. CRC Press, Boca Raton, Florida.

Li, C., Boileau, A. J., Kung, C., and Adler, J. (1988). *Proc. Natl. Acad. Sci. U.S.A.* **85,** 9451–9455.

MacDonald, R. E., Lanyi, J. K., and Greene, R. V. (1977). *Proc. Natl. Acad. Sci. U.S.A.* **74,** 3167–3170.

Mackie, G. A. (1980). *J. Biol. Chem.* **255,** 8928–8935.

Mackie, G. A. (1986). *Nucleic Acids Res.* **14,** 6965–6981.

Macnab, R. M., and Castle, A. M. (1987). *Biophys. J.* **52,** 637–647.

Maloy, S. R. (1990). *In* "Bacterial Energetics" (T. A. Krulwich, ed.), pp. 203–224. Academic Press, San Diego.

McLaggan, D., Selwyn, M. Y., and Dawson, A. P. (1984). *FEBS Lett.* **165,** 254–258.

McMorrow, I., Shuman, H. A., Sze, D., Wilson, D. M., and Wilson, T. H. (1989). *Biochim. Biophys. Acta* **981,** 21–26.

Mitchell, P. (1961). *Nature (London)* **191,** 144–146.

Mitchell, P., and Moyle, J. (1967). *Biochem. J.* **105,** 1147–1162.

Mochizuki-Oda, N., and Oosawa, F. (1985). *J. Bacteriol.* **163,** 395–397.

Muller, V., and Gottschalk, G. (1992). *In* "Alkali Cation Transport Systems in Procaryotes" (E. Bakker, ed.). CRC Press, Boca Raton, Florida. In press.

Muller, V., Blaut, M., and Gottschalk, G. (1987). *Eur. J. Biochem.* **162,** 461–466.

Murakami, N., and Konishi, T. (1990). *Arch. Biochem. Biophys.* **281,** 13–20.

Nakamura, T., Hsu, C., and Rosen, B. P. (1986). *J. Biol. Chem.* **261,** 678–683.

Nakao, T., Yamato, I., and Anraku, Y. (1987). *Mol. Gen. Genet.* **208,** 70–75.

Nelson, N., and Taiz, L. (1989). *Trends Biochem. Sci.* **14,** 113–116.

Niiya, S., Yamasaki, K., Wilson, T. H., and Tsuchiya, T. (1982). *J. Biol. Chem.* **257,** 8902–8906.

Oesterhelt, D., Hartmann, R., Michel, H., and Wagner, G. (1978). *In* "Energy Conservation in Biological Membranes" (G. Schager and M. Klingenberg, eds.), pp. 140–151. Springer-Verlag, Berlin.

Olsnes, S., Tonnessen, T. I., and Sandvig, K. (1986). *J. Cell Biol.* **102,** 967–971.

Olsnes, S., Tonnessen, T. I., Ludt, J ., and Sandvig, K. (1987). *Biochemistry* **26,** 2778–2785.

Padan, E., and Schuldiner, S. (1986). *Methods Enzymol.* **125,** 337–352.

Padan, E., and Schuldiner, S. (1987). *J. Membr. Biol.* **95,** 189–198.

Padan, E., and Schuldiner, S. (1992). *In* "Alkali Cation Transport Systems in Procaryotes" (E. Bakker, ed.). CRC Press, Boca Raton, Florida. In press.

Padan, E., Zilberstein, D., and Rottenberg, H. (1976). *Eur. J. Biochem.* **63,** 533–541.

Padan, E., Zilberstein, D., and Schuldiner, S. (1981). *Biochim. Biophys. Acta* **650,** 151–166.

Padan, E., Maisler, N., Taglicht, D., Karpel, R., and Schuldiner, S. (1989). *J. Biol. Chem.* **264,** 20297–20302.

Pan, J. W., and Macnab, R. M. (1990). *J. Biol. Chem.* **265,** 9247–9250.

Pestka, S. (1972). *Proc. Natl. Acad. Sci. U.S.A.* **69,** 624–628.

Pinner, E., Carmel, O., Bercovier, H., Sela, S., Padan, E., and Schuldiner, S. (1992a). *Arch. Microbiol.* **157,** 323–328.

Pinner, E., Padan, E., and Schuldiner, S. (1992b). *J. Biol. Chem.* **267,** 11064–11068.

Pinner, E., Kotler, Y., Padan, E., and Schuldiner, S. (1993). *J. Biol. Chem.* (in press).

Poulsen, L. K., Larsen, N. W., Molin, S., and Andersson, P. (1989). *Mol. Microbiol.* **3,** 1463–1472.

Pouyssegur, J., Sardet, C., Franchi, A., L'Allemain, G., and Paris, S. (1984). *Proc. Natl. Acad. Sci. U.S.A.* **81,** 4833–4837.

Pouyssegur, J., Franchi, A., Lagarde, A., and Sardet, C. (1988). *In* "Na^+/H^+ Exchange" (S. Grinstein, ed.), pp. 337–347. CRC Press, Boca Raton, Florida.

Qi, Y., and Adler, J. (1989). *Proc. Natl. Acad. Sci. U.S.A.* **86,** 8358–8362.

Rahav-Manor, O., Carmel, O., Karpel, R., Taglicht, D., Glaser, G., Schuldiner, S., and Padan, E. (1992). *J. Biol. Chem.* **267,** 10433–10438.

Raley-Susman, K. M., Cragoe, E. J., Jr., Sapolsky, R. M., and Kopito, R. R. (1991). *J. Biol. Chem.* **266,** 2739.
Reed, R. H . (1986). *In* "Microbes in Extreme Environments" (R. A. Herbert and G. A. Codd, eds.), pp. 55–81. Academic Press, Orlando, Florida.
Reenstra, W. W., Patel, L., Rottenberg, H ., and Kaback, H. R. (1980). *Biochemistry* **19,** 1–9.
Reizer, J., Reizer, A., and Saier, M. H., Jr. (1990). *Res. Microbiol.* **141,** 1069–1072.
Rosen, B. P. (1986). *Annu. Rev. Microbiol.* **40,** 263–286.
Rosenbusch, J. P. (1974). *J. Biol. Chem.* **249,** 8019–8029.
Rowland, G. C., Giffard, P. M., and Booth, I. R. (1984). *FEBS Lett.* **173,** 295–300.
Sakai, Y., Moritani, C., Tsuda, M., and Tsuchiya, T. (1989). *Biochim. Biophys. Acta* **973,** 450–456.
Sakai-Tomita, Y., Tsuda, M., and Tsuchiya, T. (1992). *Biochem. Biophys. Res. Commun.* (in press).
Sardet, C., Franchi, A., and Pouyssegur, J. (1989). *Cell* **56;,** 271–280.
Sardet, C., Counillon, L., Franchi, A., and Pouyssegur, J. (1990). *Science* **247,** 723–726.
Schonheit, P. (1992). *In* "Alkali Cation Transport Systems in Procaryotes" (E. Bakker, ed.). CRC Press, Boca Raton, Florida. In press.
Schuldiner, S., and Fishkes, H. (1978). *Biochemistry* **17,** 706–710.
Schuldiner, S., and Padan, E. (1992). *In* "Alkali Cation Transport Systems in Procaryotes" (E. Bakker, ed.) CRC Press, Boca Raton, Florida. In press.
Schuldiner, S., Agmon, V., Brandsma, J., Cohen, A., Friedman, E., and Padan, E. (1986). *J. Bacteriol.* **168,** 936–939.
Silver, S., and Walderhaug, M. (1992). *Microbiol. Rev.* **56,** 195–228.
Skou, J. C., Norby, J. G., Maunsbach, A. B., and Esmann, M. (1988). "The Na^+,K^+-Pump, Part A: Molecular Aspects." Alan R. Liss, New York.
Skulachev, V. P. (1987). *In* "Ion Transport in Procaryotes" (B. P. Rosen and S. Silver, eds), pp. 131–164. Academic Press, San Diego.
Skulachev, V. P. (1988). *In* "Membrane Bioenergetics" (V. P. Skulachev, ed.), pp. 293–326. Springer-Verlag, Berlin.
Slonczewski, J. L., Rosen, B. P., Alger, S. R., and Macnab, R. M. (1981). *Proc. Natl. Acad. Sci. U.S.A.* **78,** 6271–6275.
Stewart, L. M. W., and Booth, I. R. (1983). *FEMS Microbiol. Lett.* **19,** 161–164.
Storz, G., Tartaglia, L. A., and Ames, B. N. (1990). *Science* **248,** 189–194.
Sugiyama, S. H., Matsukura, H., and Imae, Y. (1985). FEBS Lett. **182,** 265–268.
Tabor, S., and Richardson, C. C. (1985). *Proc. Natl. Acad. Sci. U.S.A.* **82,** 1074–1078.
Taglicht, D. (1992). Ph.D. Thesis, Hebrew Univ. Jerusalem, Jerusalem.
Taglicht, D., Padan, E., and Schuldiner, S. (1991). *J. Biol. Chem.* **266,** 11289–11294.
Thelen, P., Tsuchiya, T., and Goldberg, E. B. (1991). *J. Bacteriol.* **173,** 6553–6557.
Thomas, M. S., and Glass, R. E. (1991). *Mol. Microbiol.* **5,** 2719–2725.
Tokuda, H. (1989). *J. Bioenerg. Biomembr.* **21,** 693–704.
Tokuda, H. (1992). *In* "Alkali Cation Transport Systems in Procaryotes" (E. Bakker, ed.). CRC Press, Boca Raton, Florida. In press.
Tokuda, H., Kim, Y. J., and Mizushima, S. (1990). *FEBS Lett.* **264,** 10–12.
Tse, C. M., Ma, A. I., Yang, V. W., Watson, A. J. M., Levine, S., Montrose, M. H., Potter, J., Sardet, C., Pouyssegur, J., and Donowitz, M. (1991). *EMBO J.* **10,** 1957.
Tsuchiya, T., and Takeda, K. (1979). *J. Biochem.* (*Tokyo*) **86,** 225–230.
Tsuchiya, T., Raven, J., and Wilson, T. H. (1977). *Biochem. Biophys. Res. Commun.* **76,** 26–31.
Tsuchyia, T., Misawa, A., Miyaka, Y., Yamasaki, K., and Niiya, S. (1982). *FEBS Lett.* **142,** 231–234.

Umeda, K., Shiota, S., Futai, M., and Tsuchiya, T. (1984). *J. Bacteriol.* **160,** 812–814.
Unemoto, T., Tokuda, H., and Hayashi, M. (1990). *In* "Bacterial Energetics" (T. A. Krulwich, ed.), pp. 33–54. Academic Press, San Diego.
Walderhaug, M. O., Dosch, D. C., and Epstein, W. (1987). *In* "Ion Transport in Prokaryotes" (B. P. Rosen and S. Silver, eds.), pp. 85–130. Academic Press, San Diego.
Waser, M., Hess-Bienz, D., Davies, K., and Solioz, M. (1992). *J. Biol. Chem.* **267,** 5396–5400.
West, I. C., and Mitchell, P. (1974). *Biochem. J.* **144,** 87–90.
Yazyu, H., Shiota-Niiya, S., Shimamoto, T., Kanazawa, H., Futai, M., and Tsuchiya, T. (1984). *J. Biol. Chem.* **259,** 4320–4326.
Yazyu, H., Shiota, S., Futai, M., and Tsuchiya, T. (1985). *J. Bacteriol.* **162,** 933–937.
Zilberstein, D., Schuldiner, S., and Padan, E. (1979). *Biochemistry* **18,** 669–673.
Zilberstein, D., Padan, E., and Schuldiner, S. (1980). *FEBS Lett.* **116,** 177–180.
Zoratti, M. (1992). *In* "Alkali Cation Transport Systems in Procaryotes" (E. Bakker, ed.). CRC Press, Boca Raton, Florida. In press.

Index

D

E

F

G

I

K

L

M

N

O

P

R

S

T

U

V

X

Y

ISBN 0-12-364539-5